Automotive Suspension and Steering
Theory and Service

Herbert E. Ellinger
Richard B. Hathaway

Prentice Hall Career & Technology, Englewood Cliffs, NJ 07632

LIBRARY OF CONGRESS
Library of Congress Cataloging-in-Publication Data

Ellinger, Herbert E.
Automotive suspension and steering : theory and service / Herbert E. Ellinger, Richard B. Hathaway.
p. cm.
Includes index.
ISBN 0-13-054123-0
1. Automobiles--Springs and suspension. 2. Automobiles--Springs and suspension--Maintenance and repair. 3. Automobiles--Steering-gear. 4. Automobiles--Steering-gear--Maintenance and repair. 5. Automobiles--Tires. 6. Automobiles--Tires--Repairing. I. Hathaway, Richard B. II. Title.
TL257.E43 1989
629.2'43--dc19 88-12428
CIP

Editorial/production supervision and
interior design: WordCrafters Editorial Services, Inc.
Cover design: Joel Mitnick Design, Inc.
Manufacturing buyer: Robert Anderson
Page layout: Meg Van Arsdale

Prentice-Hall, Inc.
A Paramount Communications Company
Englewood Cliffs, New Jersey 07632

Printed in the United States of America

10 9 8 7 6 5 4 3 2

ISBN 0-13-054123-0

Prentice-Hall International (UK) Limited, *London*
Prentice-Hall of Australia Pty. Limited, *Sydney*
Prentice-Hall Canada Inc., *Toronto*
Prentice-Hall Hispanoamericana, S.A., *Mexico*
Prentice-Hall of India Private Limited, *New Delhi*
Prentice-Hall of Japan, Inc., *Tokyo*
Simon & Schuster Asia Pte. Ltd., *Singapore*
Editora Prentice-Hall do Brasil, Ltda., *Rio de Janeiro*

Contents

Preface

In automechanics courses, the majority of the course is devoted to the engine and its operating accessories. The drive line and chassis are studied last. Students often forget that it is more important for safe vehicle operation to be able to control the vehicle on all types of roads and in traffic, and to be able to bring the vehicle to a safe, controlled stop, than it is to accelerate rapidly.

This book is written to help the reader understand how properly designed and correctly serviced suspension and steering systems contribute to vehicle safety. Routine service operations are discussed in the early chapters. Specific suspension and steering service operations are described in later chapters.

Automotive enthusiasts often overlook the fact that all vehicle control passes through the tire footprint. Tires are discussed in detail to help the reader understand how different types of tire construction help in vehicle control, how tires are matched to the vehicle, and how tires are serviced.

Wheels and hubs must be removed to perform most suspension service. Bearings and seals are open to the air when the hubs are removed; this provides a good opportunity to service the bearings, as well as to replace the seals. Bearing and seal service is discussed in early chapters so the reader will know how to service them while the suspension is being serviced.

The suspension is a major contributor to vehicle safety and control. Its maintenance includes routine service, problem diagnosis, inspection, and repair of the cause of any problem found. Repairs must be performed carefully and systematically; this is the only way the technician can be sure the vehicle will operate safely and handle correctly.

Suspension system requirements and design are discussed in the next section of the book. Included is a discussion of the principles used in four-wheel steering. A discussion of alignment principles and methods follows to give the student the information needed to align the vehicle after the suspension is serviced. Front and rear suspension service are the next topic. The section on steering systems begins with a discussion of the steering linkage. This is followed by a discussion of the operation and service of manual and power steering gears. The last chapter of the book is on elementary vehicle dynamics and includes many sample problems with calculations.

The text is general so it applies to all types of suspension and steering systems. It covers the material that the service manuals assume the technician already knows. There are very few references to specific types of vehicles in this book. It would be impossible to discuss the service details of all types, makes, and models of vehicle suspension and steering systems because the material would only be up to date at the time it was written. As a result, you should plan to use service manuals along with this book as you service suspension and steering systems.

ACKNOWLEDGMENTS

The authors wish to thank the people and organizations that have provided material for use in this book. Constructive criticism has been incorporated in the book wherever possible, especially the specific suggestions from Randall Bennett of Stark Technical College, North Canton, Ohio.

The authors wish to thank the following schools for the opportunity to photograph equipment and teaching aids for use as illustrations in this book:

Cincinnati Technical College

Kalamazoo Valley Community College

Macomb County Community College

Western Michigan University

The authors wish to thank the following organizations for their assistance with reference material and illustrations:

Firestone Tire and Rubber Company

Ford Division, Ford Motor Company

TRW Replacement Division

Chrysler Corporation

General Motors Corporation
- Buick Motor Division
- Chevrolet Motor Division
- Cadillac Motor Car Division

Uniroyal Goodrich Tire Company

1

System Introduction and Safety

The modern vehicle is a complex assembly created to meet a variety of design objectives, including fuel economy, handling, performance, safety, passenger comfort, driveability, and ride quality. No single vehicle design can meet all of these objectives. As a result, manufacturers make compromises to produce vehicles that emphasize some specific design objectives more than others. A performance vehicle designed for power and handling will compromise ride and fuel economy.

No matter what the designer wishes, the vehicle must have room for the driver, passengers, and load. It must have enough engine power to move the load at the desired speed and enough braking power to stop safely. The moving vehicle must be controllable so that it can be driven to follow changes in the roadway and traffic conditions.

The following discussion introduces vehicle parts that affect vehicle control, ride, and handling. The design, maintenance, inspection, and service of the suspension and steering parts are covered in more detail in later chapters.

1-1 FRAME DESIGNS

The driver and passengers expect a comfortable ride to minimize fatigue. The seats must be firm and give good support. The driver must be able to see in all directions. Noise level and vibration should be as low as possible. The vehicle frame is the basic platform to which all suspension and steering linkage parts attach. A vehicle will neither steer nor handle well if the frame is too flexible. A rigid frame structure may pass unnecessary vibrations into the passenger compartment. The frame and suspension design will affect the ride quality, handling, and durability, as well as the levels of both noise and vibration.

Manufacturers use several different types of construction on their vehicles as shown in Figure 1-1. Of these, separate body and frame construction was the most common through the 1970's. It is still used in large vans, pickups, and trucks. In this type of construction, the engine, drive line, running gear, and body mount to the frame through insulators. *Insulators* are synthetic rubber pads that keep road and engine noise and vibration from going into the passenger compartment.

A second type of construction is the unitized body. This design is by far the most popular in modern vehicles. The unitized design has a lightweight structure with the required strength. In this type of construction, the frame is welded into the body as part of the body structure. Body panels add strength to the frame pieces. The running gear and drive line are mounted to the unitized body through large, soft synthetic rubber insulators. The insulators minimize the transfer of noise and vibration. If the insulators are too soft they will allow too much running gear and drive line movement. This movement,

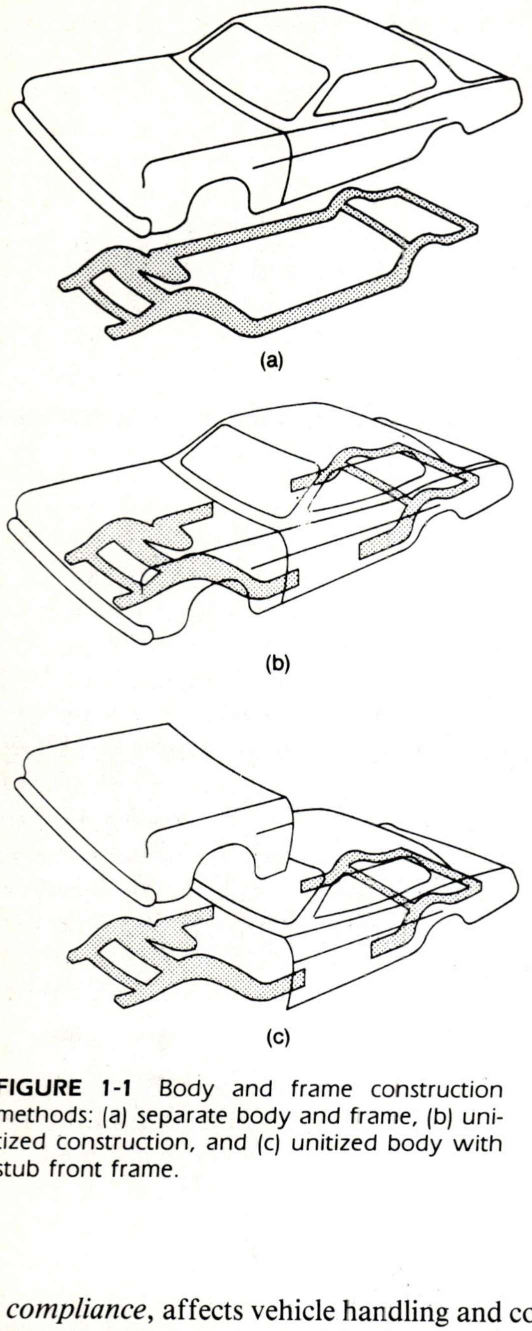

FIGURE 1-1 Body and frame construction methods: (a) separate body and frame, (b) unitized construction, and (c) unitized body with stub front frame.

called *compliance*, affects vehicle handling and control. If the insulators are too hard, they will not insulate noise and vibration as they should. The manufacturer carefully designs the insulators and puts them where they will produce a vehicle with low noise and vibration transmission that still has proper handling and feel. Insulator properties change with age, changing original characteristics as the vehicle becomes older.

A third type of construction combines the features of the first and second types. It uses a stub frame from the bulkhead forward and a unitized body from the bulkhead back. The unitized part is very rigid, while the stub frame provides a place for good insulation.

Manufacturers select the type of construction that is most economical to build, while providing the noise, vibration, and ride and handling characteristics they want in the vehicle. Large older vehicles, vans, and trucks generally use separate body and frame construction. The newer, smaller vehicles generally use unitized construction.

1-2 BRAKE SYSTEMS

Service brakes must be able to stop the vehicle, prevent excess speed when coasting, and hold the vehicle in position while it is stopped on grades. They are designed so the driver can adjust the braking effort to maintain vehicle control. Vehicle control is influenced by brakes as well as the suspension and steering systems. Faults in the brake system can lead to wheel pull during braking. To repair suspension systems, parts of the brake system may require disassembly. For these reasons, the brake system will be discussed briefly in this text.

The brake system must provide smooth stopping power that can be controlled by the driver. The force required on the brake pedal must not be so high that the wheels cannot be locked. To meet these braking requirements, minimum braking standards have been set for vehicle brakes.

The driver controls the braking force through mechanical, vacuum, and hydraulic mechanisms. The amount of braking increases as more force is placed on the brake pedal. This force is transferred through the brake system to push stationary brake linings against the rotating brake surface. This slows the vehicle as it turns kinetic energy (energy of motion) into thermal energy (heat). Maximum braking occurs just before the wheels lock to cause the tires to slide on the road surface. Maximum braking, therefore, depends on the adhesion between the tire and the road surface. When the tire slides on the road, braking effect is reduced and directional control of the vehicle may be lost.

The stationary parts of the front brake assemblies are mounted on the steering knuckle of the front suspension (Figure 1-2). In the rear, they

FIGURE 1-2 Typical disc brake mounted on the steering knuckle.

are mounted on the axle housing or the rear spindle assembly (Figure 1-3). The cast- iron brake drum or disc rotates with the wheel.

Disc Brakes. Disc brakes have discs that rotate with the wheel. The brake disc is usually called a *brake rotor*. A hydraulically operated piston in a stationary caliper is used to force the lining of the brake pad against the braking surface of the rotor. The friction between the lining and rotor is used to slow or stop wheel rotation. The stationary caliper housing keeps the pads from rotating when they are being forced against the rotating brake disc. A typical disc brake is pictured in Figure 1-2.

Disc brake pads move perpendicular to the face of the brake rotor. In this way, they clamp on the rotor to slow the vehicle motion. The clamping force is proportional to the force the driver puts on the brake pedal.

Drum Brakes. Drum brakes use stationary, internal expanding brake shoes with linings. They are mounted inside a rotating brake drum. The brake drum is fastened between the wheel-tire assembly and the hub assembly or the axle flange. A typical drum brake assembly is shown in Figure 1-4. The brake shoes slow drum rotation when the diameter of the shoes is expanded to bring the lining in contact with the brake surface. This is done by a hydraulically operated wheel cylinder. Fluid pressure from the master cylinder is forced into the wheel cylinders, expanding them. The expansion of the wheel cylinder moves the brake shoe through mechanical linkage to press the linings against the rotating brake drum. This provides braking action as it slows the rotation of the drum.

FIGURE 1-3 Typical drum brake mounted on the rear of a front-drive vehicle.

FIGURE 1-4 Typical drum brake with the drum being removed.

1-3 DRIVE LINES

The drive line includes all the parts from the engine to the drive wheels. The transmission and final drive carry the torque from the engine. Other parts form the link from one part to the other. The engine torque during acceleration and the torque during braking place loads on the suspension parts.

During suspension repair, it may be necessary to disassemble parts of the drive line. Noises produced when the suspension moves may originate from drive line parts. A basic understanding of different drive line assemblies is presented here to give you a working knowledge so that you can do suspension repair.

Drive lines with front-wheel drive often combine the transmission and the final drive into one assembly. This is also true of mid- and rear-engine vehicles. The assembly is called a *transaxle*. Short *half-shafts* with universal joints at each end connect between the transaxle and the wheels. These shafts carry power from the final drive to the wheels even when the suspension moves and steers. Figure 1-5 shows a typical transaxle drive arrangement with half-shafts.

A *differential* in the final drive splits incoming power, sending half to each drive wheel. This allows the drive wheels to turn at different speeds

FIGURE 1-5 Typical transaxle drive arrangement with MacPherson struts.

FIGURE 1-6 Differential action while cornering.

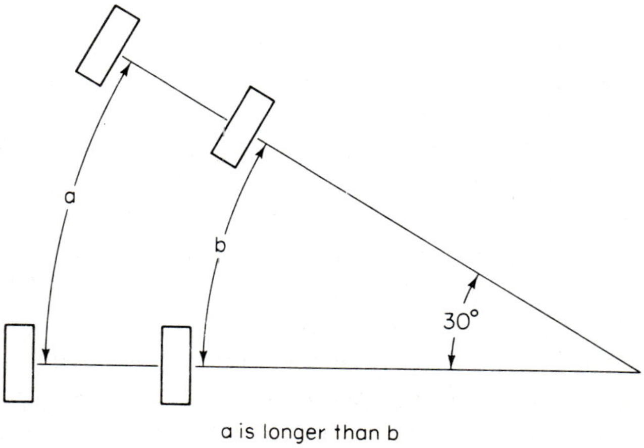

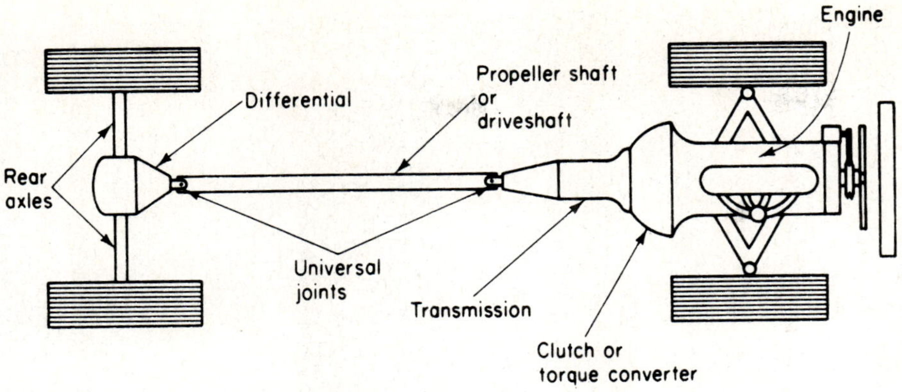

FIGURE 1-7 Typical front-engine, rear-wheel drive.

while rounding corners. Figure 1-6 shows why differential action is necessary while cornering.

In front-engine, rear-wheel drive vehicles (Figure 1-7), the transmission is located under the front floor of the passenger compartment. A *drive shaft* is used to carry engine power to the rear axle. The drive shaft has a *universal joint* at each end. It carries power through the changing drive line angles as the suspension moves.

A vehicle with independent suspension at the drive wheels has the final drive attached rigidly to the vehicle frame or the engine. This drive arrangement produces forces, without any torques, on the suspension parts during acceleration. If the brakes are mounted inboard so the caliper mounts to a frame piece and not to a suspension, the brake will also not produce a torque on the suspension. A suspension designed to handle only acceleration and braking torques can be designed differently than one that must handle both suspension forces and torques.

1-4 SUSPENSION SYSTEMS

The suspension includes springs, shock absorbers, and control linkages. It must be strong enough to support the vehicle body and load. The suspension must also resist engine and brake reactions. The most important job of the suspension is to keep the tires in contact with the road as much of the time as possible. This is done while supporting the vehicle and its load, even while traveling over rough roads. The four tire footprints are the only place the vehicle touches the road. All of the engine power, steering, and braking forces operate through the tire-to-road footprints. Control of the vehicle (power, steering, and braking) is reduced or lost any time a tire does not stay on the road or when skidding begins.

The vehicle body is supported by springs. The springs can be of the coil, leaf, torsion bar, or pneumatic type; these are shown in Figure 1-8. Coil springs are the most popular design used in the modern automobile. Coil, torsion bar, and pneumatic springs all require links and arms to hold the wheel in position. Leaf springs provide lateral and longitudinal control to prevent unwanted wheel motions. They are commonly found on vans and trucks.

Suspension systems have been changed and refined as the passenger automobile has developed. Design objectives differ between luxury sedans, performance vehicles, small compact vehicles, and light trucks. Tire improvements, along with improvements in shock absorbers, steering systems, and suspension control devices, have continually upgraded vehicle handling characteristics.

Tire-to-road contact is needed for safe, positive vehicle control under all operating conditions. Keep in mind that all four tires must stay in contact with the road at all times for maximum vehicle control. Compromises are made in handling response, tire wear, driver comfort, and ride harshness to achieve positive vehicle control.

Front Suspension. Front suspension designs have developed from relatively rugged solid-axle designs to the modern lightweight, high-strength, strut-type independent designs. These have been upgraded with added linkage. Suspension design improvements have followed improvements in roadways and driver expectations.

Dependent-type, front-suspension systems with solid axles were used on early domestic vehicles. On

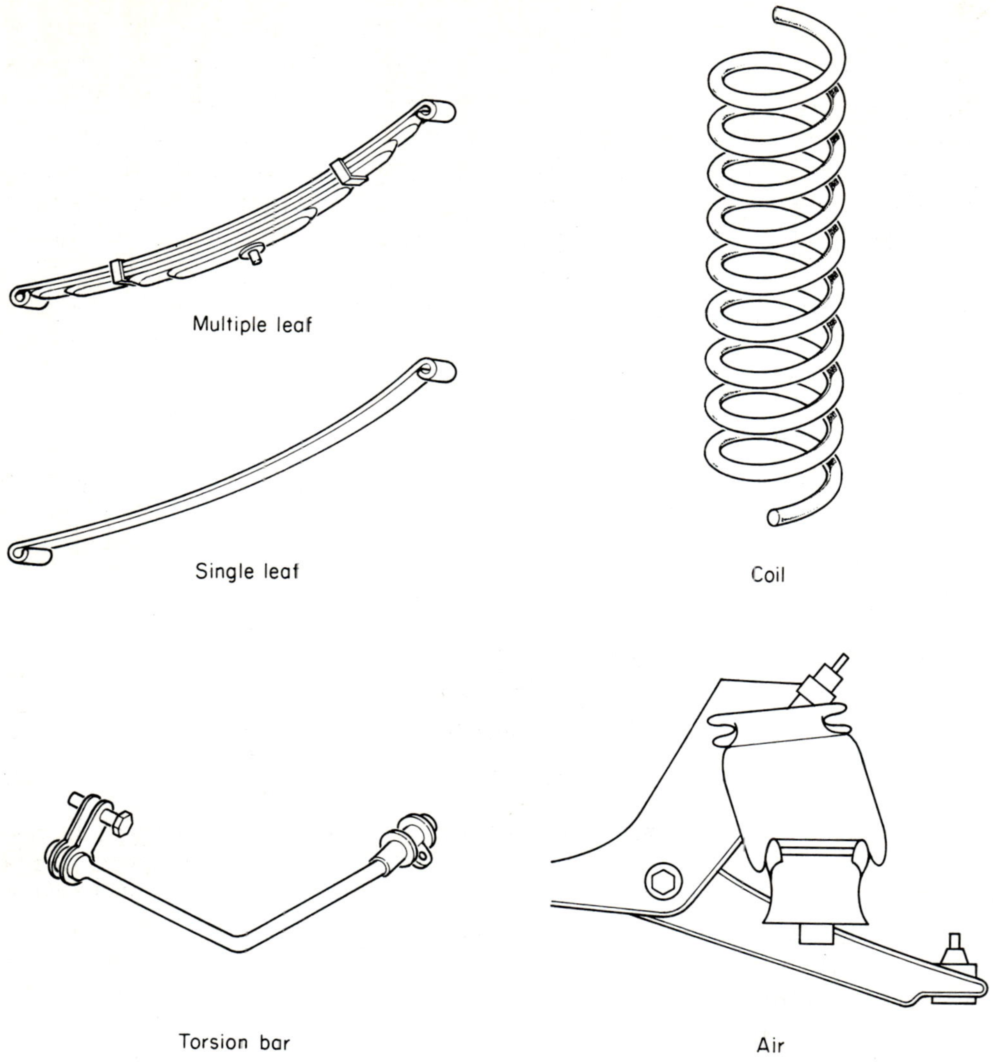

FIGURE 1-8 Types of suspension springs.

a dependent suspension, motion at one wheel causes movement at the other wheel. Dependent-type, solid-axle suspension is still in use on some heavy-duty trucks, farm vehicles, and motor homes. The dependent suspension, shown in Figure 1-9, has an I- beam or solid axle that connects the left and right front-wheel knuckles, holds them in position, and connects them to the front springs. The knuckles pivot on a shaft, called a *kingpin*, which fits into the end of the axle. The wheel hub and bearings are mounted on the outer part of the knuckle, called the *spindle*. The steering linkage is connected to an arm that is fastened to the knuckle. When required, link and lever control devices connect between the axle and frame. These linkages absorb braking and turning forces as well as improve ride quality.

The solid-axle front suspension is used on vehicles that go over rough roads at low speeds. The solid axle is used extensively on large trucks and off-the-road vehicles because it is the strongest type of front suspension system.

The independent front suspension system grew from the operator's demand for improved ride quality and improved handling at increased highway speeds. In independent suspensions, motion of one wheel does not cause movement at the other wheel.

Modern independent short-long arm suspension systems have control arms, steering knuckles, ball joints, springs, stabilizer bars, brake struts, and shock absorbers. You can see most of these parts in Figure 1-10. Selected combinations of these parts provide the required support for braking and turning forces, steering, and vertical wheel movement while maintaining vehicle control.

Today's smaller, lightweight, front-engine, front-drive vehicles require careful packaging to maximize the interior dimensions of the vehicle. The strut-type suspension (Figure 1-11) is a lightweight durable design that requires less room in the engine compartment. This design, called a *MacPherson strut*, uses fewer parts than the short-long arm suspension and it still gives the vehicle excellent handling. The basic parts of this suspension include lower control arms, lower ball joints, struts with integral shock absorbers, springs, a stabilizer bar, and, on some vehicles, control linkages.

A variation of the independent front suspension design, called the *twin I-beam suspension*, is used on some trucks (Figure 1-12). This design uses two long members called *I-beams*. It has kingpins

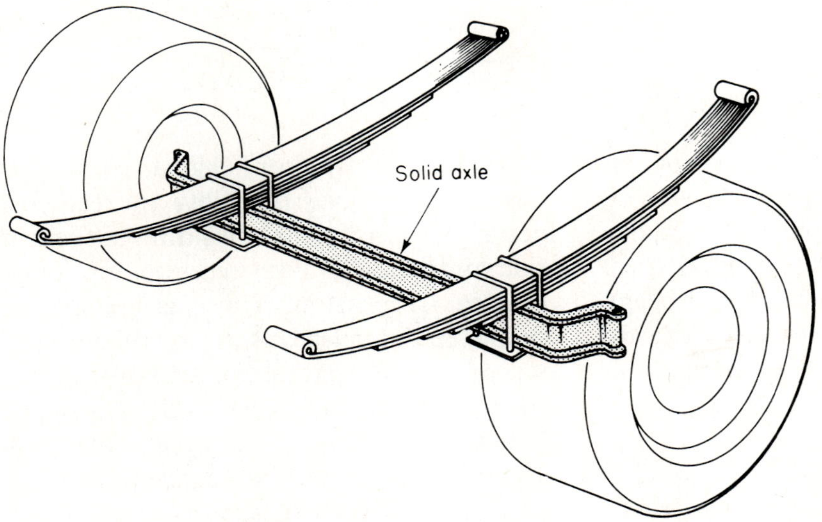

FIGURE 1-9 Typical dependent-type front suspension with a solid axle.

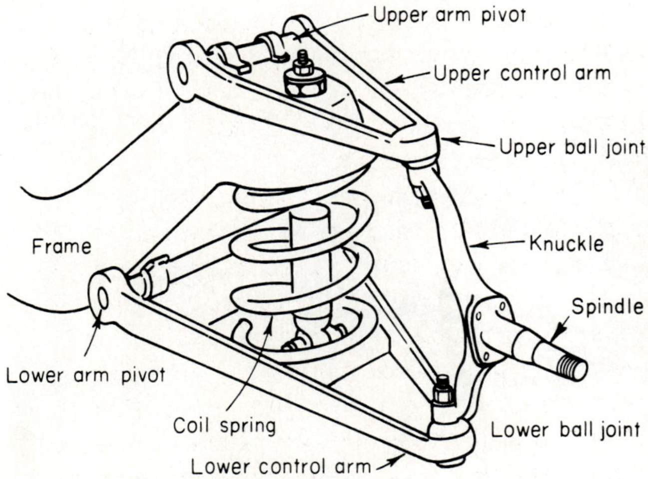

FIGURE 1-10 Modern independent long-short arm front suspension.

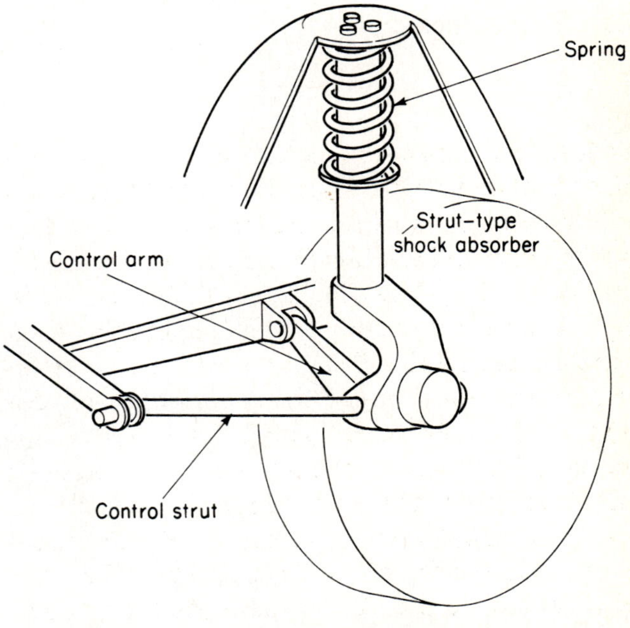

FIGURE 1-11 Typical strut-type front suspension.

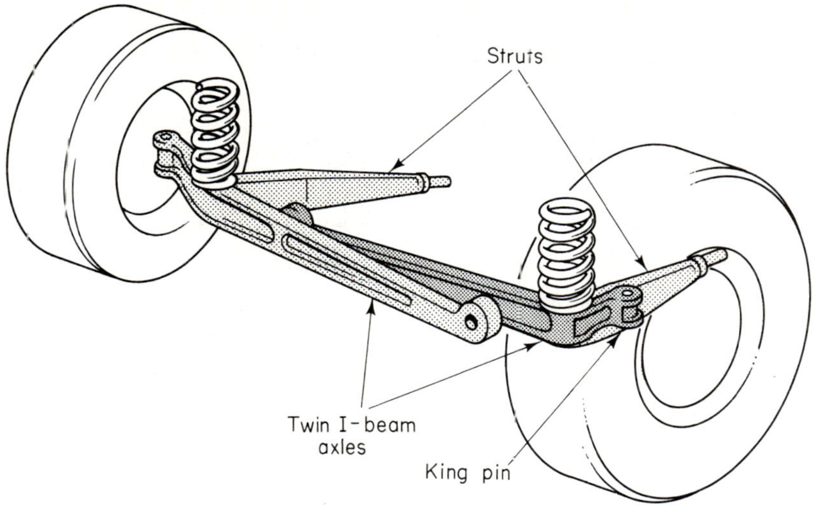

FIGURE 1-12 Twin I-beam front suspension.

on the outer ends to allow the wheels to pivot for turning. A pivot connects the I-beam to the vehicle frame at the end opposite the kingpin. Struts on the I-beam absorb braking forces. Coil springs are used to support the vehicle weight. The twin I-beam design provides a compromise between the independent suspension ride quality and the ruggedness of the solid-axle dependent suspension.

Rear Suspension. Most front-engine, rear-wheel-drive vehicles use a simple dependent rear suspension (Figure 1-13). Rear-wheel-drive independent suspension (Figure 1-14) is much more complex and expensive. As a result, it is only used on a few passenger vehicles.

The move to front-engine, front-wheel drive has allowed more flexibility in rear suspension design. By moving the drive train to the front, only ride control and braking reactions are controlled by the rear suspension. This has led to the use of simplified dependent suspension (Figure 1-15), semi-independent suspension (Figure 1-16), and independent rear suspension (Figure 1-17). The improvement in independent suspension designs allows its use in a larger number of new vehicle designs.

Shock absorbers, springs of all types, control linkages, control arms, and either an axle or spindle are the primary rear suspension parts. Struts, trailing arms, and/or lower control arms are used to control both vertical and front-rear movement of the wheel.

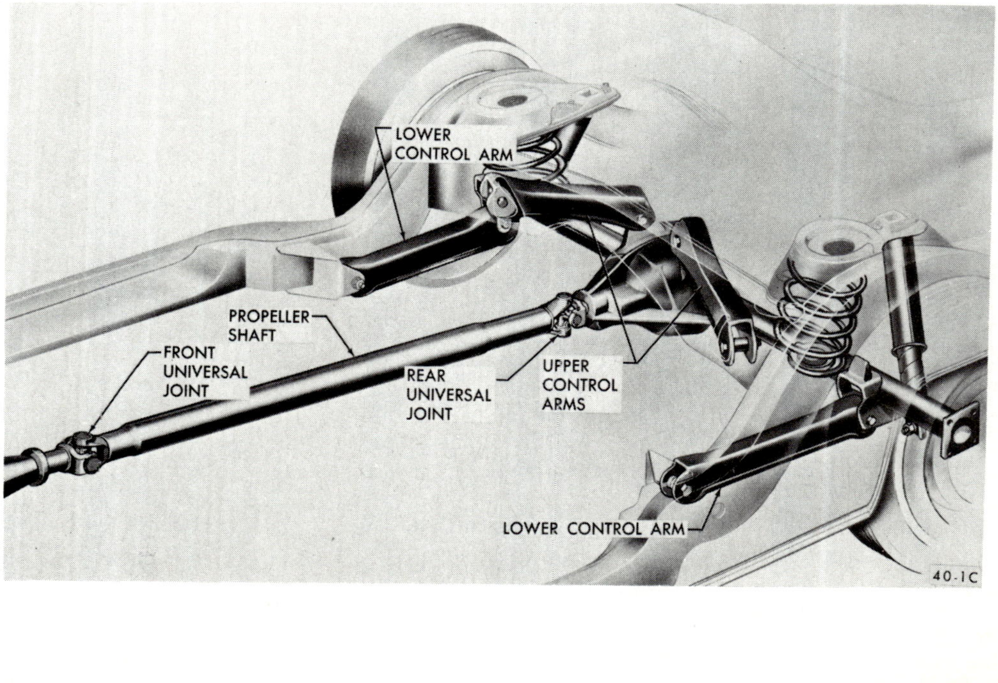

FIGURE 1-13 Typical dependent rear suspension used with front-engine, rear-wheel drive.

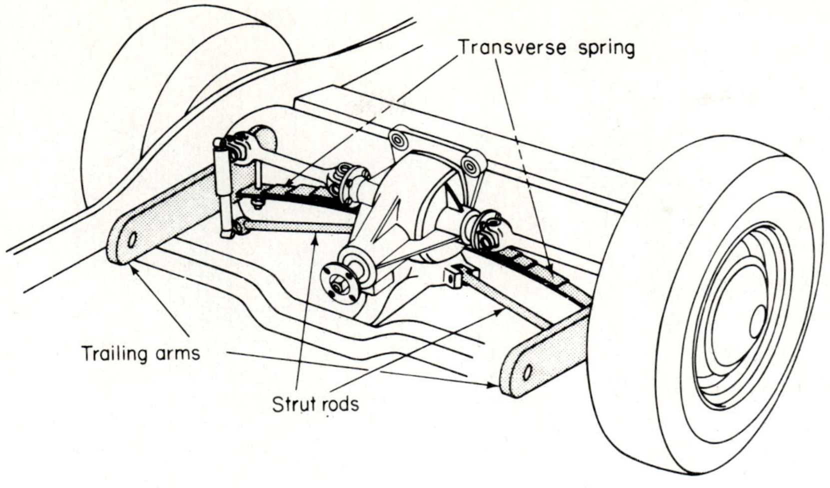

FIGURE 1-14 Typical independent rear suspension used with front-engine, rear-wheel drive.

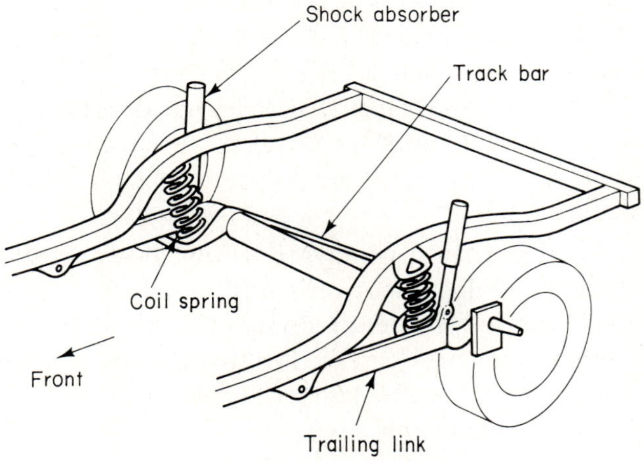

FIGURE 1-15 Typical dependent rear suspension used with a front-wheel drive.

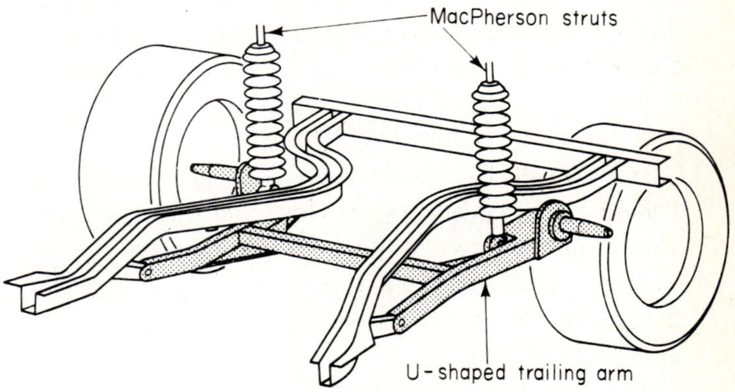

FIGURE 1-16 Typical semi-independent rear suspension used with a front-wheel drive.

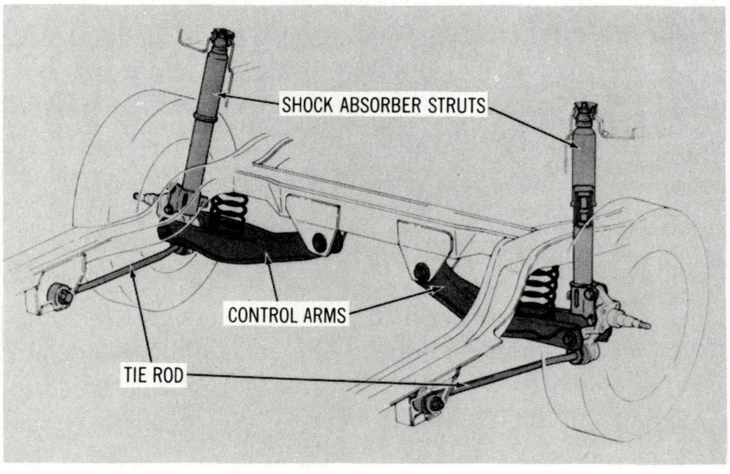

FIGURE 1-17 Typical independent suspension used with a front-wheel drive.

1-5 STEERING SYSTEMS

The driver controls the direction of the front wheels of the vehicle through the steering gear. Modern steering gears have two major units, a steering column and a gear unit. The steering column has a supported shaft that connects the driver's steering wheel to the gear unit. The gear unit multiplies the driver's steering effort to move the steering linkage.

The front wheels of rear-wheel-drive vehicles rotate on a spindle. The spindle is part of the steering knuckle (Figure 1-18). The knuckle is connected to the front suspension members with ball joints. The ball joints allow for steering as the suspension moves up and down. The wheel hubs on front-wheel-drive vehicles rotate on hollow axle stub shafts inside bearings within the steering knuckles (Figure 1-19).

The steering wheel controls the steering gear assembly. This, in turn, moves the knuckle through the steering linkage. Two steering gear designs are in use today, the rack and pinion (Figure 1-20) and the recirculating ball (Figure 1-21).

Today's vehicles are designed with responsive steering. As a result, more effort is needed to steer the vehicle when it is moving slowly. Power steering supplies this effort on many vehicles.

FIGURE 1-18 Typical steering knuckle and spindle on a long-short arm front suspension.

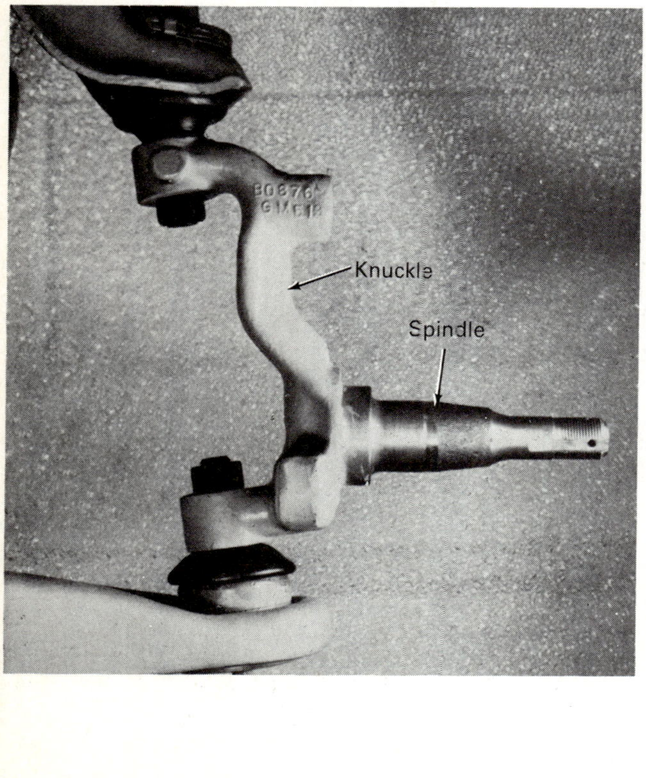

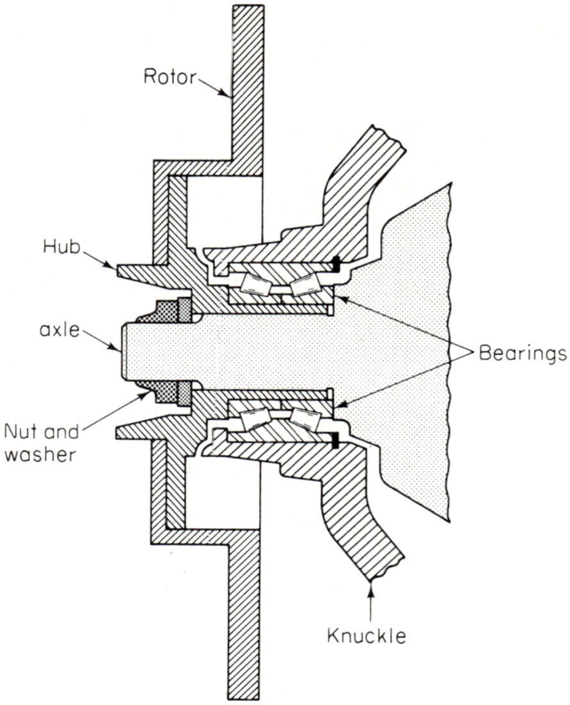

FIGURE 1-19 Typical steering knuckle used with a front-wheel drive.

With power steering doing most of the work, steering ratios are decreased so that the vehicle can be steered with small steering wheel movements. The power steering gear is similar to the standard steering gear. It includes surfaces upon which fluid pressure is applied to aid the driver's steering effort. Both rack and pinion and recirculating ball gears may have power assist.

Power for the steering gear is provided by an engine-driven pump (Figure 1-22). The pump forces power steering fluid through a system controlled by a valve. This control valve can sense the driver's steering effort. It puts fluid pressure against a pressure surface in the steering system. This fluid pressure takes over some of the effort needed to steer the vehicle.

The steering column in the modern vehicle has many parts. It is designed to collapse or fold in a vehicle collision to protect the driver. In some installations, it may be tilted and telescoped to adjust the position of the steering wheel for the comfort of the driver. To reduce the chance of theft, it also has a steering gear lock. On many vehicles, it has

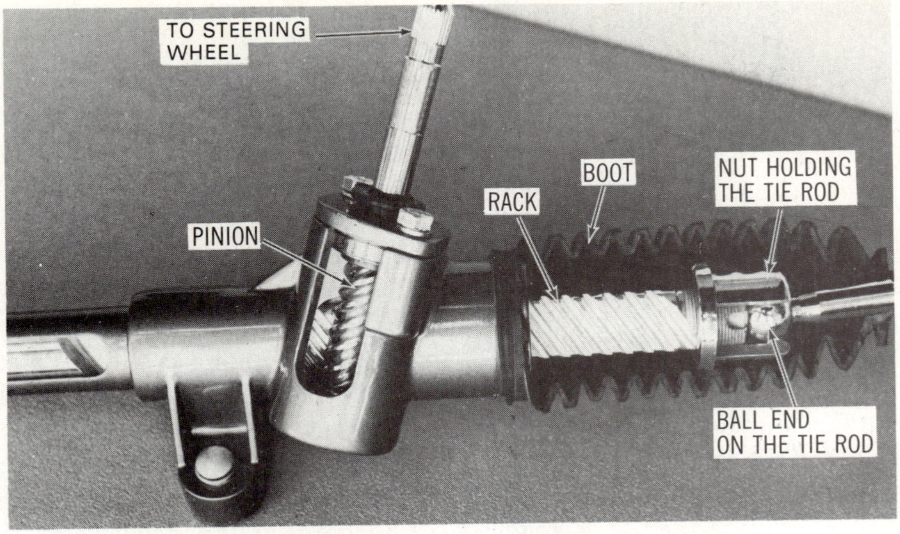

FIGURE 1-20 Major parts of a typical manual rack and pinion steering gear.

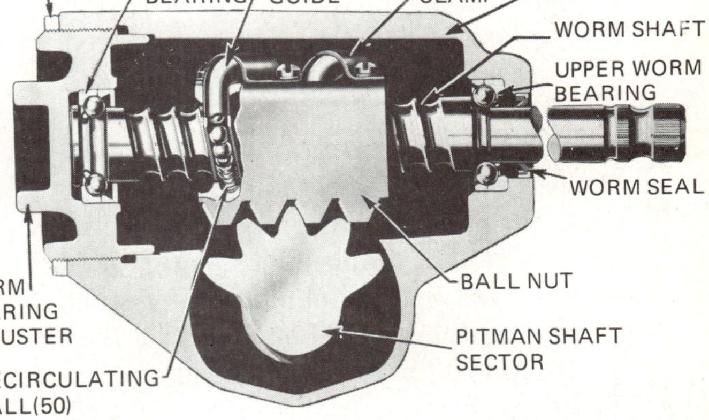

FIGURE 1-21 Major parts of a manual recirculating ball steering gear.

FIGURE 1-22 Schematic drawing showing the main parts of a typical ball-type power steering gear system.

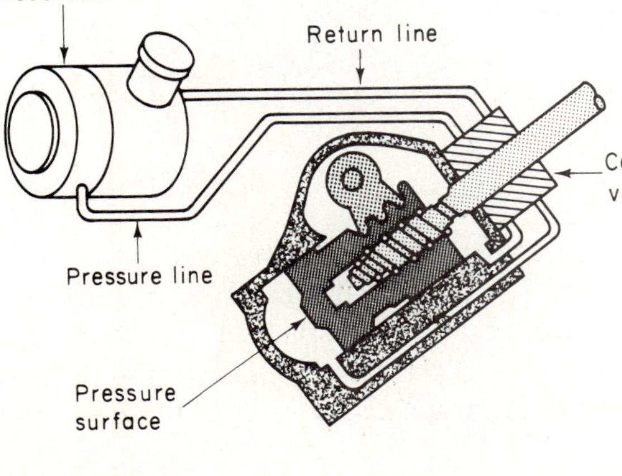

a transmission lock. Because it is within easy reach of the driver, the steering column may carry the transmission shift control lever, turn signal switch, headlight and dimmer switches, wiper switch, emergency flasher switch, and speed control.

1-6 TIRES

Tires and wheels affect vehicle ride, braking, handling, and fuel economy. The tires perform three important functions. First, the tires support the vehicle weight. The vehicle weight on the tires is called the *radial load*. Second, the tires control the vehicle

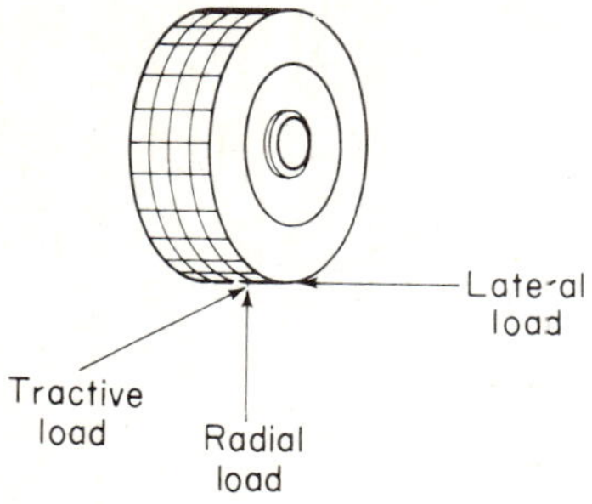

FIGURE 1-23 Direction of loads on a tire.

direction while it is in motion. The side load on the tires when the vehicle is steered is called the *lateral load*. If the tire could not control a lateral load, the vehicle would go straight ahead, even though the wheels are turned at an angle. Third, the tires and wheels transfer engine and braking torque to the road surface for accelerating, driving, and stopping the vehicle. These forces are called the *tractive load* on the tire. Figure 1-23 shows these loads. When the tire has less traction than the tractive load caused by vehicle acceleration or braking, the tire will slip on the road surface.

The frictional force or *traction* of the tire on the road surface is approximately equal in all directions (Figure 1-24). Therefore, the maximum lateral and tractive loads do not simply add. Lateral loads placed on the tire reduce the amount of frictional force remaining for tractive load on the tire at that same instant (frictional force = lateral load + tractive load). The reverse is also true; any tractive load on the tire will reduce the amount of lateral load the tire can handle at the same instant without slippage. Only a given amount of combined steering and braking can occur at any instant.

FIGURE 1-24 Traction forces on a tire are equal in all directions.

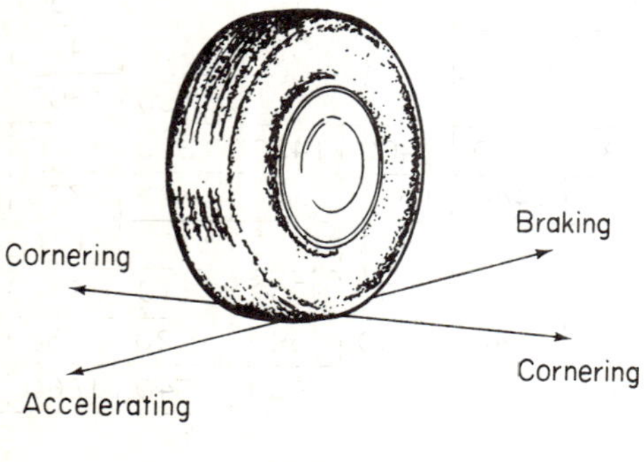

Tire Characteristics. The operator expects maximum mileage from each tire, along with desirable vehicle handling characteristics. Maximum tire mileage with a given tire type is greatest with a tread rubber compound that is relatively tough and hard. Maximum frictional force for maximum slip resistance and traction is best with a relatively soft tread rubber compound. Tread rubber compound is one place where a compromise is made so that tires will have the desired operational characteristics.

Tires are designed with a tread that is a compromise between dry road handling, wet roadability, and noise. As the tread wears in service, the performance characteristics gradually change. The water channeling, that is so important on wet roadways, is reduced considerably as the tread wears. This means that, for normal tire traction, a vehicle with worn tire tread must be driven slower on wet roads to reduce the chance of hydroplaning.

Tire Design. Three basic types of tires are used on passenger vehicles: bias ply, bias belted, and radial ply. Each type has slightly different operating characteristics.

The bias ply tire, as shown in Figure 1-25, has two or more layers of cord material running at an angle from bead to bead. The cord angle of each ply runs opposite to the angle of the cords in the next ply. The plies cross the tread centerline at an angle of 25 to 40 degrees, depending on the tire design.

The bias belted tire, as shown in Figure 1-26,

FIGURE 1-25 Bias ply tire construction. (Courtesy of Uniroyal Incorporated)

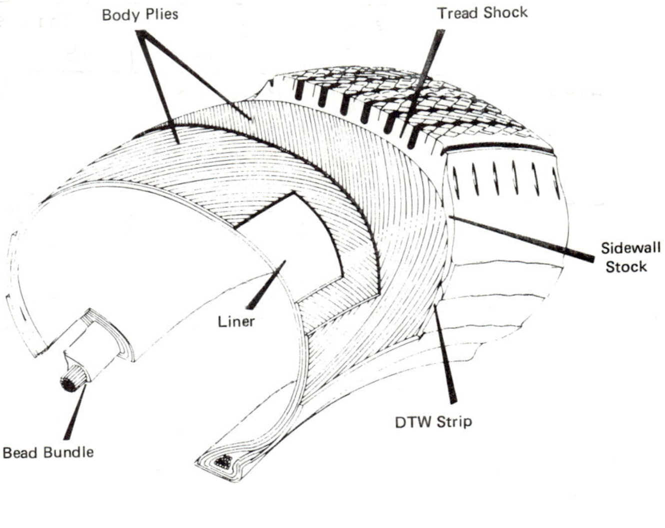

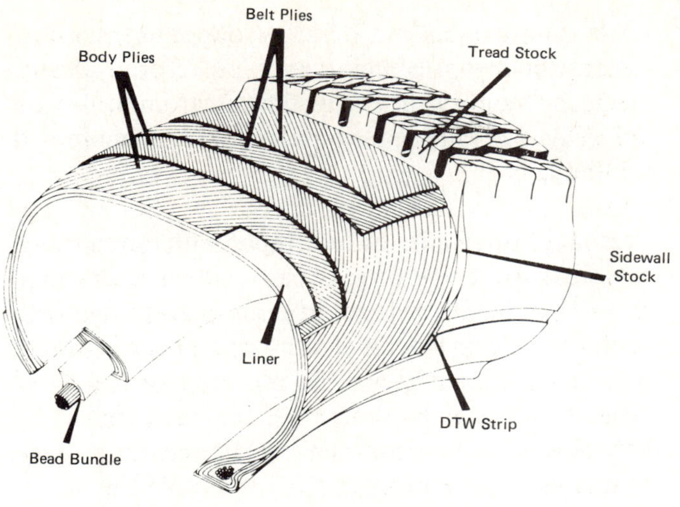

FIGURE 1-26 Bias belted tire construction. (Courtesy of Uniroyal Incorporated)

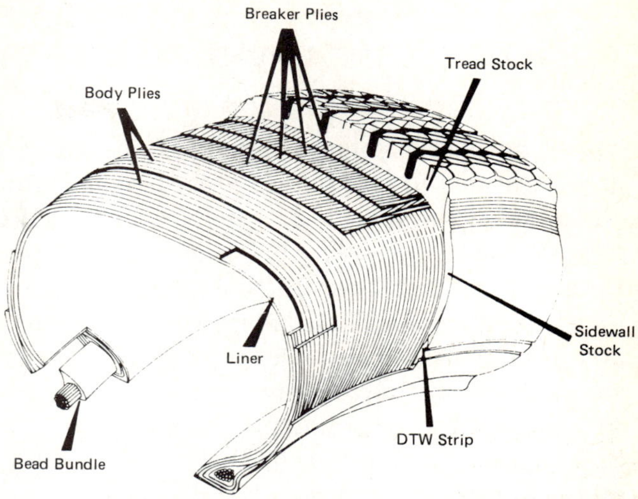

FIGURE 1-27 Radial ply tire construction. (Courtesy of Uniroyal Incorporated)

is made much like the bias ply tire, with strengthening belts added in the tread area. These belts are wrapped around the tire underneath the tread rubber. The cords in the belts run at an angle of 20 to 35 degrees to the centerline of the tire tread. The belts add to the strength and rigidity of the tire in the tread area. This improves tire tread mileage.

As illustrated in Figure 1-27, the radial ply tire differs in construction from the bias ply tire. The cord material in the body plies runs from bead to bead. It intersects the centerline of the tread at approximately 90 degrees. There is no criss-crossing cord material to strengthen the carcass of the radial ply tire. Strength is added to the tread area with breaker plies or belts. The cords of these belts run at a low angle, usually 10 to 30 degrees to the centerline. The type of tire to be used can only be properly selected when the vehicle requirements are known.

Tire Specifications. Tire manufacturers use a series of numbers and letters to identify tire size, load capacity, inflation limits, and type of service for which the tire is intended. The table in Figure 1-28 shows the popular tire sizes.

FIGURE 1-28 Load-carrying capacities of a number of popular tire sizes.

Tire load range table

Bias and bias-belted series 78	70	60	Metric series	Radial series 78	70	Cold inflation pressure (psi) 24	26	28	30	32 (Load range B)	36 (Load range C)	40 (Load range D)
B78-13	B70-13		P175/75R13	BR 78-13	BR 70-13	980	1030	1070	1110	1150	1230	1390
D78-14	D70-14			DR 78-14	DR 70-14	1120	1170	1220	1270	1320	1410	1490
			P185/75R14			1160	1210	1260	1310	1360	1450	1540
E78-14	E70-14			ER 78-14	ER 70-14	1190	1240	1300	1350	1400	1490	1580
			P195/75R14			1270	1330	1390	1440	1500	1600	1690
F78-14	F70-14			FR 78-14	FR 70-14	1280	1340	1400	1450	1500	1610	1700
G78-14	G70-14		P205/70R14	GR 78-14	GR 70-14	1380	1440	1500	1560	1620	1730	1830
F78-15	F70-15	F60-15		FR 78-15	FR 70-15	1280	1340	1400	1450	1500	1610	1700
			P205/75R15			1370	1430	1490	1550	1610	1720	1820
G78-15	G70-15	G60-15		GR 78-15	GR 70-15	1380	1440	1500	1560	1620	1730	1830
			P215/75R15			1480	1550	1620	1680	1740	1860	1970
H78-15	H70-15			HR 78-15	HR 70-15	1510	1580	1650	1770	1770	1890	2010

Maximum pound load limit per tire at the cold inflation pressure

1-7 SAFETY IN THE SERVICE AREA

Safety is always of primary importance in the work area. Learning proper service procedures must include the development of safe working habits. Properly developed safety habits are not time consuming, as some believe. If they are used regularly they will save time. These habits will not only help you as an automotive service technician, but they will reflect positively on coworkers as well.

Some unsafe working practices may not only be dangerous to you, but they may be a danger to others working around you. Vehicle owners prefer to have repairs done by a service technician who works safely and is careful in performing vehicle service. An automotive service technician who is careless about personal safety will generally be careless about service procedures and the protection of the vehicle being serviced. Vehicle protection in the work area includes the use of fender and seat covers, as well as the proper use of safety equipment.

Safety takes on many forms in the service area. All have the same goal: preventing harm to you, the equipment, and others in the service area. Using safety stands, as shown in Figure 1-29, will protect you while you work under a vehicle. Connecting the shop exhaust hose to the engine tail pipe before running an engine removes the engine exhaust from the building. This safety precaution affects everyone in the building.

FIGURE 1-29 Safety stand under a front suspension.

Personal Protection. Safety begins with the clothing you wear while working. Excessively loose clothing should not be worn. It could catch in rotating parts of the vehicle or machinery and cause injury. Good work shoes should be worn to protect your feet from objects on the floor that could puncture thin rubber soles or fabric shoes. Heavy shoes offer protection from spills and from parts you might drop on them. The best foot protection is offered by work shoes with a steel reinforcement in the toe area.

Eye protection is of primary importance. *Always wear safety glasses when you are in the work area.* Although you may not be doing a job that would normally cause eye injury, the person in the next work area may cause a small object to fly toward you. Always purchase the best eye protection possible; it will usually be more comfortable to wear.

Head protection is recommended whenever work is being done under the vehicle or under the hood. A hard hat will give good head protection. Even if you do not wear a hard hat at all times, one should be within reach when you work under a vehicle. Anyone who has serviced vehicles raised on hoists will be able to describe the need for hard hats to prevent head injury.

Breathing protection is required when working in a dusty area. For example, in the brake service area the dust may contain asbestos. It can get into the air when you remove the brake drums, clean brake parts, and fit brake shoes. Since asbestos is harmful to your health, brake dust should never be blown free with an air hose. It will enter the air and travel throughout the shop to affect everyone. The best method of reducing airborne dust is to loosen brake contamination with a brush or scraper and remove it with a vacuum cleaner. Remaining dust can be removed with a damp cloth.

Gloves should be included in your toolbox. They are used when removing parts of the vehicle that are hot or have rough metal edges or surfaces.

Remember, injuries are not only uncomfortable and time consuming, but they can also disfigure and cause permanent injury. The time spent putting on protective clothing and safety glasses is an investment in your future. Its rewards are always greater than the investment.

Safety Through Use of Proper Equipment. Safety can even reflect back to the type of equipment chosen to do a service operation. The condition of the equipment, its use, its cleanliness, and its maintenance all contribute to safety. In addition, shop training and management have an influence on safety.

The equipment selected should always be the best suited for the job, not just the most readily available. Time taken to locate the correct tool or piece of equipment will usually result in saving time to complete the job safely. The condition of the tools and equipment is just as important as proper selection. Screwdrivers with broken blades, for example, should be replaced immediately. Screwdrivers are not intended to be pry bars or chisels. They are rarely broken when used properly.

Cleanliness plays an important part in safety. A clean tool provides a good grip, prevents slippage, and reduces the chance of injury. In addition to cleaning, the tool itself should be properly maintained.

Training in the correct use of tools and equipment is important. Service technicians have been injured because they failed to get instruction on the correct use of service equipment. Instruction on proper use applies to small hand tools as well as complex shop equipment.

Chisels are dangerous if they are not properly maintained or if they are not used correctly. Chisel holders, similar to the one shown in Figure 1-30, should be used with hand chisels. Chisels will not cut parts that are harder than the chisel itself. This includes hard steel bearings. Eye protection *must* always be worn when using a chisel.

Correct use of drills is important to safety in the service area. Make sure the drill bit is properly sharpened. It must be tight in the drill chuck before drilling. Always keep a firm grip on the drill motor to hold it from twisting when the drill bit suddenly breaks through the metal. Whenever possible, use drill handles to increase your holding leverage. When using a drill press, always clamp the part in a vise or hold it with a C-clamp. Be certain the vise is securely fastened to the drill press table.

When using a bench grinder, always keep the protective shields in place. Wear eye protection and keep a firm grip on the part while you are grinding. If the piece is small, clamp it firmly in locking pliers to hold it while grinding. Never grind on the side of the grinding wheel. This will cause unnatural loading which could cause the grinding wheel to fly apart.

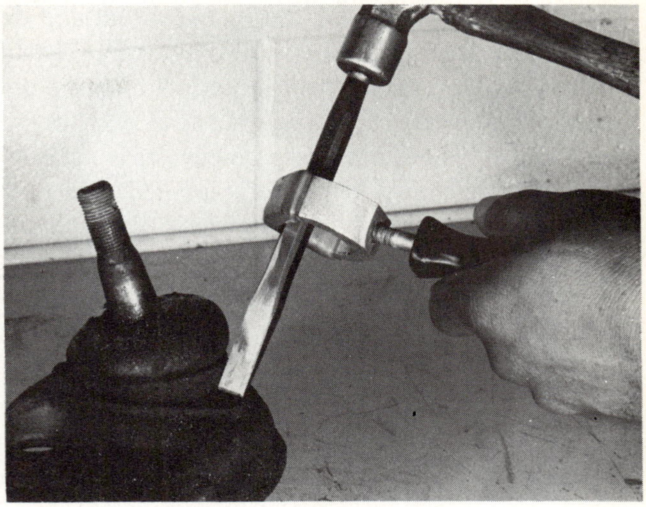

FIGURE 1-30 Chisel holder in use.

The proper use of tools, equipment, and safety precautions will reduce the chance of personal injury. Whenever you use a tool or piece of equipment, ask yourself how it should be used to prevent damage or injury. Then, adjust your position or your work methods so that safe working conditions result.

1-8 HAZARDOUS WASTE

Handling of hazardous waste has become a new concern in the automotive service shop. A new hazardous waste law came into effect in August 1985. It requires all companies that generate (produce) small quantities of hazardous waste to handle it the same way large companies do. This law applies to companies that generate from 220 to 2200 pounds (100 to 1000 kg) of hazardous waste a month. This is about one-half to five 55-gallon drums. Most automotive service shops generate this much waste in a month.

Automotive repair shops generate most of their hazardous waste in spent (used and dirty) painting and cleaning chemicals. These include used solvents, thinners, and reducers. Used rust removal and degreasing acids and alkalies are also hazardous waste. Under this law, these spent materials must be disposed of properly. It is required that each of the waste materials be kept in separate waste containers. The containers must be properly labeled to identify the waste material and the shop where the waste was generated. The waste must be accompanied by a manifest. This is an Environmental Protection Agency form that must be filled out and a copy kept by the shop. The waste must be carried to a hazardous waste disposal facility by a transporter that is properly licensed.

It is the responsibility of the shop to dispose of the waste properly. As an automotive service technician, you must be aware of hazardous waste disposal. You can do your part by placing the spent waste material in the proper storage container.

STUDY QUESTIONS

Section 1-1

1. How does frame flexibility affect the vehicle?
2. Name three types of body and frame construction.

Section 1-2

3. Why are brakes discussed in a book on suspension and steering?

Section 1-3

4. Why is the drive line discussed in a book on suspension and steering?

Section 1-4

5. What parts are included in suspension systems?
6. What is the most important job of the suspension?
7. Name four types of suspension springs.
8. How does a dependent suspension differ from an independent suspension?
9. What is the advantage of a strut-type front suspension?
10. What is the advantage of a twin I-beam front suspension?
11. Name three types of rear suspensions used with front-wheel-drive vehicles.

Section 1-5

12. How does the steering knuckle used with front-wheel drive differ from the knuckle used with rear-wheel drive?
13. Name two types of steering gears.
14. What controls can be found on the steering column?

Section 1-6

15. What three important functions do the tires perform?
16. What are the three basic types of construction used in passenger car tires?
17. What do belts do for a tire?

Section 1-7

18. Why is it important to develop safe working habits?
19. What should you wear to work safely in the auto shop?
20. How should dust be cleaned up to minimize the amount of airborne dust?

21. Give two examples of how tool selection can improve shop safety.

22. Give two examples of how tools should be used to improve shop safety.

Section 1-8

23. Who is responsible for disposing of hazardous waste?

24. How can you help to properly dispose of hazardous waste?

2

Routine Maintenance, Lubricants, and Fluids

Three things are needed for proper automotive service. First, you must have the tools to make the repair. Second, the parts must be available. Third, you must know what is to be done and how to do it. The following chapters are designed to help you master the knowledge and skill necessary to service suspension and steering systems and to identify and correct problems in related systems.

2-1 AUTOMOTIVE SERVICE

There are only three reasons to service a vehicle: (1) when preventive maintenance is due, (2) when the owner wants a change in performance, appearance, or comfort items, and (3) when there is a specific problem to be corrected.

Preventive maintenance is done following a time and/or mileage periodic maintenance schedule. It is done to extend the life of a vehicle and to prevent problems on the road. Typical examples of preventive maintenance include lubrication, brake inspection, wheel alignment, and tire rotation. Preventive maintenance is routine when standard servicing procedures are used.

Changes in economy, performance, appearance, and comfort usually involve installing manufactured parts such as air-leveling shock absorbers or special wheels and tires. Parts may need to be replaced or added to make a performance change. This type of service could include a change in tire and wheel assemblies, springs, or stabilizer bars. Here again, the service procedures are specified. You know exactly what the customer wants and you have specific instructions telling you how it is to be done.

In most vehicle service work, you will repair or replace the part causing a problem. This kind of work starts with a clear understanding of the customer's complaint. The customer wants three things: (1) to have the vehicle repaired so it will operate correctly, (2) to have it repaired at minimum cost, and (3) to have it done as soon as possible.

In general, problems can be corrected by straightforward service procedures. The first step is to verify the complaint so you correctly understand the problem. Check the system to identify the location of the problem. The system is then tested to pinpoint the cause of the problem. Required repairs are made and the system is rechecked to make sure the problem has been corrected. In some cases, the cause of the problem is extremely difficult to pinpoint. When this happens, you must have a detailed knowledge of the part's function and how the parts work together. This, along with the ability to use test equipment, will allow you to rapidly and accurately correct vehicle problems.

An operator will get maximum service life from a vehicle when the vehicle is given scheduled periodic maintenance. This service includes lubrication of some moving parts, fluid-level inspections, fluid and oil changes, and an overall visual inspection. Properly performed periodic maintenance and inspection will result in reduced total operating costs to the vehicle owner during the useful life of the vehicle. Preventive maintenance will extend the service life of each part or system. Periodic inspections will reveal worn or otherwise faulty parts before they cause a problem, become dangerous, or damage other parts. As an automotive service technician, you must not only know how each part or system works, but you must also know the correct service procedures.

Periodic lubrication is the most common preventive maintenance service operation. It is important to select the proper lubricant for each part or assembly. The wrong lubricant can shorten the service life of a part as rapidly as no lubrication at all.

2-2 LUBRICANTS

Lubrication increases the service life of parts by reducing friction, heat, and wear. In some cases, the lubricant keeps contaminants from reaching the lubricated surfaces. Many different types of lubricants are used in vehicles. Each has a specific purpose and special characteristics. If a lubricant is used in the wrong place or in the wrong quantity, it will not give the required protection.

Common lubricants used in modern vehicles include engine oil, gear lubricants, and greases. Some lubricants also transfer motion and force. These are used in the brakes, power steering, and automatic transmission.

Other special purpose lubricants are used around the shop. Some examples are penetrating oils, silicone lubricants, and antisieze lubricants that make disassembly and reassembly of vehicle parts easier.

Gear Lubricants. Gear oils are special lubricants used where there is high friction and high pressure between parts. Gear oils are used in manual transmissions and rear-drive axles. These gear oils are specified by API-GL (American Petroleum Institute-Gear Lubricants) service classification and an SAE viscosity number. Some manual transmissions and many transaxles use engine oil or automatic transmission fluid as a lubricant. Always consult the appropriate shop manual to determine the type recommended. Do not mix lubricant types.

A limited slip differential requires a lubricant with special properties. It has the same API-GL and SAE numbers as conventional gear oils plus additives to make it compatible with the limited slip differential clutches. Gear oil for limited slip differentials, although more expensive, will work satisfactorily in conventional differentials and in manual transmissions requiring the same gear oil numbers.

Greases. Grease is used as a lubricant in suspension and steering parts. Lubricating grease is solid or semisolid. It is made by combining a thickening agent and a liquid lubricant. Lubricating greases are classified by the NLGI (National Lubricating Grease Institute) number, the type of thickener used, and the texture. The NLGI number describes the relative hardness or consistency of the grease. This is the most important property of a lubricating grease. The type of thickener (aluminum, barium, calcium, sodium, or lithium) and the amount used in the grease determines its general properties. These properties include melting point, texture (smooth, fibrous, stringy, or long fiber), and water resistance. Additives are put in the grease to improve some of its properties. Fillers are also added to grease to improve the load-carrying ability. Some of the more commonly used fillers are graphite and molybdenum disulfide, usually called *moly*.

The most common type of grease used in automotive service has a *lithium* base. Lithium grease has a high melting temperature. This makes it usable as a wheel bearing lubricant. It also has excellent water resistance, making it a good chassis lubricant. As a result of its ability to provide satisfactory lubrication for many applications, it is called a *multipurpose grease*. Moly grease is usually lithium grease with molybdenum disulfide used as a filler.

In automotive applications, NLGI #2 multipurpose grease is recommended for wheel bearing and chassis lubrication. Often, a soft chassis grease is pumped through overhead grease systems in a lubrication bay. This grease is not NLGI #2. It should not be used for wheel bearing applications where NLGI #2 grease is recommended. If the NLGI number or the filler type of a grease is not known, or if it is not labeled for wheel bearing use, do not use it to repack wheel bearings. Grease is packaged specifically for wheel bearing lubrication. Wheel bearing greases may be barium, sodium, or lithium based. These greases do not intermix; therefore, grease should never be added to the used grease remaining in the wheel bearings. All of the used grease must be thoroughly cleaned from the bearings before they are repacked with fresh wheel bearing grease.

2-3 FLUIDS

The fluids described in this section are liquids used to transfer motion or force as well as to lubricate brake and steering parts. Depending on their application, additional properties such as the boiling point, effect on seals, and effect on other parts in the operational system may be important.

Power Steering Fluid. Power steering fluid, although similar to automatic transmission fluid, is not the same. Automatic transmission fluid is colored red to identify leaks from the transmission. Power steering fluid has no coloring added. Fluids used in power steering systems do not need dynamic friction properties like those used in automatic transmissions. Power steering fluids need high-temperature oxidation properties and antifoaming properties. The two types of fluids should not be used interchangeably. Power steering systems are designed to operate with fluids specifically made for their use.

Brake Fluid. Brake fluid is one of the brake system's most important elements. It must meet government and industry quality standards. Brake fluid should be handled with care and stored properly so it will remain a high-quality product.

Brake fluid is a nonpetroleum, glycol-based fluid that is compatible with the rubber and metal parts in the brake system. Many requirements are placed on brake fluids by DOT (Department of Transportation). These include minimum boiling point, viscosity, and effect of the fluid on the rubber parts. Brake fluid is *hygroscopic*, which means it will absorb moisture. It will even pick up moisture from the atmosphere if the fluid can or brake fluid reservoir is left open. Moisture in brake fluid will lower the fluid boiling point and cause corrosion in the system.

DOT3 brake fluid is recommended for most brake systems. Where high brake temperatures are expected, some manufacturers recommend high boiling temperature DOT4 brake fluid.

Silicone-based brake fluid DOT5 is used in some vehicles (Figure 2-1). This fluid has a number of advantages over the glycol-based fluid and it has a higher price. The boiling temperature of silicone brake fluid is considerably higher than the best glycol-based fluid and, for this reason, the silicone fluid has gained popularity. The silicone fluid retains its viscosity at extremely low temperatures. This makes it an excellent fluid at extreme temperatures. Silicone brake fluid has little tendency to absorb moisture. As an added advantage, the fluid does not attack painted surfaces as glycol-based fluids will. General Motors cautions against using silicone brake fluid because it tends to cause excessive wear in the brake cylinders.

Always observe all of the following handling and storage precautions when working with brake fluids:

1. Store brake fluid in its original container to be sure it is properly identified.

FIGURE 2-1 Standard DOT3 and Silicone DOT5 brake fluid containers.

2. The container should always remain tightly capped and free from dirt to prevent contamination.
3. Never re-use brake fluid that has been used for bleeding brakes. Always pour it in the proper waste container when the brake repair is completed.
4. Brake fluid will damage painted surfaces. If it is accidentally spilled, immediately flush it off with water.
5. Never use petroleum products, such as naphtha, to clean brake system parts.

Cleanliness practiced when working on brake systems will always extend the service life of the brake repair being made.

2-4 LIFTING AND SUPPORTING VEHICLES

You will need to lift and support the vehicle so the suspension is free when you do service work on the suspension and steering system. Whenever you raise the vehicle with a jack for service work, **you must support it with safety stands before you work beneath it**. Technicians have been severely injured when a vehicle fell that was not properly supported. Be careful not to jack or place safety stands on half-shafts going from the transaxle to the wheel.

Lifting devices range from the simple one-post emergency jack that is furnished with all new vehicles to the heavy-duty, in-floor, twin-post hoists used in the service area. Each is designed for a specific purpose. Do not use them for purposes for which they are not made.

Safety Stands. Safety stands or jack stands are small supports placed under a raised vehicle as a safety measure. Safety stands, such as the one shown in Figure 1-29, are available in a number of different designs. They each have a different load-carrying capacity and extension length, a large base, and a positive lock. Safety stands are designed to support the vehicle after the jack has been removed.

Hydraulic Jack. The hydraulic extension jack, as shown in Figure 2-2, is used in the service area. It is sometimes carried in the vehicle. This type of jack is handy because of its small size; it can be placed directly under the suspension part to be lifted. As a result, you have to reach under the vehicle to position and operate the jack. The wheel will lift as the jack begins to extend. When using some other types of jacks, the body must be lifted far enough to fully extend the shock absorber before the wheel will lift. The hydraulic extension jack is stable, but you should still use a safety stand when you need to work under the vehicle.

FIGURE 2-2 Hydraulic extension jack under a front suspension.

Hydraulic Bumper Jack. A second type of jack, one that is common in service stations, is the pneumatic, or hydraulic, bumper jack shown in Figure 2-3. It is used to lift the entire end of the vehicle far enough to raise the wheels from the floor. This type of jack should not be used on some vehicles because it will twist the body so much that the glass will break. A hydraulic bumper jack is often used when servicing tires. It has a relatively large base to keep it from tipping. There is still danger that the vehicle could slip from the jack or that the bumper support could fail. For this reason, make sure you always place safety stands under the vehicle when you need to work under it. The bumper jack will not move on the floor when it is supporting weight. It should only be used on a level floor surface so the vehicle can move toward the jack as it is being raised.

FIGURE 2-3 Hydraulic bumper jack.

Hydraulic Floor Jack. One of the most popular service jacks is the hydraulic floor jack shown in Figure 2-4. This type of jack is used on all types of vehicles. It comes in different sizes, depending on the lifting capacity of the jack. The hydraulic floor jack is on wheels and has a large base to give stable support. You can position and operate the jack using the long handle without having to reach under the vehicle. The wheels of the jack allow it to move toward the vehicle as it is raised. This is necessary because the jack lifting pad moves in an arc as the jack is raised. The hydraulic floor jack can be used to lift a single wheel or half of the vehicle. When lifting from the side, the suspension will fully extend for tire removal. Some tire stores use four hydraulic floor jacks to raise all four wheels when replacing a full set of tires. It is important to position the floor jack only at reinforced lifting points under the vehicle. Even with this heavy jack, safety stands should be used for support whenever you have to go underneath the vehicle to make repairs.

FIGURE 2-4 Hydraulic floor jack.

Hoists. In the modern service area, vehicles are most often raised with a hoist. It will lift the vehicle quickly and safely, high enough so you can walk under it. A typical hoist is pictured in Figure 2-5. Hoists may lift on the tires, the rear tires and the front suspension, rear suspension and front suspension, or on the frame. There are advantages to each type. The type of service work to be done determines the best type of hoist to use. Hoists usually have lifting posts beneath the floor, although above-ground hoists are becoming more popular.

Below-the-floor hoists have one or two hydraulically operated cylinders. Shop air is used to pressurize the hydraulic oil. This is called *air over oil hydraulic operation*. The air pressure forces the

FIGURE 2-5 Dual-post hoist supporting an entire vehicle.

oil into the hoist cylinders for lifting the vehicle. Above-ground hoists usually operate electrically. An electric motor may drive a hydraulic pump. All single-post hoists have a mechanical safety lock. The lock engages when the hoist is raised to its maximum height. This keeps the hoist from falling if pressure is lost. Due to their design, dual-post hoists do not require a mechanical lock. One post of dual-post hoists will move toward or away from the other post. This allows the post to be moved to match the wheelbase of the vehicle being lifted.

The vehicle must be positioned over the center of the hoist, side to side. When using a frame contact hoist, the front-to-rear balance point should be close to the lifting post. Always position the lifting arms of the hoist as far apart as possible for the greatest stability. Shop manuals indicate the location of the reinforced frame contact points you should use with the frame contact hoist. Vehicles placed improperly on a hoist have been known to fall off, doing extensive damage to the vehicle, shop, and equipment.

2-5 TIRE AND WHEEL INSPECTION

Frequent inspection of the tires and wheels on vehicles will provide the vehicle operator with greater safety, reliability, and tire mileage. Abnormal conditions can be corrected before they become a problem. At the same time, the technician has the satisfaction of knowing that the customers are operating safe vehicles. Inspection is often done as routine preventive maintenance while the vehicle is raised on a hoist for lubrication or other service. Abnormal tire wear is one of the first indications of a tire or wheel fault. Because all of the braking and handling reactions pass through the tire footprint, the condition of the tires affects steering response and handling. An inspection may also be the result of a customer complaint about the ride or handling of the vehicle. A normal preventive maintenance inspection is a visual inspection and adjustment of the tire air pressure. Diagnosis of a customer complaint also starts with a visual inspection. Diagnosis often leads to the use of measuring tools that help determine if the tire is the source of the problem. If the problem is tire related, the service technician has to find which one of the tires is at fault.

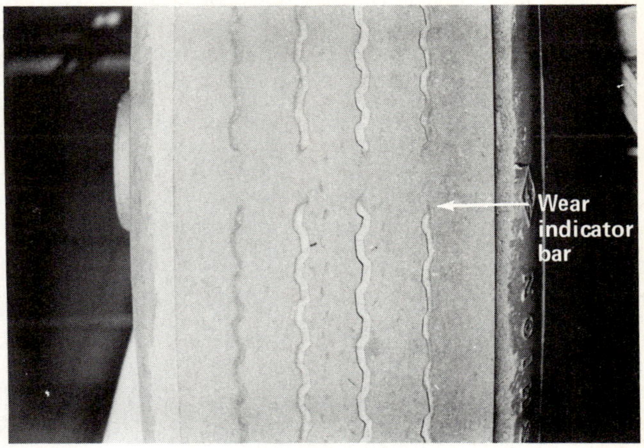

FIGURE 2-6 Tire worn to the tread wear indicator bar.

Poor driving habits, improper maintenance, mechanical problems, or a possible tire fault often show up as abnormal tread wear patterns. Frequent inspection will give the technician an opportunity to find and correct a fault before it becomes a problem so serious that the tire must be replaced. Figures 5-2 through 5-5 in Chapter 5 show abnormal tire tread wear conditions and the faults that usually cause them. Tires with tread worn to the wear indicator bars (Figure 2-6) must be replaced for safe operation.

2-6 TIRE MAINTENANCE

Tire maintenance consists of visual inspection, checking inflation, tire rotation, and wheel balance. Tires should be given a visual inspection at required intervals. Tire inflation is checked and air is added as needed. Tire rotation and balance will be covered in detail later in the text.

Tire Inflation. Tire manufacturers determine recommended inflation pressure for a tire by measuring the tread temperature immediately after the vehicle stops. The tire should have the same temperature across the tread. Tire pressure is based on a pressure that will provide the vehicle with satisfactory ride, fuel economy, tire life, and handling under normal driving conditions. The pressure is marked on a decal on the driver's door or on the door frame.

The tire size is selected to carry the expected load with normal tire pressure. Recommended tire pressures are determined by the tire and vehicle design. Different tire pressures may be recommended for different vehicle loads. It may also be different if the vehicle is driven under other-than-normal conditions. Maximum inflation pressure should never exceed the pressure marked on the tire. It should never be less than the minimum pressure recommended by the vehicle manufacturer. Operating a tire at less than the minimum pressure can cause excessive flexing. This produces heat within the tire that causes tire failure. Overinflation can cause abnormal tire wear and possible ply separation that leads to tire failure.

Tire pressures should be adjusted to the normal pressure during seasonal temperature extremes. Air expands when it is heated as the tire rolls, so the pressure may increase as much as 6 psi when the tire is hot. The pressure should be checked when the tire is cold and the vehicle has been driven less than 1 mile after standing for over 3 hours. In addition, pressure should be checked and adjusted before taking long trips. Proper tire maintenance will add to the life of the tires, provide safety for the operator with normal handling and braking, and improve the vehicle's fuel economy.

Tire Rotation. Automobile manufacturers recommend that tires be periodically rotated to ensure even tire wear and maximum mileage from a set of tires. It is normal practice to rebalance the tire and wheel assembly when the tires are rotated. The recommended rotation period is usually 10,000 miles (16,000 km) or less. It usually occurs at an oil change interval. More even wear results from the fact that rear tires have different operating conditions than front tires, especially on front-wheel-drive vehicles. This is due to differences in suspension design, load on the tires, and driving and steering conditions. As a result, rotation tends to correct slight abnormal wear conditions that are starting on the tire tread.

The recommended tire rotation sequence differs among manufacturers. All manufacturers recommend that bias ply tires be switched to the opposite side of the vehicle as they are exchanged between the front and rear. A radial tire is said to develop a "set" after operating on one side of the vehicle and rolling in one direction. Placing a tire on the opposite side of the vehicle will cause the tire to rotate in the opposite direction. Changing the radial tires from side to side on a rear-wheel-drive vehicle may adversely affect either tire life or handling. As front-wheel-drive vehicles have become more popular, some tire manufacturers have found that cross-rotation, like that used for bias ply tires, improves the wear and life of radial tires. The rotation sequence patterns are illustrated in Figure 2-7. Bias belted tires are rotated in the same sequence as bias ply tires.

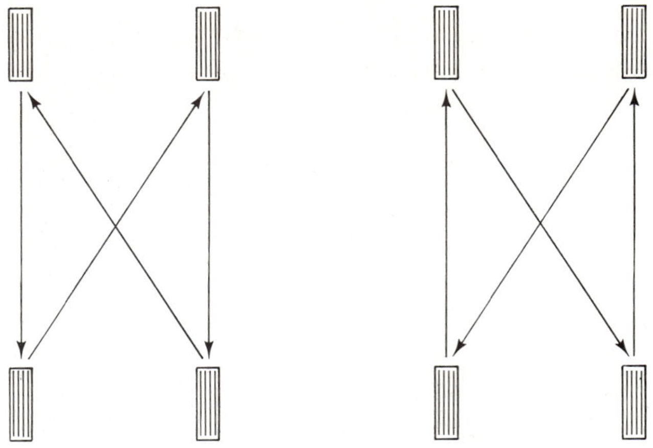

FIGURE 2-7 Two recommended tire rotation sequences.

Tires are serviced when there is a problem. Service includes checking radial and lateral runout, balancing, repairing, replacing, and, in some instances, truing or trimming tires. Always follow a logical sequence to analyze a tire problem.

2-7 CHASSIS LUBRICATION

Periodic inspection is done when the chassis is lubricated. The inspection will uncover faults that lead to steering and suspension problems. Chassis lubrication is based on both time and mileage intervals. These intervals vary, depending on the make of the vehicle and the operating conditions. They range from 1 year or 12,000 miles (19,300 km) to 3 years or 35,000 miles (56,300 km), whichever comes first—years or mileage. When operating under severe conditions, such as dusty air, stop-and-go driving, or fleet service operations, the vehicle

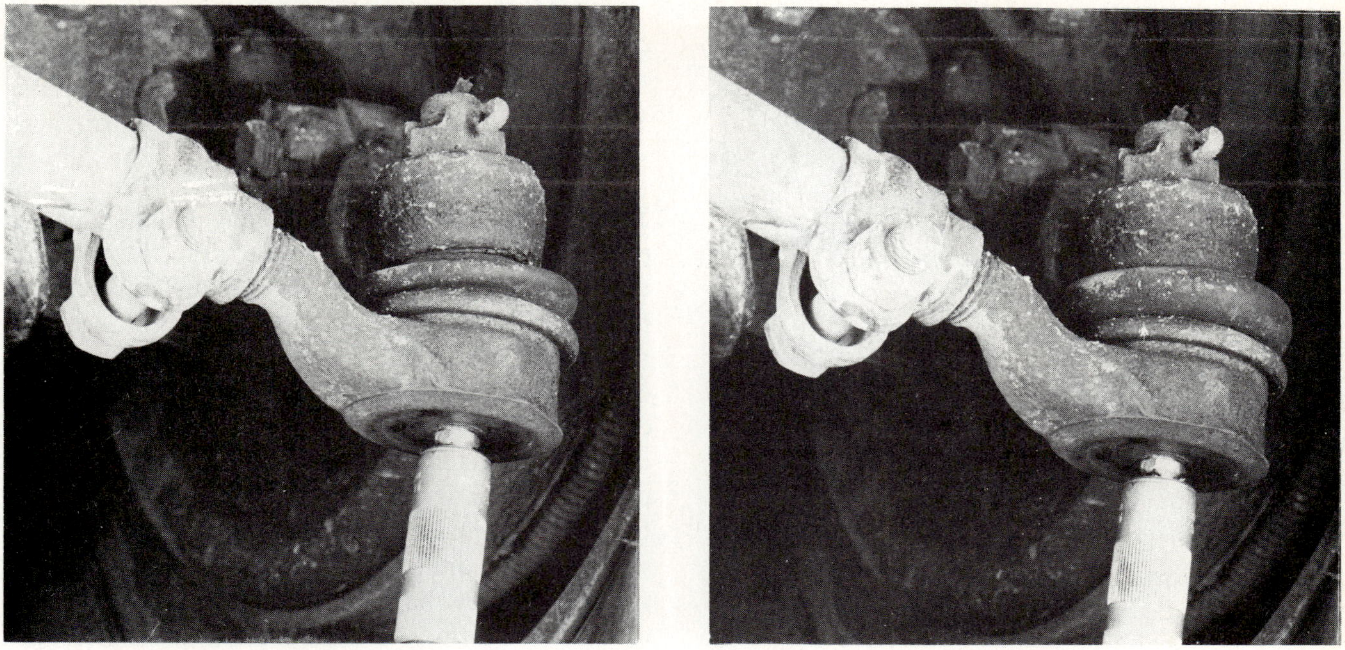

(a) (b)

FIGURE 2-8 Greasing a tie rod end: (a) collapsed boot seal; (b) inflated boot seal as new grease causes a slight swelling.

will require lubrication more often. Some suspension parts are sealed at the factory with enough lubrication for the life of the part. There is no way to add lubricant to these parts. Suspension and steering joints with rubber bushings are never lubricated.

The manufacturer cannot schedule exact chassis lubrication requirements. These requirements can only be determined by careful inspection of each part requiring periodic lubrication. The boot or seal around a ball joint or tie rod end will collapse when it needs lubrication. If the boot is slightly swollen or inflated, the boot seal is in good condition and it has enough grease. An example of this condition is shown in Figure 2-8. The ball joint or tie rod ends must have enough lubricant to hold the seal slightly swollen. This will effectively seal the joint against moisture and contamination.

Lubrication begins with raising and supporting the vehicle either with a hoist or on safety stands. While the vehicle is raised for lubrication, you have an excellent opportunity to perform other service operations. These include an inspection of all underbody parts, tires, suspension parts, steering linkages, brake lines and hoses, and fluid levels. A systemic approach to lubrication and inspection should be used. It will speed the inspection as well as eliminate any chance of overlooking a lubrication or inspection point.

Lubrication Procedure. The first step in chassis lubrication is to look at the lubrication handbook to make sure you use the correct lubricant. You also need to identify the location and number of lubrication fittings. There will be one of the following at each point in the suspension where relative motion occurs: a rubber bushing, a sealed joint, or a joint requiring lubrication. Lubrication fittings are usually found at each tie rod end, on both ends of the center steering link, and at the upper and lower ball joints (Figure 2-9). Other lubrication points are sometimes found at the universal joints and on the clutch linkage. Grease is also applied to the shift linkage and parking brake cables. Each lubrication fitting is wiped clean before it is lubricated. Carefully in-

FIGURE 2-9 Typical lubrication points.

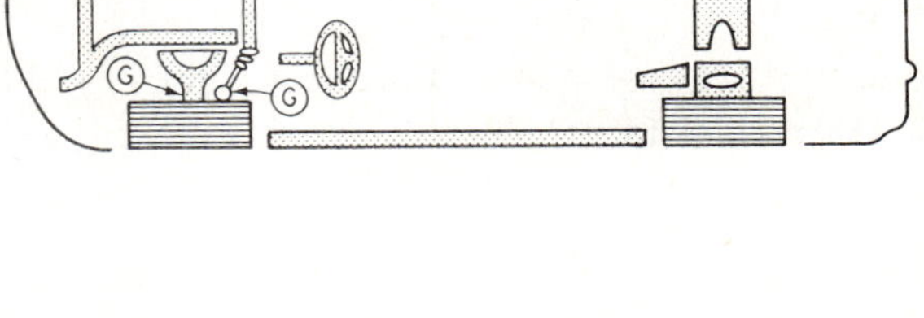

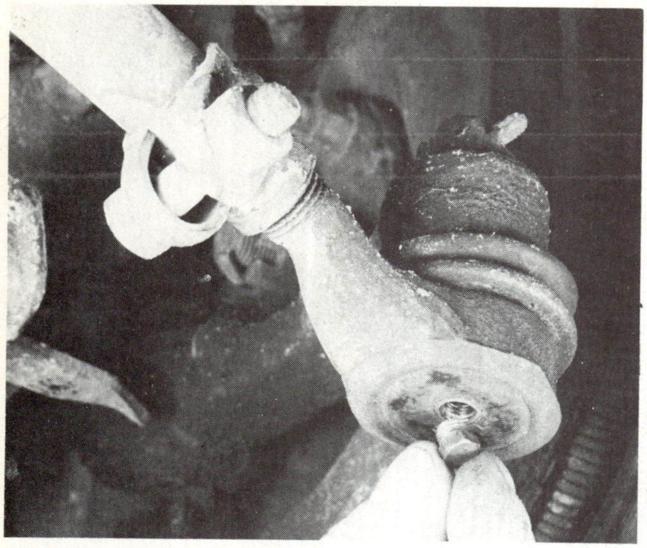

FIGURE 2-10 Removing a plug from a tie rod end.

spect the seal at each suspension joint. Some vehicles have a small plug located in place of the fitting (Figure 2-10). You may have to remove the plug and temporarily install a lubricating fitting to lubricate the joint (Figure 2-11). The plug may be reinstalled after lubrication. Each fitting is lubricated until the seal starts to balloon or grease begins to seep from around the base of the boot. Always wipe excess grease from the area. This procedure is continued until grease is put in all of the lubrication fittings. Watch for excessive movement of the parts each time you grease a joint. Movement indicates a worn joint.

FIGURE 2-11 Installing a grease fitting in place of the plug.

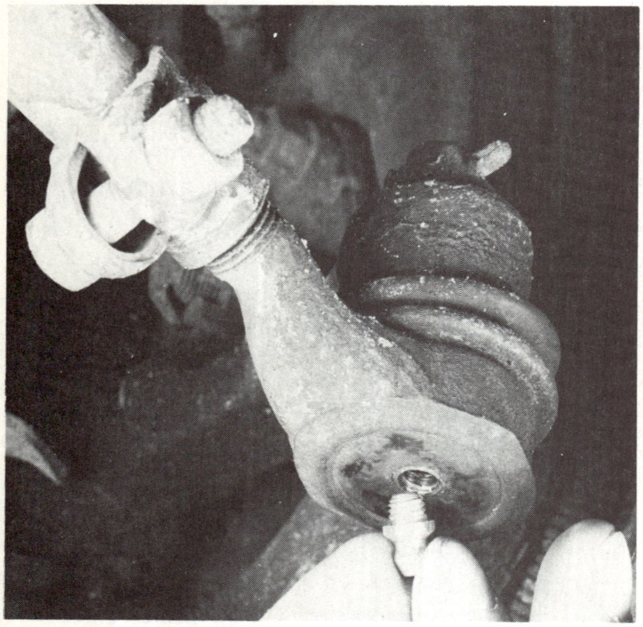

Maintenance Inspection. If an engine oil change is part of the periodic service, an inspection can be done while the oil is draining. The differential and manual transmission oil levels should be checked and refilled as required. Make a visual inspection of the exhaust system and rear suspension. Any leakage, or any loose or broken parts, should be noted. The tires are inspected for wear, damage, and inflation pressures. Any abnormal wear should be noted and the cause for the wear determined. The front suspension bushings are inspected. Look for a condition where the rubber has relaxed so the pivot is allowed to move off center as shown in Figure 2-12. The engine mounts are inspected for signs of fatigue or failure. Check the condition of the brake lines, brake hoses, and parking brake cables and linkages. If the oil has been changed, the drain plug and filter are installed.

The vehicle is lowered and the lubrication is completed under the hood and on the body. Oil should be put in the engine and the engine started to establish oil pressure; then check for leaks. The coolant and battery fluid levels are inspected. Fluid is added if it is needed. Check the fluid level in the master cylinder. Low fluid indicates an external leak. Leaks must be found and repaired. All hoses are in-

FIGURE 2-12 Relaxed and deteriorated suspension bushing.

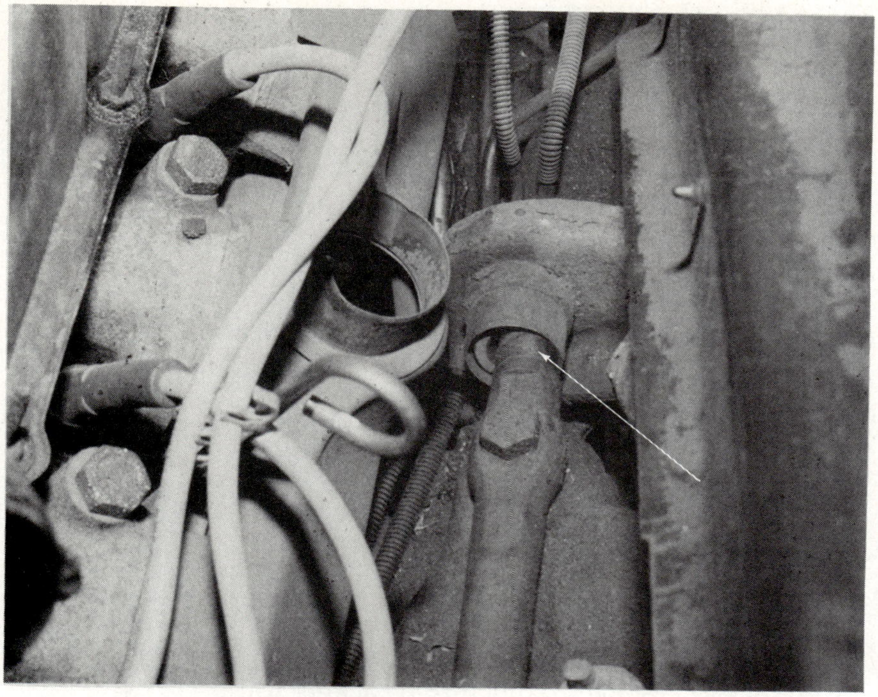

FIGURE 2-13 Failed upper suspension bushing seen from under the hood.

spected for signs of deterioration. The upper suspension bushing condition is usually inspected from under the hood. Figure 2-13 shows a failed upper suspension bushing.

The accessory drive belts are inspected for looseness, wear, and cracks. The automatic transmission fluid and power steering fluid levels are checked and adjusted as necessary. White grease is applied sparingly to hood and deck latches, and stick grease is applied to the door latches. Put a few drops of engine oil on the hinge pins. Lubricant should never be used on any carburetor linkage, rubber bushing, rear springs, sealed bearings, or drive belts.

Mark the lubrication sticker and place it on the door jamb of the vehicle. Safety inspections should also be made on the lights and windshield wiper/washer.

2-8 TOWING THE VEHICLE

When a vehicle cannot be driven for servicing, it must be towed. Professional towing companies and many service stations are equipped to tow vehicles.

The faulty end of the disabled vehicle has to be raised from the road surface to prevent further damage. One concern when towing a vehicle is the drive line. If the drive wheels can be lifted, towing will be no problem. Precautions will have to be taken if the vehicle must be towed with the drive wheels on the ground. When the drive wheels rotate, they will rotate the axles, differential, drive shaft or half-shafts, and the transmission main shaft. The axle, differential, and drive shaft or half-shafts work in the normal manner. The transmission does not. Both manual and automatic transmissions are not properly lubricated when only the main shaft turns. Severe transmission damage may result if the vehicle is towed for any distance without normal lubrication. The vehicle may be towed only for a short distance at a low speed when the drive wheels are on the ground.

The alternative is to have the tow rig operator place the drive wheels on a dolly. The dolly carries the vehicle wheels off the ground. The vehicle can be moved, for any distance, without damage to the transmission when the wheels are cradled on the dolly. If dolly wheels are not available, the shaft to the drive wheels can be disconnected. This will allow the vehicle to be towed with the drive wheels on the road with no transmission damage. Professional tow rigs have large rubber belts that prevent sheet metal damage as the vehicle is raised. Do not attempt to tow vehicles without the proper equipment to avoid doing more damage.

STUDY QUESTIONS

Introduction

1. What three things are needed for proper automotive service?

Section 2-1

2. For what three reasons does a vehicle need service?
3. Why is preventive maintenance done on a vehicle?
4. What three things does a customer want when the vehicle has a problem?
5. What steps are taken to correct a problem?
6. As a service technician, what must you know to correct a problem?

Section 2-2

7. What could be the result of using the wrong lubricant?
8. What is used to identify gear oils?
9. What is used to identify greases?
10. When is a grease a multipurpose grease?
11. Why should the old grease always be removed before repacking wheel bearings?

Section 2-3

12. How does power steering fluid differ from automatic transmission fluid?
13. How does brake fluid differ from power steering fluid?
14. What is a hygroscopic fluid?
15. How does DOT5 brake fluid differ from DOT3 brake fluid?
16. What are the advantages and disadvantages of DOT5 fluid?
17. What precautions should be taken with brake fluid?

Section 2-4

18. What are the advantages of a hydraulic extension jack?
19. When should you NOT use a bumper jack?
20. What is one of the most popular jacks?
21. What precautions should be taken when jacking a vehicle with a hydraulic floor jack?
22. What are the advantages of using a hoist to lift a vehicle?
23. What care should be taken when lifting a vehicle with a host?

Section 2-5

24. How can a preventive maintenance inspection improve tire mileage?

Section 2-6

25. What does tire maintenance include?
26. What would be the result of operating a tire with low air pressure?
27. What would be the result of operating a tire with too much air pressure?
28. Why are tires rotated to a different location on the vehicle?

Section 2-7

29. What is done at each lubrication period?
30. How can you tell if a suspension or steering joint has enough grease?
31. What suspension and steering inspections should you make when the oil is changed?

Section 2-8

32. What precautions should be taken when towing a vehicle with the drive wheels on the ground?

3

Bearings and Seals

Bearings are used in vehicles wherever there is motion between two parts. Bearings come in a number of shapes, sizes, and types. Bearings on nondriving axles are called *wheel bearings*. Bearings on driving axles are called *axle bearings*. In operation, the bearings must be properly supported, lubricated, and sealed. Seals are used to keep contaminants out and lubricants in the bearing. Seals must be replaced when they allow lubricant to get on the brake linings. When you remove steering and suspension parts, you often have to remove brake and hub parts along with disassembly of some bearings. To do the job properly, you, as a technician, must fully understand all parts of the system. This chapter presents bearings, bearing service, and seals commonly found when servicing steering and suspension systems.

3-1 BEARINGS

Bearings hold shafts and they minimize friction and wear where there is movement between two surfaces. Rolling contact bearings are used where there are high loads and relatively high rotating speeds. These bearings operate with low friction and little clearance while using a minimum amount of lubrication. There are three rolling contact bearing designs. They are described by the shape of the rolling element within the bearing. These are ball, straight roller, and tapered roller designs as shown in Figure 3-1.

Ball Bearings. Ball bearings are used in many places on vehicles. They are used because they can operate at high speed with minimum friction. The balls within the bearing assembly support the load. This load is called a *radial load*. A spot on each ball makes a theoretical point contact where it meets the race, as shown in Figure 3-2. This point contact can cause the ball bearing to fatigue early when it is overloaded. Ball bearings can support some (side) *axial loads* as well as radial loads. This is shown in Figure 3-3. They can do this because the balls roll in grooves called *raceways*. Ball bearing assemblies are easy to install in a vehicle because they are complete assemblies. They have been used as front-wheel bearings on rear-drive cars. They are used as axle bearings on some vehicles, although roller bearings are being used in new axle design. Double row ball bearings are used on the front axles (stub shafts) of many front-drive and four-wheel-drive cars (Figure 3-4). Ball bearings also are used within transmissions, in the steering gear, and in engine accessories.

Straight Roller Bearings. Straight roller bearings are used in a number of places in vehicles. Straight roller bearings can carry a greater radial load than ball bearings. Instead of a point contact, they make a line contact between the roller and the surface of the race. This is illustrated in Figure 3-5. The line contact has more rolling friction than the point contact of a ball bearing. As a result, roller bearings are limited to low- and moderate-speed applications.

FIGURE 3-1 Rolling contact bearings: from left to right–ball, roller, needle, and taper roller.

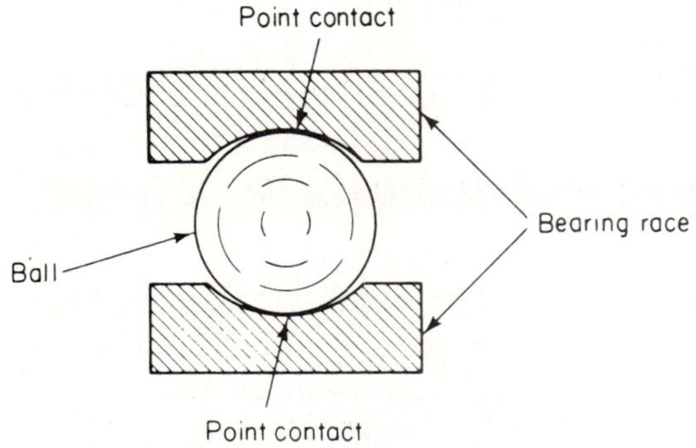

FIGURE 3-2 Ball bearing point contact.

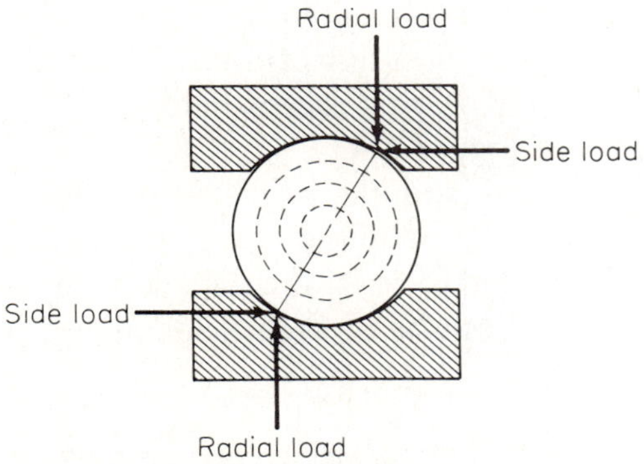

FIGURE 3-3 Ball bearings can support axial side loads and radial loads.

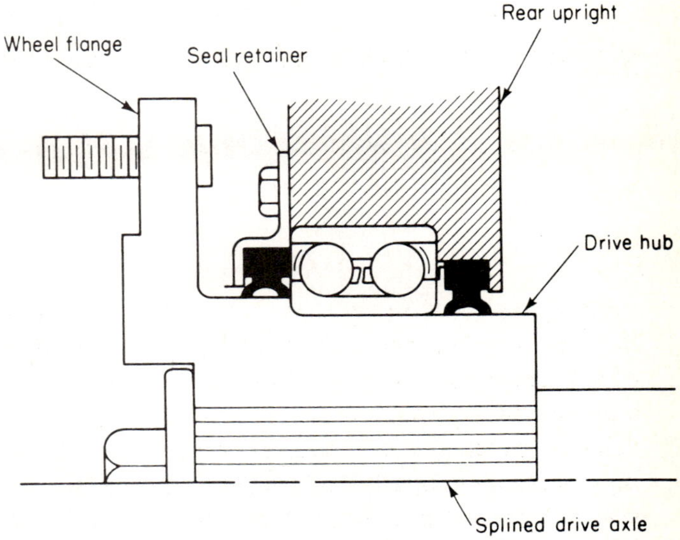

FIGURE 3-4 Double-row ball bearings on a front-wheel-drive hub.

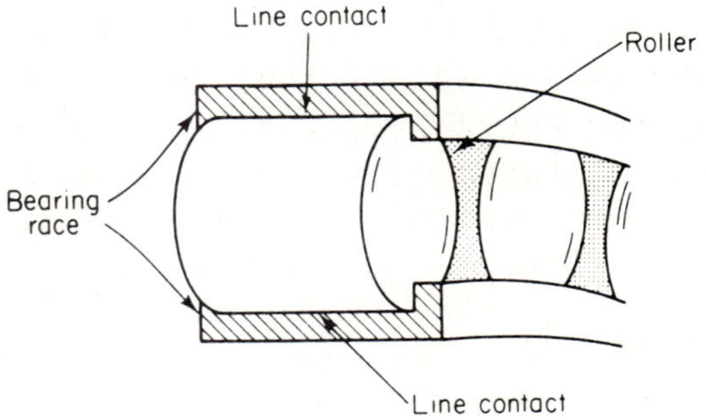

FIGURE 3-5 Roller bearing line contact.

Roller bearings are more expensive than ball bearings. They can be designed to operate directly on a journal surface that is ground on a shaft. This eliminates the need for a separate bearing inner race. The roller bearing has a small diameter so it is used where space is limited. Roller bearings have one major drawback. They cannot support an axial (side) load. Axial loads are held by the differential case side bearings when straight roller bearings are used on the outer end of the axle in a dependent rear-drive vehicle.

Tapered Roller Bearings. Tapered roller bearings are being continually used in new applications on vehicles. The inner race, rollers, and cage assembly of the bearing is called a *cone*. The outer race is called a *cup*. They have some of the advantages of both ball and straight roller bearings. Tapered roller bearings will support a radial load. They also sup-

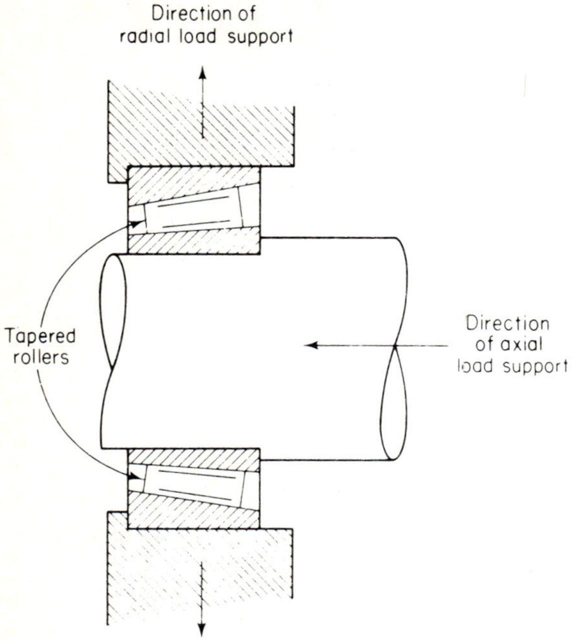

FIGURE 3-6 A tapered roller bearing will support a radial load and an axial load in only one direction.

port an axial load in one direction. This can be seen in Figure 3-6. As a result, they are used in pairs. The rolling element and the race form a line contact. As a result, it has a high load-carrying capacity and fatigue resistance like straight roller bearings. The tapered roller bearing is also limited to low and moderate speeds.

Tapered roller bearings are used in differentials as case support bearings and pinion bearings. They have replaced ball bearings for use as wheel bearings on the nondriving axles. Some front-wheel-drive vehicles also use tapered roller bearings. Tapered roller bearings are used as axle bearings on some rear-drive vehicles. Their superior fatigue resistance makes them a good bearing design for these uses.

Bearing Support. Bearing support, for this discussion, includes axle housings, knuckles, axle shafts, wheel spindles, and wheel hubs. A bearing is designed to provide low friction while maintaining the required internal radial clearance. The correct bearing clearance depends on its fit in the support as well as on the clearances within the bearing. It is important to understand the factors that influence bearing fit and resulting bearing life.

The driving axle bearing, whether it be front- or rear-wheel drive, will have a stationary outer race in the support housing. The inner race will rotate with the axle shaft. If the bearing has straight rollers, the axle shaft itself often has a ground surface to act as the inner race for the bearing.

Some nondriving rear wheels of front-wheel-drive vehicles have a sealed nonserviceable hub-bearing assembly. The nondriving axle of most passenger vehicles has the outer bearing race (cup) pressed into the wheel hub. The inner wheel bearing race (cone assembly) has a slip fit over the wheel spindle. The outer bearing race rotates with the wheel and the inner bearing race remains nearly stationary on the spindle.

Wheel Spindles. The nondriving wheel hub and bearings fit on a spindle (Figure 3-7). The spindle is a nonrotating axle that supports the wheel bearings and hub. Wheel spindles are a part of the steering knuckles when they are used on the front of rear-wheel-drive vehicles to allow the wheels to be steered. When the vehicle has front-wheel drive, the rear-wheel hub fits on a nonrotating spindle or stub axle. The front-wheel hubs fit in the inner bearing race. The outer bearing race is held in the steering knuckle. See Figure 3-4.

The wheel spindle is made of steel. This material has the required wear properties and toughness needed to resist the impact of the tire on rough roads. Because of high loads placed on the

FIGURE 3-7 Nondriving wheel hub and bearing fit on a spindle.

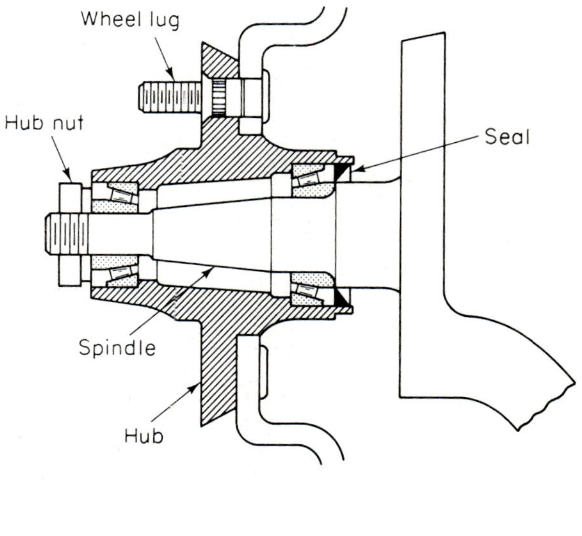

wheel spindles and the fact that they must remain perfectly aligned, no attempt should ever be made to straighten bent spindles. Bent spindles must be replaced.

The steering knuckle carries the disc brake caliper mounting and the steering arm. The steering linkage connects to the steering arm. Steering knuckles can be made from forged or cast steel. Steering knuckles made of forged steel have the steering arm and the brake caliper adapter bolted to them. Some vehicles have the steering knuckle, caliper mounting, and steering arm cast as one piece. A steel spindle is then pressed into the cast knuckle. This unit, shown in Figure 3-8, is a rugged assembly. Strut-type suspensions may use a stamped strut assembly with the stub axle or spindle welded or bolted to it.

The spindle supports the wheel bearing. The size and finish of the spindle area on which the bearing races fit is important. The inner bearing races must rotate or creep slightly on the spindle for maximum bearing life. Creeping allows the inner bearing race to gradually rotate so the load does not stay on the same part of the race. This will minimize bearing fatigue at one spot on the inner race to increase bearing life. The creep is actually in the opposite direction to wheel rotation, as shown in Figure 3-9. Wear does not normally occur on the spindle. If the bearing seizes, the inner race will rotate on the spindle at wheel speed. This will wear a groove on the spindle under the race. If wear has occurred under the race, the spindle assembly will have to be replaced.

FIGURE 3-8 One-piece cast knuckle with a pressed-in spindle.

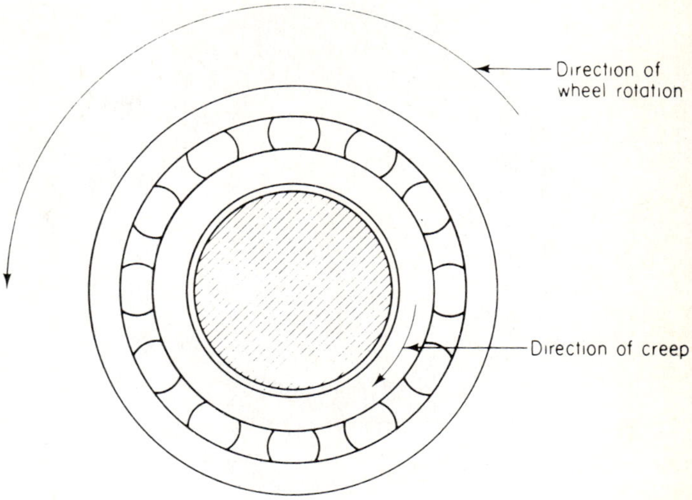

FIGURE 3-9 Wheel bearing movement to minimize fatigue.

3-2 SEALS

Seals used throughout a vehicle have different shapes and they are made from many different materials. Seal shapes used in suspension and steering systems are the radial lip seal, lathe-cut seal, and O-ring seal. These seals are usually made of some type of synthetic rubber. Gaskets are usually used where there is no movement between the surfaces to be sealed. Chemical sealers can be used as substitutes for gaskets in some places. The seal shape and material depend on the temperature range in which the seals will operate, the qualities of the lubricant, the types of lubricant, the pressures to be sealed, the type of motion that may be present, the types of contamination, and how effective the seals must be.

FIGURE 3-10 Typical radial l p seal.

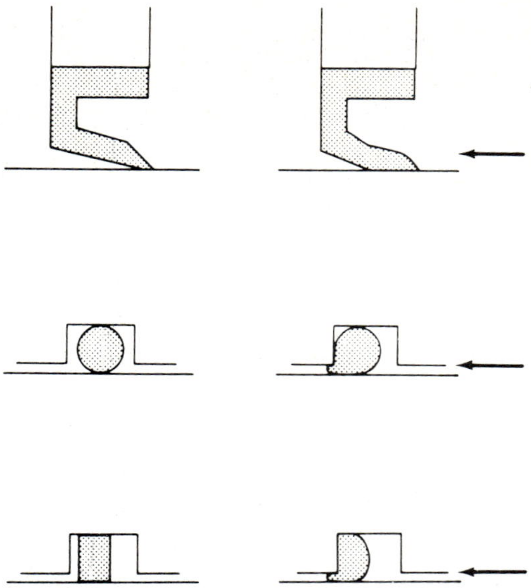

FIGURE 3-11 Shape and sealing principles of common seal types.

Lip Seal. The radial lip seal, as shown in Figure 3-10, is a dynamic seal. It is used to seal between rotating axle shafts or hubs and the stationary spindles or axle housings. Most dynamic seals have synthetic rubber sealing elements. Synthetic rubber seals usually have a spring, called a *garter spring,* to aid in holding the sealing lip in contact with the moving part. The radial lip seal is usually encased in steel. This holds the seal in the correct position. Some synthetic seals may have a double sealing lip. The secondary sealing lip keeps contamination from reaching the primary sealing lip. Lip seals are used at the wheel bearing hubs to keep bearing lubricant out of the brakes and to seal dirt out of the bearing.

O-Ring Seal. The O-ring and lathe-cut seals, as shown in Figure 3-11, are used as both dynamic and static seals. These seals are inexpensive and they work well in applications where there is axial (in and out) motion with little radial (rotational) motion. Friction caused by this type of seal is greater than friction caused by lip-type seals. O-rings make good static seals where changes from heat and vibration may make the joint difficult to seal with a gasket.

Seals that fail prematurely during service must be replaced. Seal failure is most often caused by improper installation. It is also caused by poor lubrication, very high or very low temperatures, contamination, too much movement of the parts, misalignment of the shaft and seal, or age. You will need to be aware of problems that cause seal failure so you can make the proper repair for the normal service life of the seal.

Gaskets. Gaskets are a special type of seal used where there is no movement between the surfaces to be sealed. They are used where a low-cost seal is needed. Gaskets are generally used where there is only a small pressure difference across the surface to be sealed. Sometimes the gasket is designed to hold high pressure. An example is the cylinder head gasket in the engine. Gaskets on differential covers and brake backing plates are examples of places where gaskets seal little or no pressure. They just keep the oil from leaking out.

Gaskets are made of materials that are compatible with the fluid to be sealed. They are made from a combination of asbestos, cork, cellulose, synthetic rubber, and organic and inorganic materials with binders to give them the needed properties. In high-pressure sealing and very high-temperature conditions, metal will be used in gasket construction.

Chemical Sealing. Many gaskets have been replaced by sealing the joint with chemicals. When chemical sealants are used, the joint must have a good fit. The joint flanges must be thoroughly clean and dry.

FIGURE 3-12 Comparison of four thread types.

Special precautions should be taken to remove all traces of lubricants from both surfaces. The correct amount of chemical sealant must be used. Put the sealant on one surface and assemble the joint before the sealant begins to cure. Always use the least amount of chemical you can. Excess chemical will usually squeeze inside and this can cause internal damage.

RTV silicone sealant is used as a gasket substitute between a stamping and a casting. It is put on one of the clean surfaces as a bead with about 1/8- to 3/16-inch diameter. The part is assembled before the bead cures. RTV silicone curing time is affected by moisture in the air. In humid weather, it must be assembled more quickly than it needs to be on a dry day. Wet shop towels placed around the assembled joints will speed curing of RTV silicone.

Anaerobic chemicals are also used as a sealer. They are used as a gasket substitute between two clean machined surfaces. Anaerobic chemicals do not cure while they have air around them. They cure after air (oxygen) is squeezed from the assembled joint.

3-3 FASTENERS

Before you start to do service on vehicles, it is important for you to know some basic facts about fasteners. On some vehicles, the fasteners are made to customary dimensions (inches). Other vehicles have fasteners made to metric dimensions (millimeters). You must use the correct fastener with the correct thread.

Threads. Screw threads are used on fasteners to hold parts firmly together. They are made to Society of Automotive Engineers (SAE) standards. Four thread types are commonly used: course, metric, fine, and tapered pipe (Figure 3-12). There are standard thread sizes and dimensions for each. You will need to be aware of the size and type of thread to assemble the cap screws, nuts, and fittings correctly. The most common automotive bolt and cap screws run from 1/4-inch through 9/16-inch. Metric automotive bolts and cap screws are sized from 6 mm through 16 mm. Thread sizes are compared in Figure 3-13. Tapered pipe threads used on vehicles are 1/8-inch, 1/4-inch, and 3/8-inch. In other ap-

FIGURE 3-13 Comparison of thread sizes.

Bolt size	Diameter		Threads per inch			
	Inch	(mm)	Fine	Inch	(mm)	Course
1/4	0.25	(6.35)	28			20
M6 X 1	0.2362	(6.0)		25	(1.0)	
5/16	0.3125	(7.9)	24			18
M8 X 1.25	0.3150	(8.0)		20	(1.25)	
3/8	0.375	(9.5)	24			16
M10 X 1.5	0.394	(10.0)		16.9	(1.5)	
7/16	0.4375	(11.1)	20			14
M12 X 1.75	0.472	(12.0)		14.5	(1.75)	
1/2	0.500	(12.7)	20			13
M14 X 2	0.551	(14.0)		12.7	(2.0)	
9/16	0.5625	(14.3)	18			12
M16 X 2	0.6299	(16.0)		12.7	(2.0)	

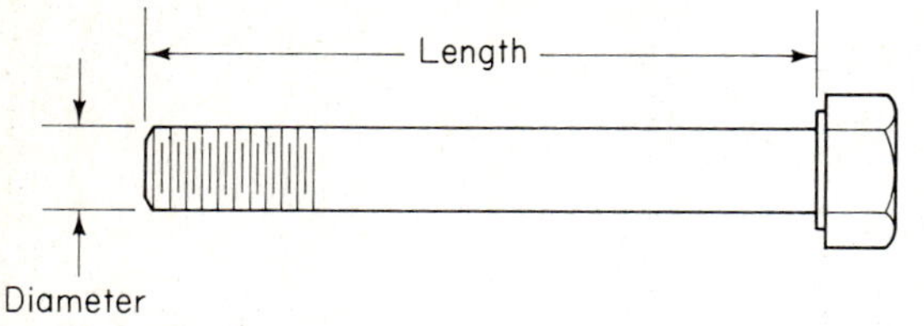

FIGURE 3-14 Cap screw size identification.

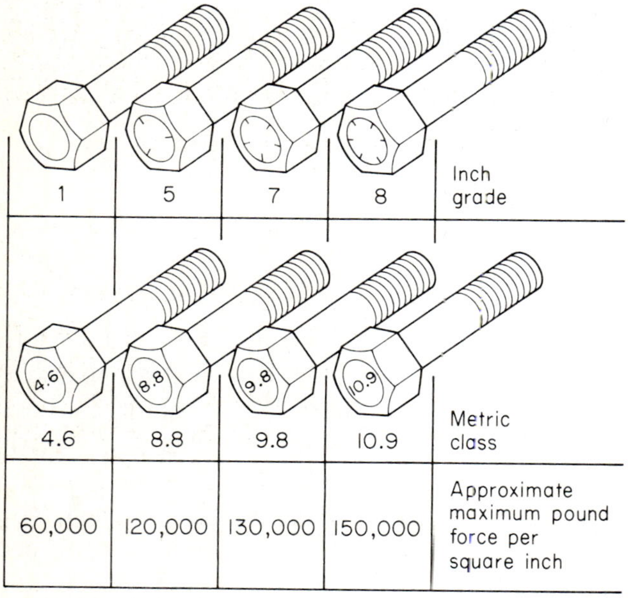

FIGURE 3-15 Typical cap screw head markings and strength.

plications, a nut is turned on the cap screw threads to hold parts together. When it is used with a nut, the cap screw is called a *bolt*. Cap screws are identified by their diameter, thread type, and length as measured from below the head end (Figure 3-14). Many different materials are used for cap screws. Inexpensive standard-grade steel cap screws are used for general assembly work. High-grade steel cap screws must be used to assemble highly stressed parts. Cap screws of different strengths are identified by head markings, as shown in Figure 3-15. Never use a cap screw of a lesser strength than the original.

Nuts. Most nuts used on cap screws have the same hex size as the cap screw head. In some specialized cases where clearance is a problem, the next smaller hex size is used. Some inexpensive nuts use a hex size larger than the normal cap screw head size. Metric nuts are often marked with dimples to show their strength. More dimples indicates stronger nuts. Many nuts used on cap screws in automotive work have a smooth ring on the fastener side, as shown in Figure 3-16.

In many applications, it is necessary to make sure a nut or cap screw does not accidentally loosen. In the past, this was done by drilling the head of the cap screw, then fastening two or more together with a wire through the holes. This took a great deal of time. Chassis nuts were held with cotter pins. Many wheel bearing nuts are still being locked with cotter pins. Lock washers with sharp edges are used under the heads or nuts to keep the nut or cap screw from accidentally loosening. In many applications, nuts and cap screws are kept from accidentally loosening by using interference fit threads. This is done by slightly distorting the shape of the nut or by deforming part of the threads. Nuts are also kept from loosening with a nylon washer fastened in the nut, or with a nylon patch or strip on the threads (Figure 3-17). Anaerobic sealers are used on the threads where the nut or cap screw must be both locked and sealed.

FIGURE 3-16 Smooth ring under the cap screw head and on the nut.

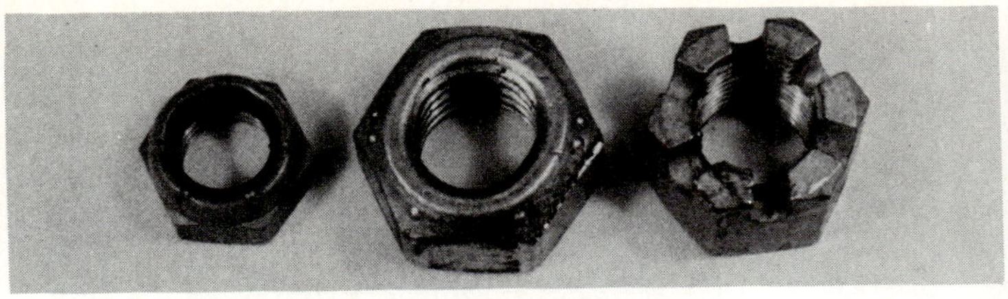

FIGURE 3-17 Nut locking methods: left, nylon ring; center, distorted shape; and right, castle.

3-4 WHEEL BEARING MAINTENANCE

Periodic maintenance is required on serviceable wheel bearings used on nondriving wheels. The condition of the lubricant in the wheel bearings should be checked every time the wheel hub is removed to allow for steering or suspension service. The bearings should be cleaned and repacked whenever there is not enough grease, or when disassembly has caused contaminants to enter the bearing area. Grease should never be added to wheel bearings. They must be thoroughly cleaned and repacked when grease is needed. New wheel bearing seals should be installed when the wheel bearings are repacked with new grease.

Disassembly. The vehicle must be raised and supported on safety stands so the wheel being serviced is off the floor. One lug bolt and its matching hole should be marked. This will allow you to replace the wheel without changing the balance of the wheel assembly. It should be noted that the brake pedal must not be pressed at any time while the brakes are disassembled. Pressing the brake pedal will push the pistons from the wheel cylinder and caliper.

If the vehicle has disc brakes, the brake caliper must be removed before the hub and rotor assembly can be removed. On a floating or sliding caliper, a C-clamp is placed between the outer pad and the cylinder to force the piston inward as pictured in Figure 3-18.

Some caliper housings are held on the adapter by pins. You can see this in Figure 3-19. The caliper floats or slides as the brakes are applied. This design is called a *floating caliper* or a *pin slider* caliper. On other caliper designs, both the caliper and adapter have machined surfaces. They slide on the machined surfaces as the brakes are applied. This design is called a *sliding caliper* (Figure 3-20).

FIGURE 3-18 C-clamp used to move the caliper piston inward to make clearance between the pads and rotor.

The pins must be removed from the pin slider caliper to free the caliper housing from the adapter. Retaining screws and retaining clips are removed from the adapter of sliding calipers to free the caliper from the adapter. The caliper can then be lifted from

FIGURE 3-19 Caliper that floats or slides on pins.

FIGURE 3-20 Caliper that slides on machined surfaces.

FIGURE 3-22 Caliper supported so that it does not hang on the brake hose.

the steering knuckle (Figure 3-21) or suspension part. On some vehicles, it is necessary to remove a brake hose clip to give enough hose movement so the caliper can be lifted from the rotor. The caliper must be supported (Figure 3-22) to prevent damage to the brake hose.

If the vehicle is equipped with drum brakes, the brake shoes may have to be retracted before the drum and hub assembly can be removed. This is caused by drum wear. The shoes are retracted by inserting a wire or screwdriver into one end of the adjusting slot to hold the automatic adjuster lever away from the star wheel. The star wheel can then be turned in the proper direction to retract the shoes. This is illustrated in Figure 3-23.

FIGURE 3-21 Lifting a caliper from a rotor.

To remove the drums or rotor from a spindle, first take off the dust cap on the wheel hub (Figure 3-24) so you can remove the cotter pin from the spindle nut. Next, remove the hub nut. The washer and bearing can then be lifted from the spindle as shown in Figure 3-25. Place them where they will not be

FIGURE 3-23 The backing plate has been cut away to show how the drum brake automatic adjuster lever is held away from the star wheel with a release tool as the star wheel is turned with a brake spoon. This creates more clearance between the drum and linings.

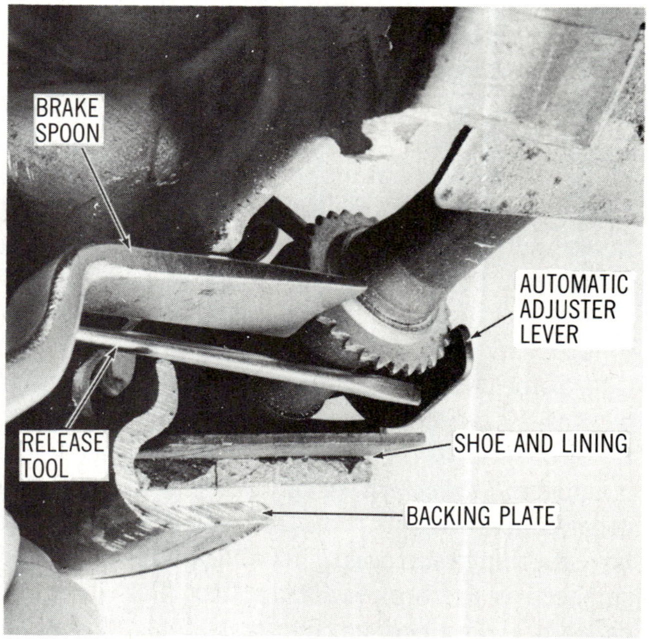

FIGURE 3-24 Removing the dust cap from a wheel hub.

mixed with parts from the other wheel hub. The dust cap is a good container for these parts. Screw the hub nut back onto the spindle while the hub is still in place. The inner bearing cone will catch on the nut as the hub is pulled over it. The inner bearing cone and seal will stay on the spindle. This procedure can be seen in Figure 3-26. The inner bearing and seal can then be taken from the spindle. If the seal and bearing cone cannot be easily removed in this way, remove the nut and hub. The inner bearing cone can be driven out of the hub with a large wooden drift and a hammer. The bearing cup will stay in the drum. Always keep the bearing cones

FIGURE 3-25 Lifting the washer and bearing from the wheel hub.

FIGURE 3-26 Pulling the inner bearing cone and seal from the hub.

separate so they can be returned to their original locations.

Routine Packing of Serviceable Wheel Bearings. As the wheel bearing cones are being cleaned, they should be inspected for any abnormal conditions. Bearings with abnormal conditions are replaced. Use a brush to thoroughly clean the bearing and hub in petroleum solvent. The bearing should be blown dry with clean compressed air to remove any contaminants that were not flushed out during cleaning. Do not 'spin dry' bearings with air pressure. Spinning the bearings with air pressure will overspeed and score them. Overspeeding may even cause the bearings to fly apart. The flying pieces could injure someone in the service area.

Not all types of wheel bearing lubricants mix together. Often, mixed lubricants will separate from each other in service. If this happens, bearing failure will result. Therefore, it is important that all of the old grease be cleaned from the hub cavity, bearing cups, and spindle. The bearing cups should be inspected to make sure they are in good condition and tight in the hub. The hub and bearing will have to be replaced if a cup is loose.

If all parts of the bearings are in good condition, the bearing cones are packed with a high-temperature NLGI #2 wheel bearing grease. Packing is a critical process and, therefore, it should be done carefully. A bearing packer (Figure 3-27) can be used or the bearing cones can be packed by hand (Figure 3-28). Packing by either method will force

FIGURE 3-27 Using a bearing packer to pack a bearing with grease.

FIGURE 3-29 Reserve grease in a hub cavity.

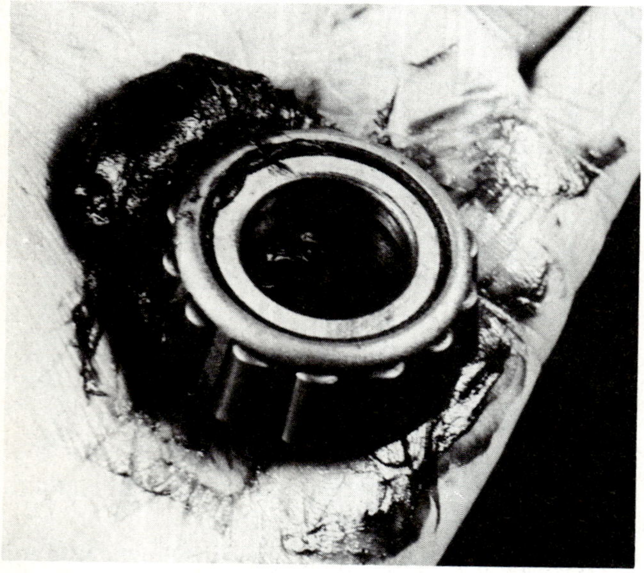

FIGURE 3-28 Packing a bearing by hand.

FIGURE 3-30 The seal is placed on a seal driver with the lip facing the bearing.

the grease around each roller, completely filling all cavities in the bearing cone. A small amount of grease should be put in the hub cavity to act as a reservoir of grease as shown in Figure 3-29. The spindle is coated with a thin film of grease to provide rust protection and lubrication. Lubrication is needed because the inner bearing race creeps on the spindle during operation.

Use a seal driver to install a new grease seal. Coat the seal with the wheel bearing grease. Put the seal on the seal driver with the seal lip toward the bearing (Figure 3-30). With the bearing cone in place, position the seal so it starts into the hub squarely. Tap the driver until the seal is seated (Figure 3-31). If a seal driver is not available, place a clean block of wood across the seal. Make sure the seal starts squarely in the bore as you tap the block to seat the seal. Never use a hammer directly on the seal because hitting will damage the seal.

The hub and drum are placed over the spindle and the outer bearing, washer, and spindle nut are installed. The nut must be tightened correctly to properly adjust the wheel bearings.

Wheel Bearing Adjustment. All tapered roller wheel bearings are adjusted in a similar way. While the drum or rotor is being rotated, the spindle nut is

FIGURE 3-31 Installing the inner seal after installing the inner wheel bearing.

FIGURE 3-32 Torquing the spindle nut to seat the bearing before final adjustment.

FIGURE 3-33 Properly installed cotter pin used to securely lock the spindle nut.

tightened to 20 to 25 ft-lbs (27 to 34 N-m) to seat the bearing (Figure 3-32). The nut is then loosened and retightened finger-tight. When a nut lock is used on a spindle nut, it is properly positioned and the cotter pin is installed (Figure 3-33). When a castle-type adjuster nut is used, the cotter pin slot in the adjuster nut may be backed off just enough to install the cotter pin. This should provide a bearing adjustment with no preload. The wheel bearing should have between 0.001 and 0.003 inch (0.02 and 0.08 mm) in and out free-play movement at the hub. If this cannot be obtained without preloading the bearing, either the bearing cup is loose or there is excessive clearance between the bearing cone and the spindle. If the bearing is preloaded, grease will not work into the bearing. The bearing will rapidly wear until it has about 0.002 inch (0.03 mm) of free play, damaging the bearing. After the cotter pin is installed, the dust caps are replaced. The disc brake calipers are reassembled. Drum brakes require preliminary manual shoe adjustment. Final adjustment is automatic. Disc brakes require resetting the pads by depressing the brake pedal several times until the pedal has normal height. Remember that you never press the brake pedal until all the brakes have been completely assembled. If you do, a wheel cylinder will separate. This will require additional labor to correct. The brake fluid reservoir should be refilled, as needed, to complete the assembly.

The wheels are reinstalled with your marks

aligned and the lug nuts tightened. It is important to use the proper lug nut tightening sequence and to tighten them to the correct torque. The best way to tighten the lug nuts is to use a torque wrench. Lug nuts are tightened by tightening every other nut in a crossing pattern. Incorrect tightening can cause brake- or wheel-related problems. Too much torque can warp the rotor or drum. It can also break the lug. Too little torque might cause the wheel to come loose. Wheel lug torque is critical in preventing loosening of the wheel. The wheel is designed so the raised part of the wheel around the lug deflects slightly (Figure 3-34), like a Belleville spring, when it is properly tightened. This holds lug tension and acts to lock the lug nuts to keep them from loosening.

FIGURE 3-34 Wheels are slightly dished, somewhat like a Belleville spring, to help hold the lug nut tension.

STUDY QUESTIONS

Introduction

1. How can you tell the difference between wheel bearings and axle bearings?

Section 3-1

2. What loads can ball bearings support that other types of bearings cannot?
3. Where are roller bearings used rather than ball bearings?
4. Why are tapered roller bearings always used in pairs?
5. What is a wheel spindle?
6. How should you repair a bent spindle?
7. Why must the inner race move on the spindle?

Section 3-2

8. When is a seal used rather than a gasket?
9. When would an O-ring seal be used rather than a lip seal?
10. What is the most common cause of seal failure?
11. In what kind of a joint is a gasket used?
12. What type of chemical sealant should be used between a stamping and a casting?
13. What type of chemical sealant should be used between machined surfaces?
14. What cures RTV silicone sealant?
15. What cures anaerobic sealant?

Section 3-3

16. Name the four fastener threads.
17. What was done to reduce the number of metric fastener sizes?

18. How does a cap screw differ from a bolt?
19. How are nuts and cap screws kept from loosening?

Section 3-4

20. What should be done to wheel bearings before packing them with grease?
21. How are the brake pads moved away from the rotor to make clearance for caliper removal?
22. What should be done to prevent brake hose damage when the caliper is removed?
23. How is the inner bearing removed from a hub?
24. What may happen to a bearing that is spin dried after cleaning?
25. What should be done when you find a loose bearing cup in a hub?
26. Where is the grease reservoir for wheel bearings?
27. What precautions should be taken when installing the seal?
28. Why is the bearing first tightened, then loosened, and again retightened by hand when adjusting wheel bearings?
29. What problems can result from incorrect lug nut tightening?

4

Tires and Wheels

All of the loads on a tire can be placed into three load groups or force components: *radial load, side load*, and *tractive load*. These were shown in Figure 1-23. The size of the load in each direction is not constant. The loads change as the vehicle is turned, accelerated, and stopped.

There is a near-zero side or lateral load on the tire when the vehicle is going on a straight road. The required tractive load is only enough to hold a constant speed, and the radial load is nearly equal to the vehicle weight or static load on the tire. As the vehicle is driven into a turn, the lateral load increases enough to cause the vehicle to go around the corner, assuming the tire does not slip sideways. The radial load on the tire is also changed while turning. Centrifugal force of the vehicle in a turn acts on the center of gravity of the vehicle. The centrifugal force causes the vehicle body to lean toward the outside of a turn. This increases the effective weight on the tires on the outside of the turn. At the same time, it decreases the effective weight on the tires on the inside of the turn. The tractive load will also be affected when the brakes are being applied or the vehicle is being accelerated.

With any given radial load on a tire, the summation (vector sum) of the tractive and lateral loads must be less than the frictional force between the tire and the road (Figure 4-1). When the vector sum of the forces is more than the frictional force, the tire will slide on the road surface. The maximum frictional force of the tire (without sliding) is usually between 60 and 70 percent of the radial load the tire has when it is on a dry asphalt surface. In some sports and racing cars, the sum of the forces is over 90 percent. This percentage is called the *tire coefficient of friction*. (See the appendix for additional discussion.) In actual use, there is a slight amount of tire slip at speeds above a walk. The angle between the direction the tire is headed and the direction it is actually going is called the *slip angle*. Slip angle is discussed in Section 4-3.

4-1 TIRE CHARACTERISTICS

Tires produce noise as the vehicle is driven on the road. A tire with a smooth tread surface will make the least noise and it has good frictional force when rolling on clean, dry paved road surfaces. On the other hand, a smooth tire would be useless for vehicle control on wet highways. The smooth tire will ride up on the surface water, a condition called *hydroplaning* (Figure 4-2). Therefore, tires are designed with a tread pattern that is a compromise between dry road handling, wet roadability, and noise. As the tread wears in service, the performance characteristics gradually change.

Tires absorb the effects of small changes in the road surface before these changes reach the suspension. To do this, the tire needs a relatively flexible sidewall and tread. The tire also carries steering

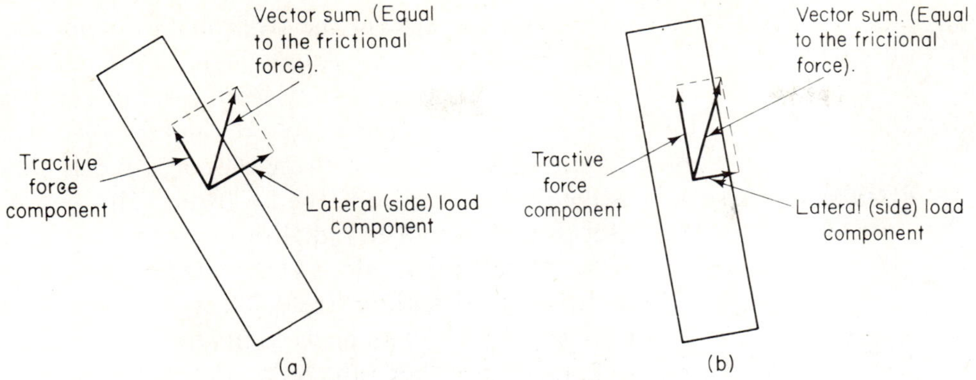

FIGURE 4-1 Maximum vector forces possible during a turn: (a) large lateral force while making a sharp turn; (b) less lateral force while making a mild turn.

forces to the road surface. To do this, the tire needs a sidewall and tread that responds quickly to driver input. Again, a compromise is made to make tires suitable for both ride quality and handling. Tires meet these requirements while rotating between 650 and 1000 rpm at 55 mph (88 km/h), depending on the tire size.

Asymmetric Tire Designs. Some special-purpose tires are designed to meet different requirements on each side of the tire. These tires are designed so that they perform best when they rotate in a specified direction. Because these special properties are built into the tires, asymmetric tires are different from side to side tires.

The asymmetric tire is specially designed to support the difference in loads across the tire while the vehicle is in a turn. It is also designed so that the tire has a more uniform contact surface on the road at high vehicle speeds. These features can only be built into a tire if the rotational direction and the mounting direction are specified. These specially designed tires have superior traction and handling when compared to a conventional (symmetric) tire. The cost, however, of an asymmetric tire is considerably greater.

4-2 TIRE CONSTRUCTION

Tires are made of many separate parts. The parts are molded together to form the complete tire structure. The parts include the bead, cords, plies, rib, tread, sidewall, and liner as shown in Figure 4-3.

The *bead* is a bundle of wires in the part of the tire that fits around the wheel rim. The inner edge of the *sidewall plies* is wrapped around the bead. The tire *carcass* is the tire without either sidewall rubber or tread rubber. Tire cords are strands of rein-

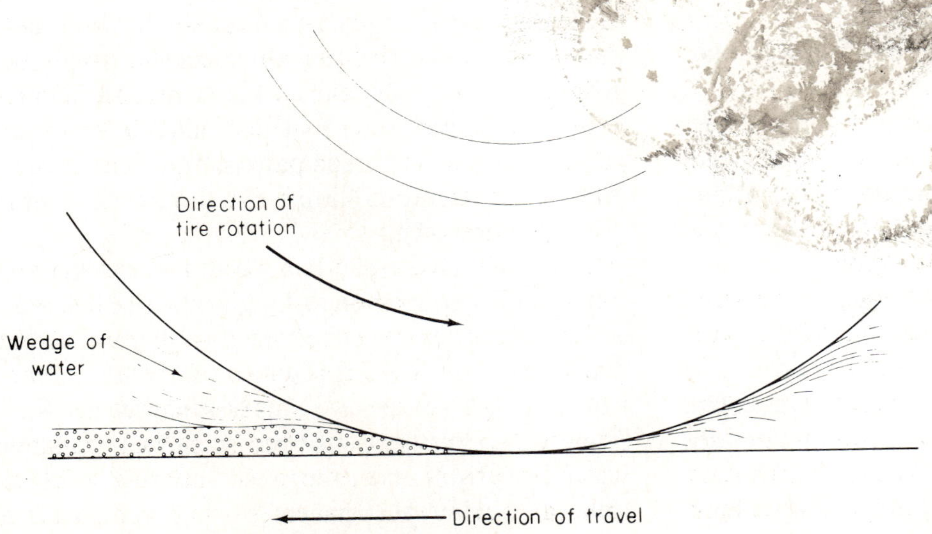

FIGURE 4-2 If water cannot get out, the tire rides up on a wedge of water, which causes the tire to hydroplane.

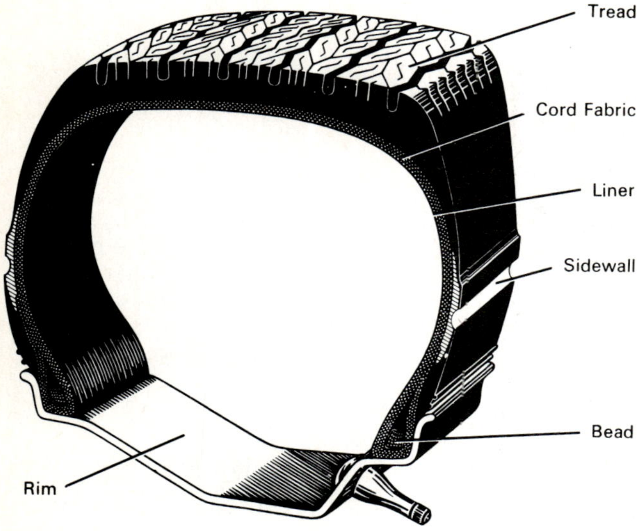

FIGURE 4-3 Section view showing the different parts of a typical tire. (Courtesy of the Firestone Tire and Rubber Company)

forced textile bonded together to form a sheet. The sheet is used to form the plies of the tire. Plies of either textiles or steel are used for *belts*. The *ribs* are continuous rows of tread rubber going around the tire. The *tread* is the part of the tire that is designed to run on the road surface. The tread rubber is grooved between the ribs with a pattern designed to provide maximum frictional force and minimum noise. When the tire is operating on a wet surface, water goes through the grooves as the tire rolls on the road surface. The *sidewall* is the flexible part of the tire that connects the bead to the tread. A thin airtight rubber *liner* covers the inner surface of the carcass.

Tire Design. Low cord angles in the carcass of bias ply tires are generally better for high-speed performance because they generate less heat and provide lateral stability. Low cord angles also increase tread wear. High cord angles in the carcass of radial ply tires, especially when they are used with belts in the tread area, will give a tire greater strength, increase its fatigue resistance, and reduce its rolling resistance. The stronger tire helps to stabilize the tires in the footprint and this reduces wear. The *footprint* is the part of the tire tread that meets the road surface.

The bias-type construction forms a sidewall that is more rigid than the radial sidewall construction. This can be seen when looking at both a bias and a radial tire from the front at the same time when both tires have equal loads and equal air pressure. The bias angle of the cord material restricts motion and this results in a more rigid tire sidewall. Flexing causes friction between the cords in bias plies as the tire rolls through the footprint. This causes heat to build up. The radial tire has no bias angle in the carcass plies, allowing the cord material in a radial tire to flex more freely. The radial cords flex with very little friction between the cords of the plies and, therefore, they run cooler and roll more freely. A tire with less friction uses less power; this results in greater fuel economy. Radial ply tires are being used on all types of vehicles, from heavy over-the-road tractor trailer trucks to very sophisticated Formula I and Indianapolis race cars.

We can see, however, that no one type of tire is best for all applications. The type of tire to be used can be properly selected when the vehicle requirements are known.

Tire Tread Patterns. You have seen many different types of tire tread designs. Some have many fine grooves while others have large grooves with a course tread, almost like having rubber blocks on the tread. Treads that have good traction are called *aggressive* treads. They become less aggressive as the tread wears.

Tire tread patterns have changed as tire materials, road surfaces, and vehicle speeds have changed. The typical rib-type tire tread has been popular for passenger cars since the late 1940s. You can see this tread design on the left side of Figure 4-4. The ribs are separated with grooves running around the tire. It is typical to have five ribs, but tires may have as many as nine ribs. On wet roads, the water under the tire flows out through the grooves so the rubber can contact the road surface. The tread has narrow cross-slots, called *sipes*, to add gripping edges. The tread pattern and sipes have an unequal pattern to minimize tire humming or siren noise (Figure 4-5).

The rib-type tread has good dry traction and fairly good wet traction as long as the tire has most of the tread rubber. As the tread rubber wears, the depth of the grooves is reduced and water cannot flow through as quickly. This lessens wet traction. Some roadways have grooved sections to channel water away from under the tires to improve wet traction. Grooving the pavement has reduced the

FIGURE 4-4 Three types of tire treads.

number of accidents on this type of road during wet weather. One problem has developed on grooved roads. The ribs on the tires tend to follow the grooves in the road. However, the grooves are not straight, and, as a result, the driver feels as though control of the vehicle has been lost as the tires follow the grooves. This is called *groove wander*. Groove wander is especially bad when riding a motorcycle.

FIGURE 4-5 Unevenly spaced sipes on the outer rib of the tread.

Rib-type tire tread operates poorly in snow, even when the tire is new. Snow tends to pack into the grooves so the tire acts like it has a smooth tread. As a result, snow tires were developed. Snow tires have an aggressive tread with a heavy lug design, like the tire tread shown on the right side of Figure 4-4. This tread design also provides good traction in mud. The large open groove design does not pick up snow or mud, so the lugs can continue to bite into the surface for good traction. This heavy lug design is very noisy when driven on dry roads.

Traction is reduced as the tread wears. In most applications, snow tires have enough acceptable snow traction for use during three winter seasons, if they are not used during the rest of the year. They wear quite rapidly on a combination of mud and dry roads. Snow tires should be replaced with conventional tires during the warm months. The heavy rubber snow tread builds up more heat than a rib-type tread. As a result, if snow tires are used in warm weather, they are likely to fail.

All-season radial tires were introduced in 1978. The tread on all-season tires is a compromise between rib-type and snow tire treads. It is more aggressive than the rib-type tread and less aggressive than the snow tread. The tread of a typical all-season radial tire is shown in the middle of Figure 4-4.

The all-season tread has an open tread pattern

to let water flow out of the sides as well as through grooves around the tire. This allows the tire tread to work well in the rain. The open tread does not clog with snow or mud, which makes it a good snow tire. These tires are usually marked "M & S" for mud and snow. As the tread wears, the wet surface and snow traction decreases. The all-season tread is computer designed and operates almost as quietly as the rib-type tread on a dry road. Also, the tread does not follow the grooves in the road surface, minimizing groove wander.

In 1983, automobile manufacturers put all-season tires on about one-third of the new automobiles. In 1984, this increased to approximately one-half of the new automobiles. All-season radial tires make up a large part of the replacement tire market.

A tread wear grade is marked on tires. It can be used by a customer when selecting replacement tires. Tread wear standards are based on a Department of Transportation standard test run under controlled conditions on a specified government test course. A tire with standard tread wear is given a grade of 100. If the tread wears 50 percent longer than the standard tread, the tire is given a tread wear grade of 150.

Tire Materials. Tires are constructed from many different materials, including strands of textile, fiberglass, and steel along with natural and synthetic rubbers. The choice of the materials used in the tire is based on material costs and on vehicle and operator requirements.

Tire Cord Materials. The cord used in the tire gives strength to the carcass. It also affects many of the tire's operating characteristics. The cord materials in the sidewall of the tire carcass must flex easily. The flexible tire is able to absorb much of the road surface roughness and provides long tire life without failure. The tread tends to lie flatter on the road surface for better traction while the vehicle is being turned when the material used in the sidewall is flexible.

Modern cord materials used for the tire carcass are rayon, nylon, and polyester. Steel is being used as a cord material in the carcasses of some truck tires. Nylon and polyester have greater strength than rayon. Nylon has a tendency to develop a "set" while the vehicle is standing. This set produces a characteristic "thump" as the tire rolls until the tire becomes warm as the vehicle is driven.

Tire cord material used for breaker ply belts is much stiffer and stronger than the cord used to make up the tire carcass. The stiff ply under the tread is helpful in preventing the tread from closing the grooves in the footprint. The greater stiffness and strength of the belt also give the tire excellent puncture resistance. The tread grooves of a bias ply tire have a greater tendency to be pushed together, or squirm, in the contact patch area than a radial ply tire. This occurs because the carcass is stiff and the tread flexes. The radial ply tire has less carcass stiffness and greater tread stiffness. Materials used for the belt cords are fiberglass, steel, or an organic fiber called Aramid. The strands in most belts run around the tire at a slight angle. As a result, there are strand ends all along the edge of the belt, as shown in Figure 4-6. If they start to loosen, the ends will come through the tread (Figure 4-7). To reduce this failure, some tires have folded belts (Figure 4-8). Armstrong introduced a belt of woven Aramid fiber (Figure 4-9). It is woven as a complete belt. This belt is placed on the tire carcass before the sidewall and the tread rubber. Experimental tires have also been made using carbon fiber belts.

A tire with minimum squirm in the footprint can use a softer rubber compound in the tread. Any squirm that takes place in the footprint will produce wear and reduce traction on wet roads.

Both the sidewall and the air pressure within

FIGURE 4-6 Cut belt with the ends of the wire cut at the edge of the belt.

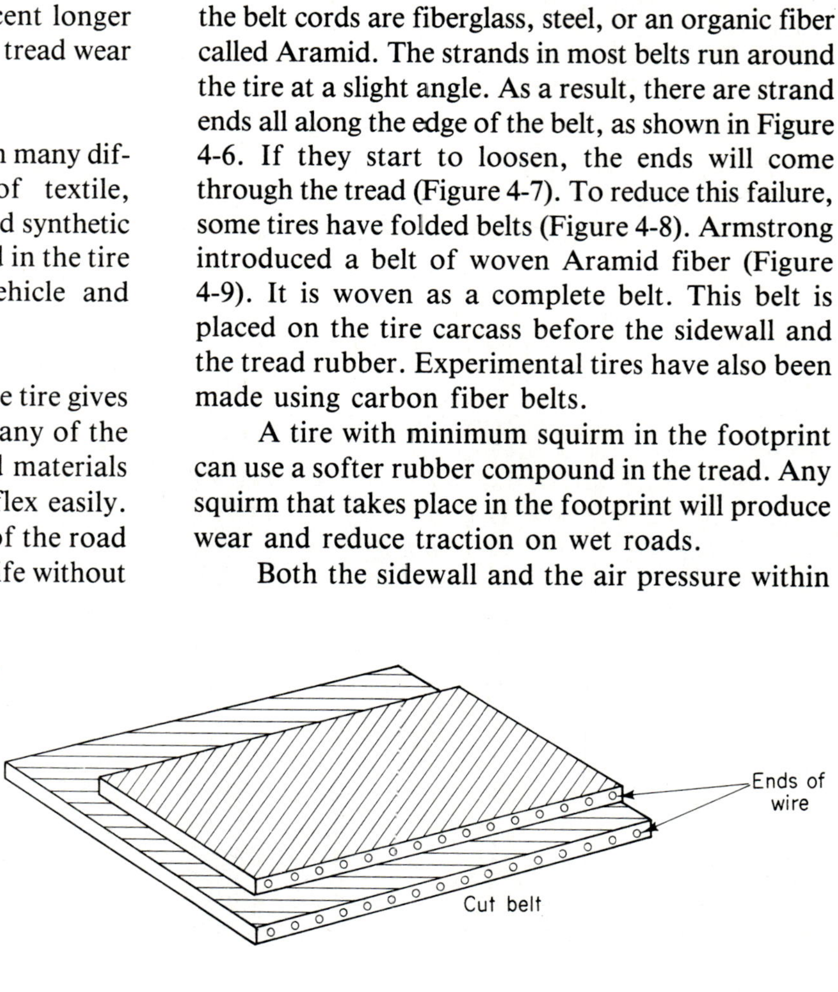

FIGURE 4-7 The belt has loosened and the edges of the belt have worn through the tread.

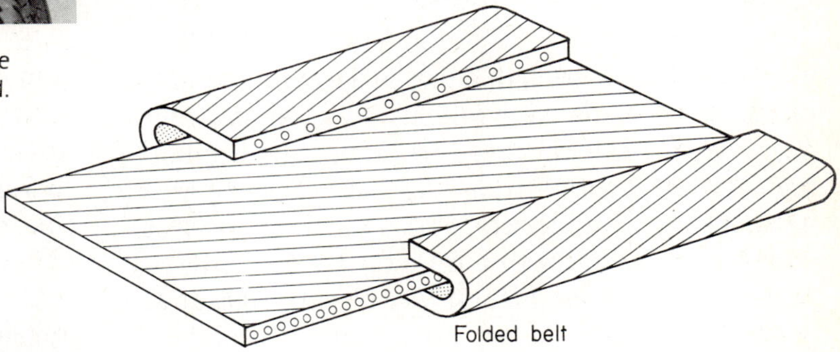

FIGURE 4-8 Inner belt folded over the outer belt.

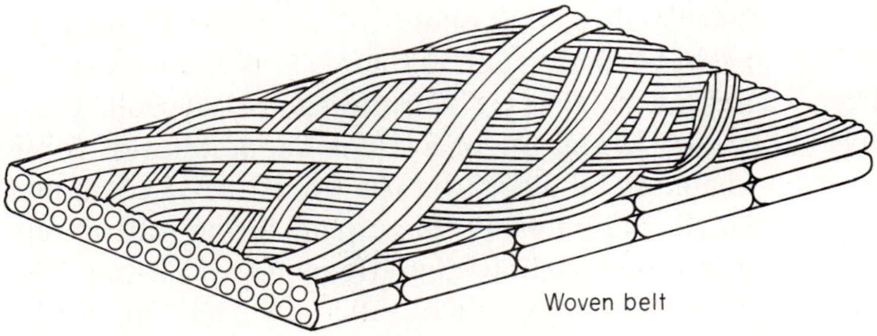

FIGURE 4-9 Woven belt design.

the tire combine to support the load. Under normal conditions, the air pressure supports up to 90 percent of the weight. As the tire sidewall is called upon to support a greater percentage of the vehicle weight, the tire tread has a tendency to push together in the tire footprint. As a greater percentage of the load is carried by the air pressure, less tire deformation occurs in the tread contact patch. A greater percentage of the weight is carried by the sidewall when the tire is underinflated. As a result, excessive heat will build up in the carcass, the tread will squirm, and the tire tread will wear rapidly.

Tire Sidewall and Tread Rubber. Tire sidewalls and treads are made using different types of rubber. A large percentage of natural rubber is used in the sidewall of the tire and in bonding the cords together that make tire plies. It flexes easily to minimize heat buildup. Synthetic rubber is used in the tire tread. It is compounded so that it is superior to natural rubber in its wear and traction characteristics.

Synthetic rubber is compounded from a hydrocarbon called butadiene. Chemicals are added to the rubber to hold the tire together. Carbon black is added to the tread rubber to provide wear resistance. Chemicals are also added to slow the aging (oxidation) process that naturally occurs in rubber materials.

White sidewalls are produced by mixing titanium dioxide with some of the sidewall rubber. The white sidewall rubber is extruded as a core in the bulk sidewall rubber before the tire is built.

The various parts of the tire are hand assembled on a tire-building machine. The assembled tire, called a *green tire*, is inspected. The green tire can be dissassembled for repair. Next, the parts are forced tightly together, called *stitching*, to interlock the parts of the tire. The green tire assembly is then cured by *vulcanizing* in a mold with heat and pressure. The black sidewall rubber over the whitewall or white letters is cut off to expose the white rubber. The finished tire is inspected again and made ready for shipment. Inspection of the finished tire determines whether it will meet original equipment specifications, replacement specifications, blemish specifications, or if it is to be scrapped.

4-3 TIRE PERFORMANCE

Tire construction and materials determine a tire's performance. Performance can be broken down into wet and dry road traction, wear characteristics, ride quality, and steering sensitivity.

Traction. The traction of a tire on a wet road results primarily from the tire's ability to channel water away from under the tread to prevent hydroplaning. The ability of a tire to stop on wet pavement when measured under controlled test conditions is marked on the tire with a traction grade: A, B, or C. Grade A has the highest wet traction; grade C has the lowest. Secondary factors in wet road performance of tires are the width of the tread and the softness of the tread rubber compound. A wide tire has a greater tendency to hydroplane than a narrow tire. The narrow tire exerts greater unit pressure, which improves the tire's adhesion to the road surface. A soft rubber compound allows the tire to interact with the road surface to improve the tractive force. On the other hand, reducing the amount of rubber in contact with the road surface (narrower tire) and using a soft compound will increase the rate of tread wear.

The dry road performance of a tire depends on the type of tire construction, rubber compound, tread type and design, as well as the cord material. Radial construction is superior for fuel economy, tread wear, and steering response.

The tread type and pattern are designed for the specific application and use of the tire. If a tire is designed for both tractive and lateral load, the tread has grooves around the tire. The tire tread grooves tend to fold under during cornering as shown in Figure 4-10. This increases the slip angle of a tire. The *slip angle* is the angle between the direction the tire is headed and the actual direction it is going. A new tire with thick tread has a greater slip angle than a used tire. A tire with little tread will, therefore, have better dry road traction than a tire that has deep tread grooves, providing the coefficient of friction of the tires is equal.

Ride Quality. The quality of the ride is affected by tire construction and material. Bias and bias belted tires have higher radial spring rates than radial tires. *Radial spring rate* of a tire is the amount of force that must be placed on the tire to deflect it down 1 inch (Figure 4-11). The radial spring rate affects the resonant (rebounding) frequency of the tire. The low radial spring rate of radial ply tires contributes to what is known as *low-speed harshness.* A bias tire has the strongest resonant frequency at about 160 hz (cycles per second). This happens when the vehicle is traveling at approximately 60 mph (96 km/h). A radial tire has a resonant frequency of about 60 hz at a vehicle speed of less than 20 mph (32 km/h). Strong resonant frequency at low speed causes noise.

FIGURE 4-10 Tire tread grooves folded under as the tire flexes during a turn.

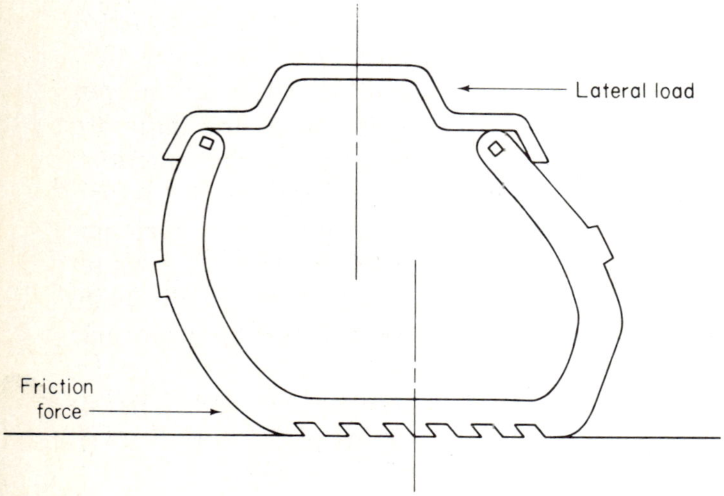

FIGURE 4-11 Radial spring rate is the vertical force that deflects the tire 1 inch.

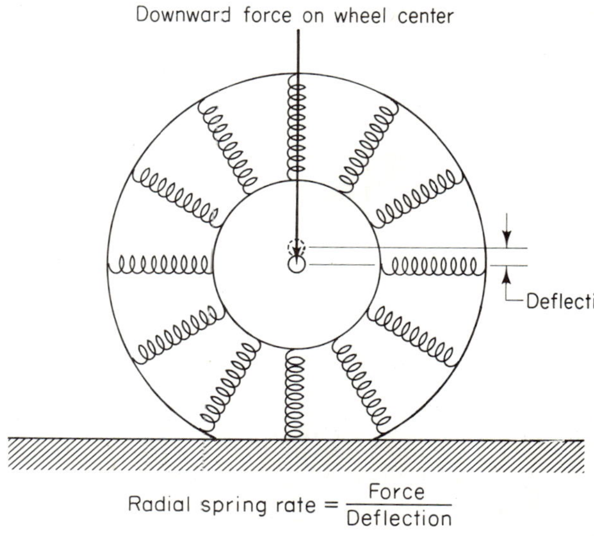

This is the major disadvantage of radial ply tires used on suspension systems designed for bias ply tires. New suspensions are tuned for radial tires to prevent low-speed harshness (see Section 7-5). Radial ply tires are more sensitive to wheel alignment settings and wheel balance than bias ply tires. This sensitivity often leads to vibration and vehicle pull in one direction.

Steering Sensitivity. Steering sensitivity is one of the greatest differences among the types of tire construction. *Steering sensitivity* is how much the steering wheel must be rotated for a given turn and how quickly the vehicle responds. A vehicle with sensitive tires tends to respond more rapidly to slight steering wheel movements. As a result, a sensitive tire requires more precise steering control. The radial ply tire has the greatest steering sensitivity.

Steering sensitivity is affected by self-aligning torque of the tire. *Self-aligning torque* is the tendency of the tire to twist between the footprint and the wheel rim. A radial tire has a low self-aligning torque when a lateral force is applied. This produces a faster steering reaction and vehicle response from the operator input. Because radial tires produce good steering sensitivity, they were first used on sports cars. Later, they were introduced on passenger vehicles, then trucks, and finally, on race cars.

Slip Angle. The slip angle of the tire is another factor affected by various types of tire construction. The *slip angle* is the angle between the direction in which the tire is heading and the direction in which it is pointed, as illustrated in Figure 4-12. The slip angle can increase until the tire is actually sliding sideways or even backwards in a skid. The amount of slip that occurs as the lateral load is increased on the tire differs among the various types of tire construction. A radial ply tire typically has a low slip angle. It remains low until the tire suddenly breaks loose and slips sideways. A bias ply tire has a slip angle that is nearly proportional to the lateral force being placed on the tire. The slip angle increases as the lateral load increases. As a result, the bias ply tire gives the driver a gradually increasing warning that the vehicle is reaching the limit of the lateral force. It gives the driver enough warning between the time the tire starts to slip and the time control is lost to make a correction before a skid begins. The slip warning is one of the reasons bias ply tires are used on many types of race cars. In racing, the vehicle is driven around curves right at the edge of the maximum lateral force, just before it starts to slide sideways. A great deal of skill is required to drive on the fine line between the start of slip and loss of control when using radial tires. The vehicle will suddenly spin out when the lateral force exceeds the traction of the tires.

FIGURE 4-12 Twist in the tire tread within the footprint causing slip angle.

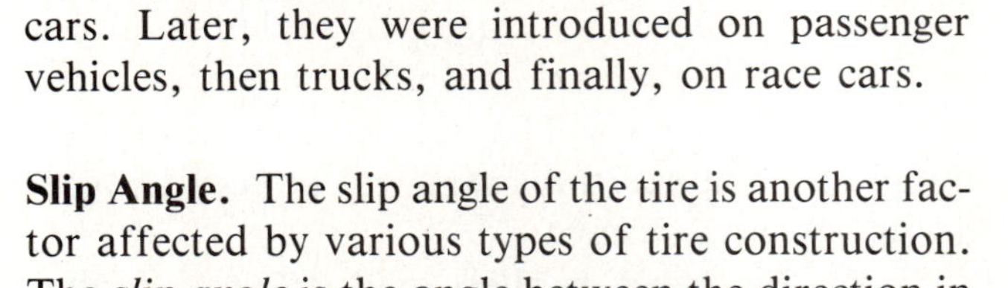

The low slip angle of the radial ply tire gives the operator more sensitive control at all speeds, until slip results in loss of control. The small difference between control and loss of control was one of the reasons manufacturers were slow in changing from bias ply to radial ply tires on automobiles for the general public. Fortunately, the average driver operates a vehicle at lateral loads well below the traction limit of the tires.

Tires with different types of ply construction should not be mixed on a vehicle. Radial ply tires have the lowest slip angle, followed by bias belted tires. Bias-ply tires have the greatest slip angle. Mixing tire constructions on one vehicle can produce unpredictable and, therefore, dangerous handling characteristics. The different slip angle characteristics of the tires can produce severe unexpected steer reactions when entering a turn, especially when the vehicle is driven near the traction limit. These conditions are even greater when driving on rain-slicked highways.

If a vehicle has radial ply tires mounted on the front and bias ply tires on the rear, the different slip angles would cause the vehicle to *oversteer*. Having bias ply tires on the front and radial ply tires on the rear would cause a vehicle to *understeer*. Assume that the steering wheel were held in a constant turn position as the speed is increased. The vehicle is said to understeer when it tries to go straighter as the speed increases. It oversteers when it tries to make a tighter turn. You can see why tires with different ply construction should not be mixed on the same vehicle. Figure 4-13 shows the understeer and over-

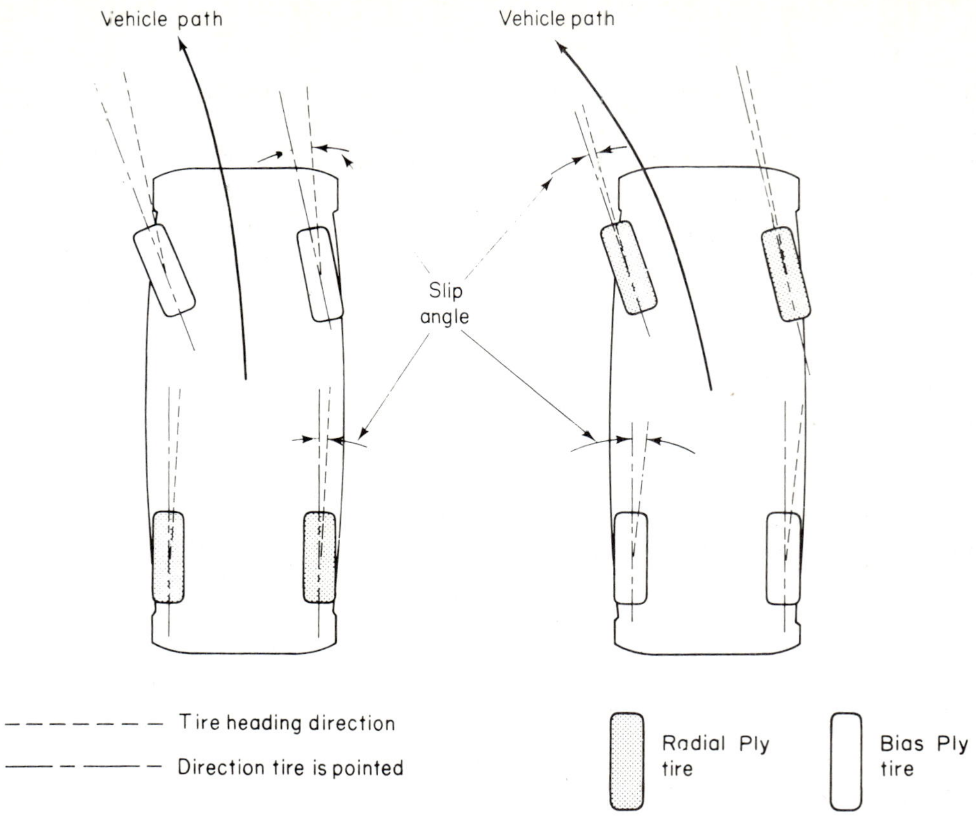

FIGURE 4-13 Effect of tire construction on vehicle steering.

steer tendency of a vehicle when tires with different types of construction are being used.

Tires and Fuel Economy. To improve fuel economy, research continues to find ways to reduce the rolling resistance of tires. This is important because, once a vehicle is moving at a constant highway speed on a level road, approximately 20 percent of the fuel energy is used to roll the tires. Most of this energy is lost in flexing the tire as it rolls through the footprint. You will remember that energy is lost from internal friction between the cords of the tire. Radial ply tires have the lowest internal friction. Their use gives vehicles the best fuel economy.

Tires are marked with a grade (A, B, or C) to indicate how well they resist heating when they are being run under standard test conditions on an indoor test wheel. All tires must meet or exceed temperature grade C. Performance tires are either grade B or the highest grade, A. Tires must be properly inflated and not overloaded to meet the temperature grade.

Tire Deterioration. We have been looking at how the different tire types and construction features affect the operation of new tires. The tire responds, as it was originally designed, for only a short time before wear occurs. Tires also deteriorate with temperature and age. The materials in the tire continue to cure as they remain in the atmosphere after manufacture, even in storage. Higher temperatures increase the rate of oxidation, the major cause of deterioration. Tires will harden faster at warm operating temperatures than they will at cool temperatures when in storage.

As the tread rubber oxidizes, the rubber becomes harder, reducing the coefficient of friction in the footprint. A tire that runs cooler on the vehicle will not deteriorate as fast as a warmer-running tire with higher internal friction. Age can also cause deterioration of the cord material in the carcass. In time, this can result in tire failure due to fatigue.

Water channeling, which is so important on wet roads, is reduced considerably as the tread wears. This means that a vehicle with worn tire tread must be driven slower for safety by reducing the chance of the tires hydroplaning.

All tires of like construction do not perform the same under similar conditions. It is difficult to judge which replacement tire is the best for a specific type of vehicle and type of driving. Performance during the life of the tread, as well as tire mileage, are methods used to evaluate tires. Most operators evaluate a tire on the price and the mileage that can be obtained without thinking about other factors. This results from the fact that most operators drive their vehicles well below the critical traction and weight limits of the tire. For satisfactory tire operation under all operating conditions, a careful balance must be made between tire-wear rate and tire performance.

4-4 TIRE TYPES AND SIZING

You will be faced with the problem of recommending a good replacement tire. Tire selection should be based on original equipment specifications, type of driving anticipated, weather conditions normally encountered, and cost of the replacement tire.

Tire Specifications. Tire manufacturers use a series of numbers and letters to identify the tire size, load capacity, inflation limits, and type of service for which the tire is intended. The table in Figure 1-28 shows the popular tire sizes. There are a number of tire properties that must also be considered when replacing tires.

When replacing tires, the first consideration is the wheel rim size. Two dimensions are considered, the diameter across the wheel where the tire bead seats and the distance between the inner and outer bead flange. These are shown in Figure 4-14. The diameter is the most important of the two dimensions. You will find 12-, 13-, 14-, 15-, 16-, and 17-inch diameter rims on passenger cars. Large wheel diameters require large tires which, in turn, will support more weight when using the same series tires. The tire must be large enough to support the weight of the vehicle and its load. Underfender clearance limits the maximum tire size that can be used on a vehicle. Oversized tires usually require wheels with wider rim widths, because the rim width must also match the tire size and design.

The aspect ratio is considered when replacing tires. The *aspect ratio,* or *tire series*, is the ratio of the tire section height to the tire section width, as shown in Figure 4-15. Manufacturers supply 75 series tires on most new vehicles. Replacement tires are available for passenger cars in 75, 70, 65, 60, and 50 series. A 50 series tire would have a section height

FIGURE 4-14 Location of wheel dimensions.

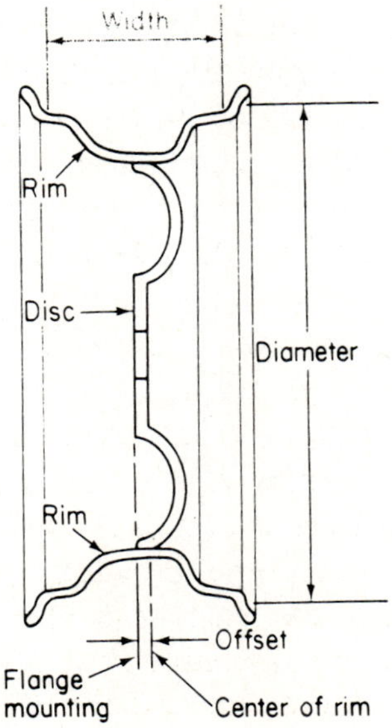

FIGURE 4-15 Meaning of tire size designations.

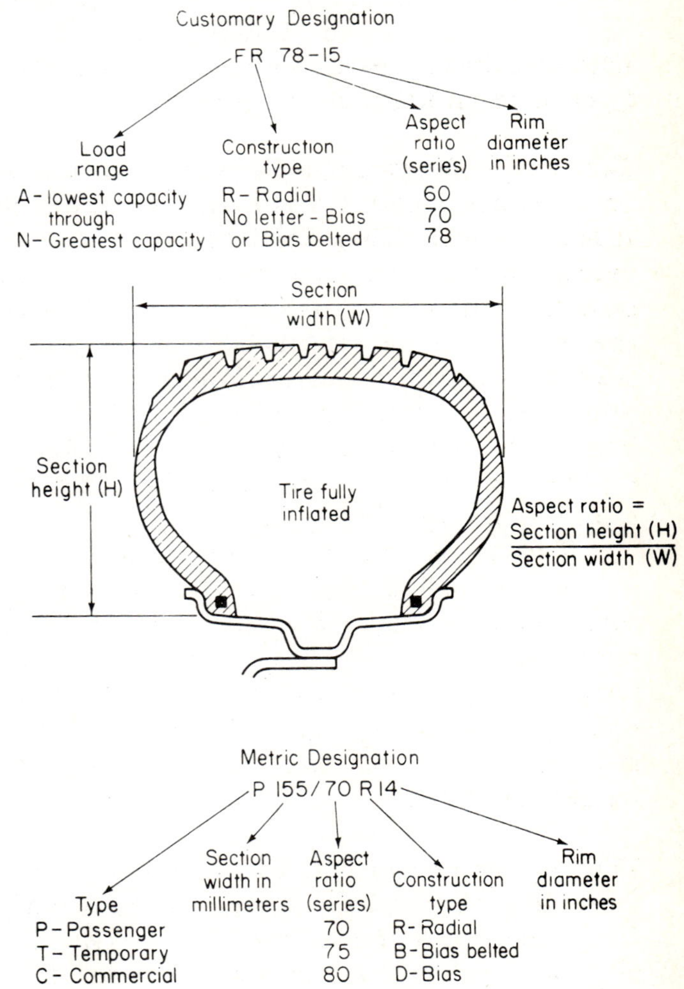

one-half as large as the section width, while a 75 series tire would have a section height approximately three-fourths as large as the section width. The low-aspect-ratio or low-series-number tires tend to have shorter sidewalls with a lower profile, giving the vehicle a sporty or performance look. They have faster steering response, increased stability, and better cornering ability. Low-aspect-ratio tires also have a wider footprint with more tread surface on the road. Race cars have aspect ratios as low as 35, which provides better handling in turns. High- aspect-ratio tires with longer sidewalls typically will have better ride characteristics with a lower handling response than low-aspect-ratio tires. Each tire series must have wheels with the correct rim width to allow the proper tire shape for maximum tire-to-road contact in the footprint.

Metric tire specifications and markings are also shown in Figure 1-28. The metric tire identifications use the section width in millimeters, "R" for radial tires, the aspect ratio to the nearest 5 percent, and the wheel size in inches. In a few cases, the wheel size has a metric diameter. The section width of the tire relates directly to the tire's load capacity. It is an accurate description of the maximum load capacity of the tire.

Some owners want to change the handling of a vehicle by changing the tire design and size. When doing this, it is important to keep the outside diameter of the tire as near to the standard tire size as possible. The outside diameter of the tire tread affects how far the vehicle will move with each revolution of the tire. A large-diameter tire will roll farther on one revolution than a small-diameter tire. A change in the tire diameter will also affect the ground clearance under the vehicle.

The speedometer is matched to the tire diameter. If the tire diameter is changed, the speedometer gear will also have to be changed to maintain correct speedometer readings. The correct vehicle speed on each tire revolution becomes critical when computers are used to control the engine. Vehicle speed is one of the input signals used by the computer. The computer will receive the wrong speed signal when the diameter of the tire is not correct. The computer will not control the ignition and fuel correctly when it receives incorrect signals.

Moderate increases in tire size will not increase the diameter of the tire when it is used on a wide wheel rim. The most effective way to improve handling is to use short tire sidewalls while maintaining the diameter of the tire. This can be done by using tires with lower aspect ratios and using larger and wider wheel rims. Basically, you would use a tire 10 mm wider for each 1-inch increase in diameter. Plus One is a wheel rim that has a 1-inch larger diameter and uses a 60 aspect ratio tire. Plus Two uses a 2-inch larger diameter and a 50 aspect ratio tire. These tires are shown in Figure 4-16. When selecting replacement tires, you must always choose a tire with load rating at least as high as the standard tire it replaces.

Standard radial tires are marked "R." Tires marked "SR" are rated for speeds up to 111 mph (177 km/h); those marked "TR" are good up to 118 mph (187 km/h). When the tire is marked "HR,"

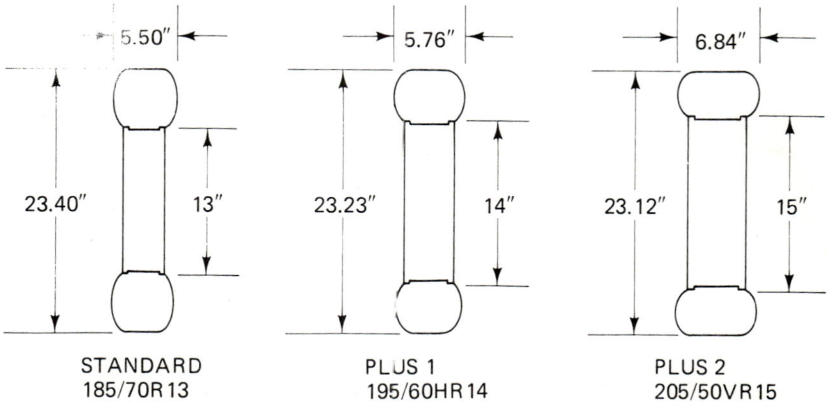

FIGURE 4-16 Standard, Plus One, and Plus Two tires compared.

FIGURE 4-17 Deflated and inflated temporary spare tires.

it can be operated at speeds up to 130 mph (208 km/h). Tires marked "VR" are rated to operate up to 149 mph (238 km/h). "ZR" tires are rated for speeds above 149 mph (238 km/h). Plus One or Plus Two performance tires should have "HR" or higher tire ratings.

Temporary tires are placed in most new vehicles for highway emergencies. These tires are designated "T" for temporary use only. Their use is limited to a maximum speed of 50 mph (80 kp/h) and a maximum range of 50 miles (80 km). The temporary tires are smaller when stored and lighter than the normal vehicle tires. As a result, they permit greater luggage space in the vehicle. Early temporary tires were carried while deflated. They were inflated with a pressure can after they were installed on the vehicle. Modern temporary tires are small and carry 60 psi inflation pressure. Both types are shown in Figure 4-17. Handling and driveability are poor when these tires are being used. Drive carefully when using temporary tires and replace them as soon as possible.

4-5 WHEELS

The wheels on a vehicle interact very closely with tire characteristics to provide ride quality, load capacity, and handling. The wheel design will also affect front suspension design as it relates to vehicle control. Most wheels used on passenger vehicles are made from stamped steel. The stamped steel is made in two pieces and then welded or riveted to form the wheel assembly. This makes a true-running, airtight, and strong tire-holding wheel. The rim is designed to provide the inner radius for the tire. The center disc is the section of the wheel that bolts to the axle or wheel hub. Some wheels are designed for appearance and/or performance. These wheels are cast in one piece. They are commonly referred to as *mag wheels*, although they may be made from either magnesium or, more commonly, an aluminum alloy. Fiber-reinforced composite wheels are starting to be used on some specialty vehicles. These wheels are lighter than metal wheels; they are accurately round and balanced; they do not corrode and they can be easily styled. As a result, they are becoming more popular.

Some heavy-duty trucks and trailers also use single-piece wheels. The rim, with the tire mounted, is bolted or clamped to the wheel hub. Other rims are designed with a removable bead seat. These wheels are used on large tires to allow easy mounting and dismounting. This type of wheel does not have a drop center and it is referred to as a *split rim*. Another wheel design is made by bolting two wheel halves together. This design would typically be used on small-diameter wheels that would otherwise make tire mounting and dismounting difficult due to the small size of the tire. Passenger car wheels use the drop center wheel rim. This type allows the tire to be mounted and removed without separating the wheel parts.

You will remember that the tire to be used determines the width of the rim. The rim diameter and offset are determined by the underfender clearance and use of the vehicle. Wheel offset is the distance between the centerline of the rim and the mounting face of the disc. It is selected to fit the requirements of the vehicle. The wheel dimension locations are shown in Figure 4-14. The offset is positive when the centerline of the rim is inboard of the mounting face and negative if outboard of the mounting face. The amount and type of offset are critical because changing the wheel offset will change the front suspension loading as well as the scrub radius as described in Section 8-1. This is important to suspension alignment and vehicle handling characteristics.

STUDY QUESTIONS

Introduction

1. When does a tire slide on the road surface?
2. Describe tire coefficient of friction.

Section 4-1

3. What type of tire tread makes the least noise?
4. What compromise is made on the tire tread design?
5. What is special about asymmetric tires?

Section 4-2

6. How does water get out from under a rolling tire?
7. What tire design has the strongest sidewall?
8. What tire design gives the best mileage?
9. Why do radial ply tires operate cooler than bias ply tires?
10. Why do tires have unequal spacing between sipes?
11. What are the advantages and disadvantages of a grooved road surface?
12. What is special about the tread on all-weather tires?
13. What is the purpose of the tread wear grade marking on the tire sidewall?
14. What causes tire tread squirm?
15. How has squirm been eliminated from tires?
16. How do sidewall rubber and tread rubber differ?
17. What is the last point in tire manufacturing that still allows the tire to be disassembled for repair?

Section 4-3

18. What tire design characteristics improve wet traction?
19. What tire design characteristics improve dry traction?
20. What tire design characteristics causes the least slip angle?
21. What is the major disadvantage of radial tires?
22. What tire construction is most sensitive to wheel alignment settings?
23. What is steering sensitivity?
24. What is self-aligning torque of a tire?
25. What is the tire slip angle?
26. How does a driver recognize understeer?
27. How does a driver recognize oversteer?
28. How does mixing tires of different constructions affect vehicle steering?
29. How can you tell how well a tire resists heating?
30. List the ways a tire will deteriorate.

Section 4-4

31. What must be taken into consideration when selecting replacement tires?
32. What dimensions must be considered when replacing tires?
33. What limits the size of the tires a vehicle can use?
34. What is the aspect ratio of a tire?
35. How does the aspect ratio of a tire affect vehicle handling?
36. How do Plus One and Plus Two tires differ from standard tires?
37. What markings on performance tires are different from the markings on standard tires?
38. Where and when should a tire marked with a "T" be used?

Section 4-5

39. From what three materials are wheels made?
40. Name three ways wheels are designed to allow the tire to be removed and installed.
41. When is wheel offset positive?

5

Tire and Wheel Service

Tires and wheels operate as an assembly. Faults in either or both can cause an abnormal ride and poor handling. Periodic inspection and preventive maintenance procedures minimize the chance of the faults causing problems. Faulty tires and wheels must be either repaired properly or replaced when an inspection shows any part of the assembly to be faulty.

5-1 INSPECTION

Poor driving habits, such as high-speed cornering, cause wear on the outer edges of the tire tread as shown in Figure 5-1. This type of wear is similar to the type of tire wear caused by underinflation. Lack of preventive maintenance allows tires to operate while underinflated and overinflated, unrotated, and out of balance. Incorrect inflation will cause predictable wear patterns, as pictured in Figure 5-2. Spotty or erratic wear patterns in the form of cupping or spot wear, of the type shown in Figure 5-3, are usually caused by unbalanced tires. This type of wear could also be caused by loose, worn, or bent suspension parts, faulty shock absorbers, or a bent wheel.

Some types of tire tread wear always indicate the need for suspension realignment. Tire wear caused by incorrect camber is shown in Figure 5-4. Caster will only produce wear when it is far out of adjustment. Toe-in, if excessive, will produce feathered edge wear on the outer edges of the tire tread ribs. Camber, caster, and toe are discussed in Section 8-1. Excessive toe-out will produce the same wear condition on the inner edge of the tire tread ribs. Wear from incorrect toe is usually on the edge of the tread, similar to that of camber wear. Toe wear is shown in Figure 5-5. The vehicle must be realigned to correct tire wear caused by out-of-specification camber and toe.

Tire balance is necessary to produce even tread wear around the tire. After a tire is reinstalled on a wheel, the assembly should be balanced to assure smooth tire operation.

Tire Failure. A faulty tire may show as ply separation, tread separation, bead separation, cord separation, or chunking. These conditions are defined as follows:

Ply separation: Parting of adjacent plies within the carcass due to failure of the rubber bond between the plies.

Tread separation: Parting of the tread rubber from the tire carcass.

Bead separation: Separation in the bead area of the tire at the wheel rim caused by a loosening of the carcass rubber and plies.

Cord separation: Failure of the rubber bond between the cord material within a ply.

Chunking: A condition where pieces of the tread rubber are thrown from the tire.

FIGURE 5-1 Wear on the edge of the tire from high-speed cornering.

FIGURE 5-2 Wear caused by improper inflation: (a) underinflation wear on the outer ribs; (b) overinflation on the inner ribs.

(a)

(b)

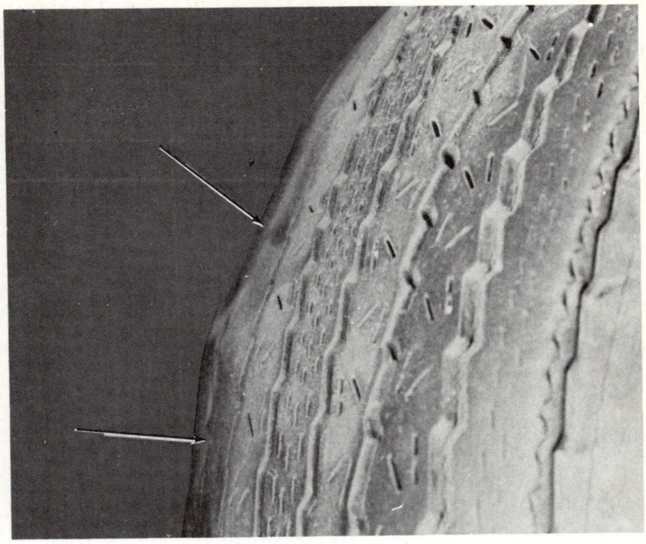

FIGURE 5-3 Cupping or spot wear.

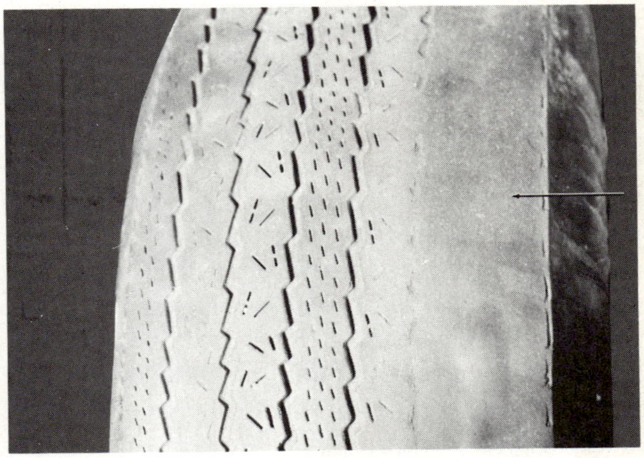

FIGURE 5-4 Wear on one side of the tread as a result of incorrect camber.

FIGURE 5-5 Feathered edge on the tread caused by incorrect toe setting.

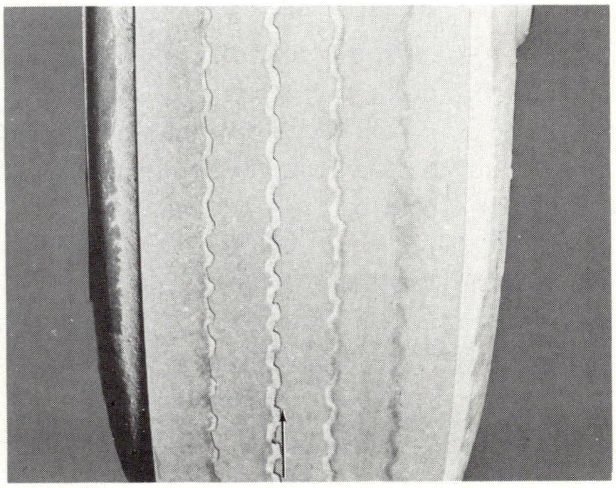

These types of tire failure are the result of excessive heat buildup. Heat buildup is caused by excessive tire flexing or by scuffing. Most often the heat buildup is caused by operating the tire with less air pressure than specified for the loads being carried by the tire. Chunking is usually caused by scuffing. This results from high heat on the tread surface. The heat releases the bond between the tread rubber and the tire carcass in localized areas. Centrifugal force of the rotating tire throws the loosened tread rubber from the tire.

Vibration and noise complaints lead to a tire inspection. Tire vibration is an out-of-balance condition causing a shake. It can be traced to one corner of the vehicle. A noisy tire could be the result of tire tread cupping. Extreme cupping will often cause vibration that feels similar to a failed wheel bearing. Separation causes a slapping noise as the loose part of the tire hits the road.

Wheel Pull. Often the wheels are realigned to correct for wheel pull caused by the tires. Wheel pull can be caused by either a tire or by faulty wheel alignment. When bias ply tires were used, faulty wheel alignment caused most wheel pull. Radial tires are more likely to cause wheel pull or lead than bias ply tires. They may also cause unusual steer characteristics. It is usually difficult to determine if the alignment settings or tires are causing the wheel pull when radial tires are used. If it is not known whether tires or alignment are at fault, the normal procedure is, first, to make sure the wheel alignment settings are correct and then check for tire problems. Sometimes, it is more convenient to check the tires for their effect on wheel pull before checking the wheel alignment settings.

If the alignment settings are known to be correct, the diagnosis of wheel pull is straightforward. It simply involves changing the suspected tires from left to right. If this change causes the vehicle to pull in the opposite direction, one of the two tires is faulty. The tires are then restored to their original locations and the faulty tire is replaced.

Wheel pull caused by tires is usually caused by the way the belt is placed on the tire carcass. If, during tire manufacture, the plies and belts are not assembled exactly, the tire may have a slight cone shape that causes *ply steer*. This results in a tire that acts somewhat like a rolling cone. Either a misplaced belt or ply steer will cause wheel pull when they are out of specification within the tire. If the ply steer is not equal around the tire, it will produce waddle. *Waddle* is a periodic pull as the tire rolls, producing a low-speed shaking of the vehicle. As a result of these problems, radial tires are more sensitive to production variations than bias ply tires.

It is possible, after alignment, to have a tire produce a directional wheel pull. This could be caused by unequal wear across the tread surface caused by the previous misalignment. With realignment, the center of pressure is no longer located in the center of the worn tread, resulting in wheel pull. This problem is discussed in more detail in Chapter 8. It could also be caused by ply and tread separation.

5-2 RUNOUT

Occasionally, tires that show no abnormal condition during routine maintenance are suspected as being the cause of a vibration. These tire and wheel assemblies should be checked for both runout and balance. Runout should be checked before balancing because it can cause vibration even when the assembly is in balance. Runout can occur in two ways: laterally and radially.

Lateral Runout. Lateral runout is the side-to-side movement as the tire and wheel assembly rotates. Excessive lateral runout will cause wheel shake. It is measured with a dial indicator positioned on the sidewall close to the tread as pictured in Figure 5-6. Acceptable lateral runout is usually no more than 0.080 inch (2.00 mm). Most tires fall within acceptable limits. Excessive lateral runout can be caused by improper tire installation, faulty wheel installation on the hub, a bent wheel, or a faulty tire.

Sometimes, excessive lateral tire runout can be corrected. First, mark the point of greatest tire runout with chalk. The lateral runout of the wheel rim is then checked (Figure 5-7) to make sure that it is within specifications. Lateral wheel runout should be less than half of the acceptable tire runout, or 0.040 inch (1.0 mm). Mark the point of greatest wheel runout. If the marks on the tire and wheel are close together, it may be possible to correct the tire runout. To do this, deflate the tire, free it from the rim, rotate it until the marks are opposite each other,

FIGURE 5-6 Dial indicator set to measure lateral runout of the tire.

FIGURE 5-7 Dial indicator set to measure lateral runout of the wheel rim.

then reinflate the tire. Mount the assembly on the vehicle and recheck the runout. If the lateral runout is increased, the tire is at fault. If rotating the tire on the wheel made no change in lateral runout, the wheel itself is at fault. If the lateral runout has decreased, it may not cause vibration even if it is still slightly out of specification. Balance the tire and test drive the vehicle.

Radial Runout. Tire runout can occur radially as well as laterally. Radial runout causes the tire tread to move away from its axis at one point and come closer to the axis at another point as the tire and wheel assembly rotates. The tread does not form a true circle around the tire.

Tires tend to develop a *set*, or flattened spots, when standing in one position as they cool. Tires must be properly inflated and should be driven enough to warm them before measuring radial runout. This will normalize the tire tread.

Radial runout is measured with the indicator pickup against the tread surface as shown in Figure 5-8. Excessive radial runout will cause enough noise and roughness to be felt by the operator. The total runout, therefore, should not exceed 0.045 inch (1.1 mm). If the radial runout is excessive, the point of greatest runout should be marked. The dial indicator is then positioned against the wheel rim, as shown in Figure 5-9. The wheel radial runout is checked and the high spot is marked. Radial runout of the wheel should not exceed 0.035 inch (0.9 mm).

FIGURE 5-8 Dial indicator set to measure radial runout of the tire.

FIGURE 5-9 Dial indicator set to measure radial runout of the wheel rim.

If the wheel radial runout is excessive, mark one lug stud and matching hole in the wheel. Then change the position of the wheel on the hub two or three bolt holes from the original location. Recheck to see if this change has corrected the runout. If the runout is now within specifications for the wheel and tire assembly, mark one lug stud and the matching lug hole in the wheel. Paint works well for this. Then, the stud can always be used to position the wheel in the same location to minimize runout. If the wheel runout is still excessive in the same direction in relation to the hub, the runout is due to a faulty hub.

Radial runout, like lateral runout, can often be reduced by rotating the tire on the rim when the hub and wheel are within specifications and the high runout marks on the tire and wheel align. Deflate the tire and rotate it on the wheel so the marks are opposite each other. After inflating the tire, reinstall the wheel. Measure the radial runout again. If the radial runout is still at the same location and the runout has not changed, the wheel is at fault. If the radial runout is greater than it was when first measured, the tire is at fault. If the radial runout decreases to an acceptable value, the tire and wheel assembly should be rebalanced and the vehicle test driven.

Loaded Runout. A tire falling within the lateral and radial runout specifications may still have runout when the tire is supporting a load. Loaded runout or tire force variation is due to the way the tire was built. It is represented as springs between the rim and tread in Figure 5-10. The tire will force the wheel spindle upwards as a stiff section within the tire rolls through its footprint. This causes up-and-down movement of the spindle with each revolution of the tire. This will be felt as a bump by the operator. Special equipment is needed to measure loaded runout. A motored drum is placed under the tire of the vehicle. The tire supports its normal vehicle weight on the drum. The drum rotates the tire while an indicator measures any up-and-down spindle movement (Figure 5-11).

When this type of equipment is not available, the cause of the shake problem can sometimes be located by overinflating the tires. All four tires are overinflated to 50 psi (345 kPa) and the vehicle is test driven. If the vehicle rides smoother, deflate one tire to the recommended air pressure and test drive the vehicle again. Repeat this procedure at each tire. When the faulty tire is returned to its original pressure, the rough ride should return. The faulty tire will have to be replaced to correct the problem.

Another way to find the cause of the shake problem is to substitute the wheel and tire assembly. Mount the spare tire or a normal tire and wheel

FIGURE 5-10 Springs used between the rim and tread to illustrate the effect of a stiff section in a tire.

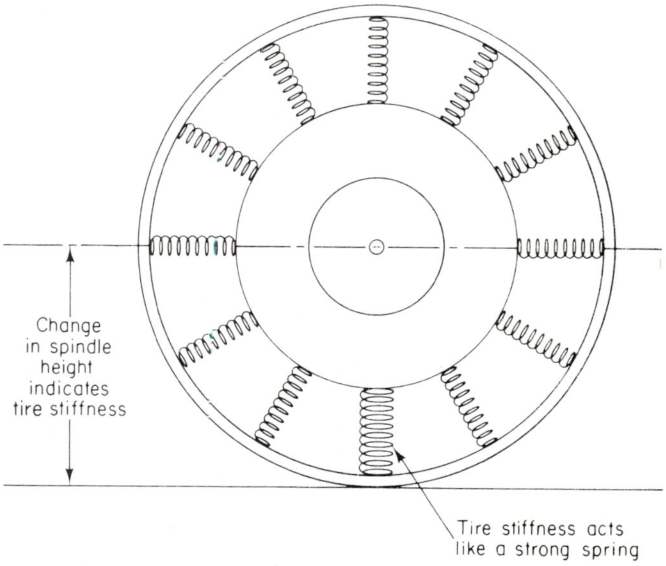

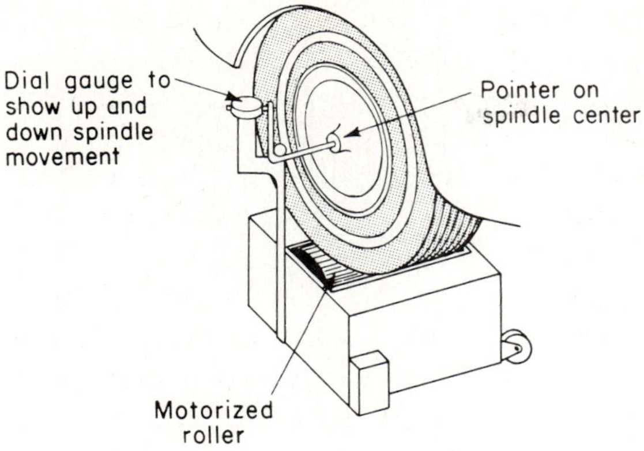

FIGURE 5-11 Measuring loaded runout with a tire problem detector as the tire and wheel assembly is rotated.

assembly in place of the assembly suspected of being faulty; then test drive the vehicle. Move the spare tire and wheel assembly to each wheel location, in turn, until the vehicle rides smoothly. The faulty tire will never roll smoothly, but it is usually good enough for a spare.

Make sure that everything affecting smooth vehicle operation is correct before condemning the tires and/or wheels.

1. Tires should always be checked for proper inflation.
2. The tires should be run enough to be warm before measuring runout: lateral, radial, and loaded.
3. The wheel bearings must be properly adjusted (as discussed in Section 3-4).
4. Make sure that there are no defective wheel or axle bearings.
5. The lug nuts must be properly tightened in the correct tightening sequence to the correct torque.

5-3 WHEEL SERVICE

Before condemning a wheel with excessive runout, double check to make sure the wheel lugs have been correctly tightened and the hub and bearings are normal. If the wheel is faulty, it must be replaced.

Wheel failure may be caused by improper maintenance, accidents, overloading, and age. Figure 5-12 shows a rim flange that was bent when the wheel went through a pothole in the road.

Improper maintenance and accidents are the major causes of wheel failures. Improper wheel maintenance will show up as wear in the bolt holes. This is generally caused by the wheel lug nuts loosening during operation. The bolt holes can also fail from fatigue in the bolt circle resulting from tightening the wheel lugs too much. Improper tightening can also damage wheel hubs and brake parts.

Overloading wheels will usually show up as a failure in the rim bead flange and seat area of the rim. This area of the wheel will fatigue under continued excessive loads and it will eventually fail. The wheels will also be overloaded when tires designed for wide rims are incorrectly placed on narrower rims. Continued overloading with side loads on the tire will sometimes result in failure of the wheel disc in the bolt circle area as shown in Figure 5-13.

Accidents are the most common cause of wheel replacement. The wheel is usually hit on the side in an accident. This will cause the rim and center disc section to bend. After an accident, the vehicle suspension and associated drive line components should also be inspected for other bent parts.

Failure of the wheel due to age can occur in several ways. Corrosion can take place inside the wheel rim. This will lead to air leaking around corroded or loose rivets or through corroded or cracked spot welds. The raised bosses and bolt holes can become worn with use so the wheel will not accurately center or the lug nuts will not remain tight.

FIGURE 5-12 Rim flange bent when the wheel went through a pothole in the road.

FIGURE 5-13 A wheel disc that failed from overloading.

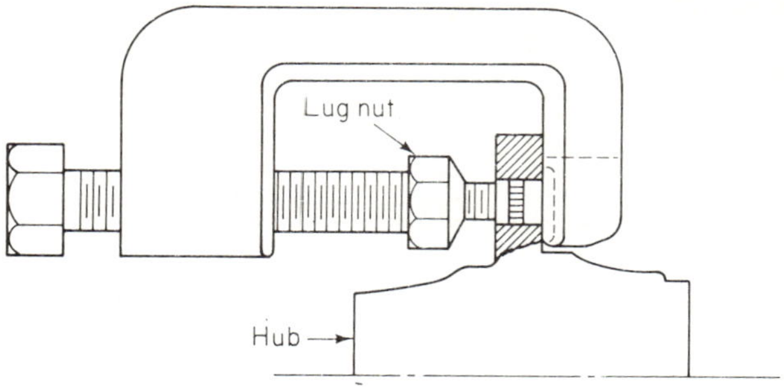

FIGURE 5-14 Removing a damaged lug bolt.

FIGURE 5-15 Oversized nut being used as a spacer to pull a new lug bolt into a flange.

Carefully inspect the wheel each time a tire is to be mounted. If the wheel is found to be damaged or defective, it should be replaced.

Lug Bolt Replacement. Occasionally, a lug bolt will break when you are removing a wheel. This can be the result of overtightening, or the bolt may have rusted in service. Broken lug bolts must be replaced to make the vehicle safe to drive. It is standard procedure to replace all the lug bolts on a wheel hub or flange if one fails. Usually, the rest of the lug bolts are in the same poor condition. The area around the broken lug bolt should be supported while the lug bolt is being removed. It must also be supported when the lug bolt is reinstalled. This will prevent damage to the flange, hub, or rotor.

Lug Bolt in Flange. Replacement is straightforward when a lug bolt breaks on a wheel axle flange. The wheel and brake drum or rotor are removed. If necessary, loosen the brake shoes following the directions given in Section 3-4.

The broken bolt can be removed from the axle flange with a puller or press (Figure 5-14). Do not use a hammer to remove lug bolts. Hammering on the flange will damage bearings. A new lug bolt is lubricated and positioned in the bolt hole. A thick washer or oversized nut is lubricated with grease and placed over the bolt to act as a pulling surface. A lug nut is threaded upside down over the new lug bolt. It is tightened until the lug bolt is pulled against the axle flange (Figure 5-15). The nut and spacer are removed. This procedure is repeated on the remaining lug bolts that need replacement. The drum or rotor and wheel can then be reinstalled.

Lug Bolt in Drum or Rotor. The brake hub and drum assembly or rotor must be removed if a lug bolt breaks on a front wheel with drum brake or one-piece rotor and hub. The lug bolt is pressed from the outside inward to remove it (Figure 5-16). The new lug bolt is pressed into place from the inside (Figure 5-17). Replace all the lug bolts in the hub using this procedure. If a press is not available, do not try to replace the lug bolts in the hub because the hub or brake surface may be damaged.

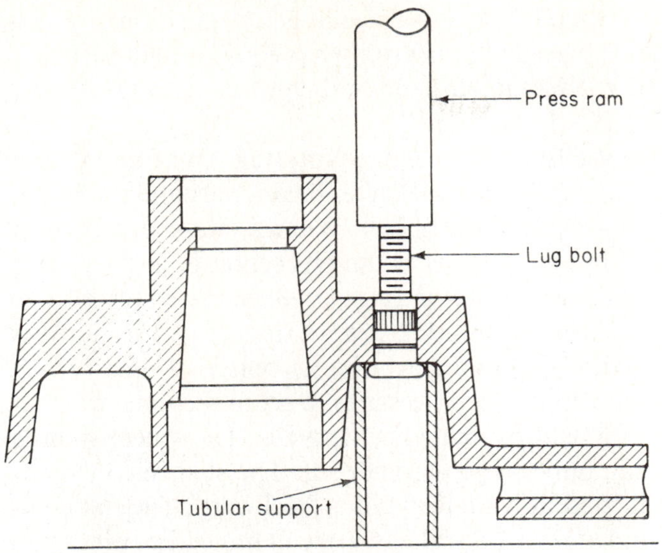

FIGURE 5-16 Pressing a lug bolt from the rotor.

5-4 BALANCE

Tire balancing is one of the most common types of tire service. Whenever a tire is mounted, the tire and wheel assembly should be balanced. The assembly can be balanced using a number of different methods, although there are only two types of imbalance or unbalance: static and dynamic.

Static Balance. The assembly is in static balance when there is equal weight distribution about the axis of rotation of the tire and wheel assembly. Therefore, static imbalance is corrected by adding enough weight to the light spot on the wheel rim to equalize the weight around the wheel center. Static imbalance can be checked with the wheel:

1. Supported in a horizontal position.
2. In the vertical plane.
3. While rotating.

The final tire balance tolerance depends on the skill of the operator and the type of balancing equipment used.

Horizontal Static Balancing. Static tire balancers that balance the tire while it is supported in a horizontal position with no rotation are usually called *bubble balancers* (Figure 5-18). The wheel fits over a hub on the balancer. Imbalance is shown by a bubble in the sight glass in the center of the hub. The bubble balancer is similar to a carpenter's level. The bubble in the sight glass shifts toward the light side of the wheel (Figure 5-19). Weights are laid on the light side of the wheel rim until the bubble is centered. Once the required location of the weights is determined, the location is marked. The tire and wheel assembly is removed from the balancer. It is customary to put half of the added weight to the inner rim and half to the outer rim, especially when 2 ounces (57 g) or more are needed. The purpose of adding weights to both inner and outer rims is to keep from changing the dynamic balance of the assembly. Weights are tapped onto steel rims or cemented on cast rims and then the balance is rechecked.

FIGURE 5-17 Pressing a lug bolt into a rotor.

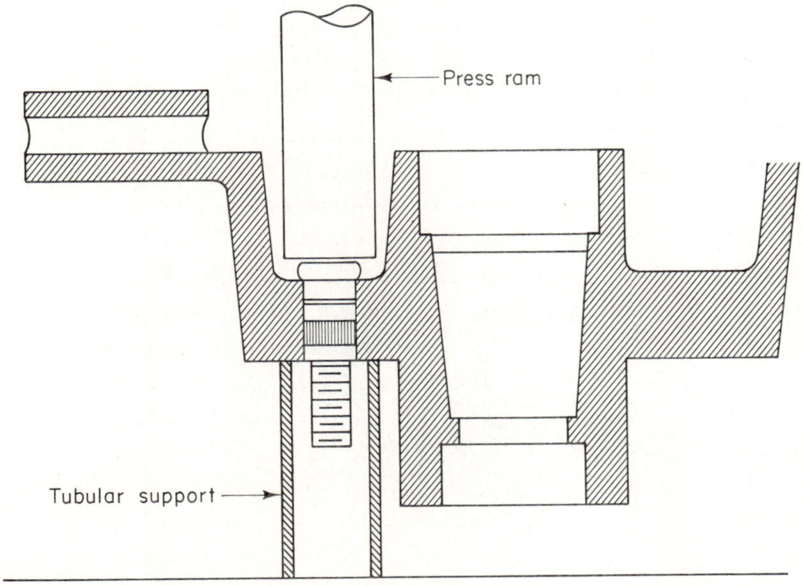

(a)

FIGURE 5-18 Static bubble balancer in the balancing position.

(b)

FIGURE 5-19 Off-center bubble indicating imbalance.

Vertical Static Balancing. Some equipment will static balance the tire and wheel assembly in a vertical position. The heaviest spot in the tire will tend to rotate toward the bottom. With this type of balancing equipment, the tire and wheel assembly is mounted on a hub. The hub arbor of the balancer is supported with antifriction bearings. The tire and wheel assembly is allowed to rotate freely until the heavy spot settles to the bottom. Weight is added at the very top of the wheel rim flange, which is the lightest part of the assembly. The tire and wheel assembly is rotated to various positions to determine if the assembly still has any tendency to rotate. Add or remove weight until there is no tendency to rotate. When the assembly remains at any position in which it is placed, the assembly has equal weight distribution about its axis and is, therefore, in static balance.

Static Imbalance While Rotating. When an object is spun in space, all forces balance around the center of gravity of the object. When a odd-shaped object is thrown, it spins around its center of gravity. You have seen a plastic funny ball that is heavy on one side. When thrown, it appears to wobble, but it is actually spinning around its center of gravity. An out-of-balance wheel tries to act in the same way but it is held by the wheel spindle. The wheel spindle can only move up or down. The suspension keeps it from moving forward or backward. The static imbalance force of the wheel will move the spindle up and down as the wheel rotates (Figure 5-20). This action is called *wheel tramp*. The effect of the shake increases four times as the speed is doubled, so imbalance is more noticeable at high speed. One type of tire balancing equipment checks imbalance while rotating the wheel on the vehicle. The tire and wheel asssembly is removed from the vehicle and mounted on the second type of balancing machine. There are advantages and disadvantages to each type.

Balancing the tire and wheel assembly while it is still mounted on the vehicle allows all of the rotating mass—tire, wheel, brake drum or rotor, and wheel cover—to be balanced at the same time. This can result in a balance with the least tolerance. The entire assembly is free of vibration. When balancing is completed with the wheel on the vehicle, one lug and a corresponding lug hole should be marked. Realigning the marks will ensure proper relocation and balance when the tire and wheel assemblies are reinstalled after being removed. Any time the tire and wheel assemblies are moved to a different wheel hub, the assembly must be rebalanced with the parts of the new hub.

FIGURE 5-20 Effect of static imbalance.

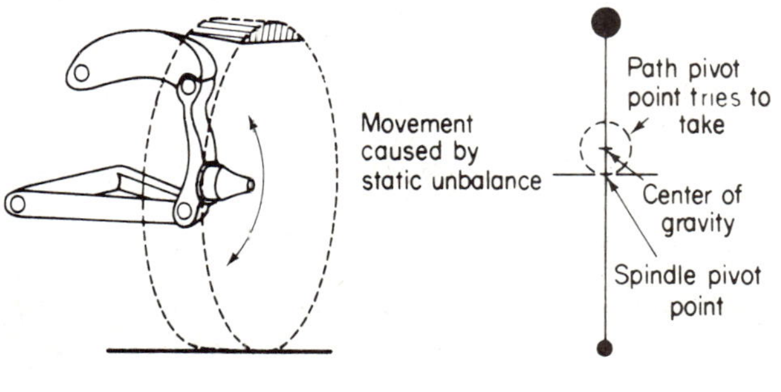

Balancing the wheel and tire assembly when it is off the vehicle on a neutral spindle allows each assembly to be balanced. As a result, the balanced assemblies can be put at any wheel location on the vehicle. If imbalance occurs after mounting the assembly, the imbalance is caused by the other rotating parts. These should then be balanced separately.

Static imbalance with the tire and wheel assembly on the vehicle relies on the operator's ''feel'' or use of an electronic readout to determine imbalance. The vehicle is jacked up so the wheel will rotate freely. The equipment, shown in Figure 5-21, has an adapter to connect the balancing head to the wheel. There are movable weights inside the balancing head. While the wheel is rotating, the weights can be moved to change the amount and position of effective weight. Nondriving wheels are rotated with a wheel spinner. Driving wheels are rotated with the engine and drive line. The weight in the balancer is adjusted as the tire is rotated with a wheel spinner until no vehicle shake can be seen or felt. Rotation is stopped and the amount of weight shown on the scale is added to the wheel rim in the position indicated. The wheel attachments are then removed and the wheel assembly is respun to check the balance. This type of balancing offers quick service while detecting vibrations that annoy the operator.

Another type of equipment that detects static imbalance while the tire and wheel assembly is on the vehicle is shown in Figure 5-22. This balancer uses a magnet to hold a vibration sensor on the bottom of a suspension part as close to the wheel as possible. The sensor detects any downward suspension movement by flashing a strobe light. A meter on the light indicates the amount of movement. The strobe flash is similar to an engine timing light. The flash will make the assembly look like it is stationary with the heavy part at the bottom.

A chalk reference mark is placed on the tire, or the valve stem is used for a reference point. The assembly is rotated with the wheel spinner or engine

FIGURE 5-21 Static balancing as the wheel is rotating.

FIGURE 5-22 Vibration sensor connected to the lower suspension arm with a magnet to sense vertical shake caused by static imbalance.

and the location of the reference mark is noted in the light of the strobe flash. When the position of the reference mark and reading of the weight indicator are noted, rotation is stopped. Do not try to stop the wheel by hand. The wheel is rotated so the reference mark is positioned as it appeared in the strobe flash. The amount of weight indicated is added to the light part of the wheel rim flange at the top. The assembly is spun again to check the balance. If, on the respin, the added weight is at the top, too little weight was added. If the weight is anywhere between top or bottom, the position of the weight is incorrect. The position is corrected by moving the weight toward the top, a small amount at a time. If, after repositioning, the weight appears at the bottom, replace the weight with a smaller weight. Add or remove weights as needed until the assembly is balanced. Strobe balancers can be used to balance any size tire, although it usually takes longer to balance tire and wheel assemblies with this equipment than with other balancing methods.

Some static wheel balancers spin balance the wheel when it is off the vehicle. They sense the force of the static imbalance about the axle of the rotating assembly. The machine indicates the amount and the location of the weight needed to balance the assembly. Weight is added to the rim; then, the assembly is respun to assure that proper balance has been attained. This type of balancer is expensive, and requires very little operator skill to properly balance tire and wheel assemblies.

Dynamic Balancing. Dynamic imbalance, unlike static imbalance, can only be detected while the wheel is rotating. The tire and wheel assembly has dynamic balance when it has equal weight distribution about the center plane. Dynamic imbalance is caused by unequal weight distribution across the center plane of rotation through the assembly as illustrated in Figure 5-23. The unequal side-to-side

FIGURE 5-23 Dynamic imbalance shown as weight on each side of the center plane.

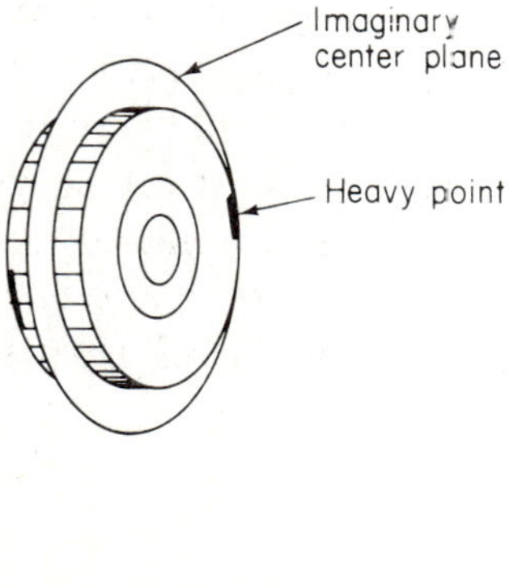

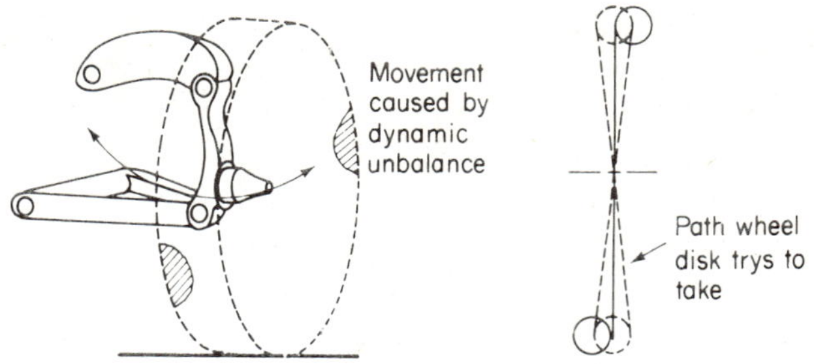

FIGURE 5-24 Effect of dynamic imbalance.

weight tries to force the tire to a new plane of rotation as the heavy section tries to move closer to the center plane. This force increases as the rotational speed is increased. Figure 5-24 schematically represents a dynamic imbalance condition. As the tire is rotated, the tire attempts to alternately turn in and out every 180 degrees of rotation. This results in wheel shimmy. It is possible to have static balance and still have dynamic imbalance. This can be shown by placing one 4-ounce balance weight on the outer rim and another 4-ounce weight on the inner rim, 180 degrees from the outer weight. The weights balance each other statically. They add together to cause dynamic imbalance. For this reason, the tire must be spun to detect dynamic imbalance. Dynamic imbalance on all balancers is corrected by placing two weights of equal amount on the wheel 180 degrees apart, one on the inner rim and one on the outer rim. The use of two equal weights 180 degrees apart does not upset static balance as it corrects dynamic imbalance.

Dynamic Balancing Equipment. One type of dynamic balancing equipment balances the tire and wheel assembly on the vehicle. A second type balances the assembly off the vehicle, combining static and dynamic balance tests. This type of balancer corrects both imbalance conditions with one test spin and two weights.

To dynamically balance the tire and wheel assembly on the vehicle, the assembly must be in static balance *before* it is spun for dynamic balance. Wheel weights, of the size indicated, are placed on the wheel rim in the position indicated by the balancer (180 degrees from each other on opposite wheel rims). The assembly is spun again to make sure that the imbalance has been corrected.

Balancers that combine static and dynamic balance of the tire and wheel assembly off the vehi-

FIGURE 5-25 Balancer that combines static and dynamic balancing. The guard is up in this illustration to show the horizontal position of the tire.

cle sense the entire weight characteristic of the tire and wheel assembly. This allows imbalance to be corrected with one weight on each side of the wheel. This type of balance is shown in Figure 5-25.

Dynamic balance may be checked and corrected on the vehicle. The strobe balancer vibration sensor is positioned against the steering arm or brake splash shield as pictured in Figure 5-26. In this position it will detect any tendency for the wheel to turn

FIGURE 5-26 Vibration sensor positioned on the steering arm to detect dynamic imbalance as the wheel is spun.

in or out as the assembly is spun. The strobe light flashes as the heavy part of the assembly moves toward the pickup. The reference mark is again used as it was for static balance. After spinning, return the wheel to the same position it was in when the strobe light flashed. For dynamic balance, one weight is added to the inside rim of the wheel at the sensor. A second weight of the same size is added 180 degrees from the sensor on the outside of the wheel. These weights tend to move the wheel away from the sensor as the wheel spins. Spin the assembly again to verify the balance. Repeat the procedure until the wheel and tire assembly is in balance.

5-5 TIRE REPAIR

Tire repair is most commonly required because the tire has been punctured. If the repair is properly performed, the tire can be placed back in service, safely and without fear of the leak reoccurring. At present, very few automobiles and light trucks use tube-type tires and their repair will not be discussed.

Tire repair begins by removing the wheel from the vehicle. Be sure to mark a wheel stud and corresponding lug hole so that the wheel can be reinstalled in the same position.

Locating the Leak. The tire is inflated to the maximum inflation pressure marked on the tire, usually 35 psi (240 kPa) (see Figure 1-28). To find leaks, first look around the tread for something still in the punctured hole. The tire and wheel assembly is then either submerged in a water tank or it is covered with a soap and water solution using a brush or sponge. The leak is directly below the bubbles when the tire is submerged in water. A tire may leak in a number of places. Each area of the tire should be individually checked. The tread is checked first. If no leaks are found, the tire should be placed flat on the floor and soapy solution applied to the sidewall and the bead areas. One side is inspected at a time. Slow leaks in these areas may not show up when the tire is in a vertical position. Leaks can occur at the valve stem and in the wheel rim. Do not overlook them.

Mark the leak with crayon so you will know where it is after the tire is removed from the wheel for repair. Use an identifying crayon mark to show the valve stem location. After repairing the tire, it can be remounted so that the original tire balance and runout is re-established.

Leak Repair. Procedures used to repair a leak depend upon the type of leak. Leak repair from outside the tire is not recommended by tire manufacturers. This type of repair is considered temporary, at best, and in some instances, it may even cause more damage to the tire. Repair from the outside is called *plugging*. An awl reamer is pushed through the punctured hole to clean it (Figure 5-27). A strip of rubber coated with tire cement is forced into the hole using a special inserting tool (Figure 5-28). When the tool is removed, the rubber remains in the hole to seal the leak. There are problems with this type of repair. The puncture may not go straight into the tire and its path cannot be followed with the awl. Forcing the awl through the tire in a straight line may actually cut part of the cord that was untouched by the original puncture. Flexing of the repaired tire during operation may loosen the plug, causing a new leak in the same location.

Repairs made inside the tire are considered permanent repairs when they are made correctly. Repairs should only be made to punctures in the tread area. A tire should be considered unrepairable if it has a leak in the sidewall or in the bead area or if the tire is worn so much that the wear indicators show, as shown in Figure 2-6. More than two punctures generally damage the tire cord enough that the tire should be considered unrepairable. These two punctures should be no closer than 15 inches from each other. Punctures larger than 1/16 inch (2 mm) are not completely safe. Never try to repair punctures larger than 1/4 inch.

FIGURE 5-27 Awl used to clean a punctured hole.

FIGURE 5-28 Inserting a rubber plug in a cleaned, punctured hole.

Leaks sometimes occur in the sidewall or bead area. Sidewall leaks may be the result of a porous tire inner liner. This could develop if the tire were run with low pressure. Excess flexing will cause enough heat to build up to damage the liner. Sometimes the liner on a new tire will leak. The specific location of liner leaks is nearly impossible to find and they are considered unrepairable. A tube may be installed in the tire for an emergency repair to prevent leakage. The tube will increase the operating temperature of a tire. It should never be used in place of a good tire repair or replacement tire.

Bead area leaks usually result from rust and corrosion in the bead area of the wheel rim. The wheel rim is cleaned with a wire brush and the bead of the tire is lubricated with nonpetroleum-based tire lubricant when the tire is remounted. This will usually stop a leak at the bead area.

Tire Dismounting. The tire can be removed easily from the wheel with a tire-changing machine. These machines use manual, electric, or pneumatic power to separate the bead from the wheel rim. Be sure to follow the procedures recommended by the equipment manufacturer.

First, the beads of the tire must be unseated from the rim. This is done by forcing the tire beads inward from the rim, as shown in Figure 5-29. After both beads are unseated, the tire is removed by placing a bar or lever under one bead to raise it above the rim while holding the opposite side of the same bead in the drop center portion of the wheel rim. The bar or lever is rotated to lift the bead above the wheel rim as shown in Figure 5-30. The process is repeated from the same side of the wheel to remove the opposite bead. The tire is then free from the wheel.

Some wheels are designed so that it is difficult to remove the tire from the front side of the wheel. Look closely at the drop center part of the wheel. The side of the wheel with the least distance between the bead seat and the lowest drop center is the side from which the tire must be removed. Carefully inspect the carcass interior. If there is no damage other than the original leak, the tire can be repaired.

Select a patch of the required size to repair the puncture. The area around the leak is cleaned with a tire buffing solvent to remove all residue. The area is then buffed with a power-driven wire brush or a rasp to remove the surface of the rubber liner so proper bonding can take place. The area is buffed slightly larger than the size of the patch selected (Figure 5-31). The rubber removed during the buffing operation is cleaned from the tire. Tire cement is applied to the buffed area. After the tire cement has dried, the protective film is removed from the patch and the patch is applied by hand. Full contact of the patch is made using a specially designed roller to ensure complete bonding. The roller has a

FIGURE 5-29 Unseating the tire bead from the rim.

FIGURE 5-30 One side of the bead is lifted over the rim while the opposite side of the bead is in the rim drop center.

FIGURE 5-31 Inner lining of the tire is buffed slightly larger than the patch to be applied.

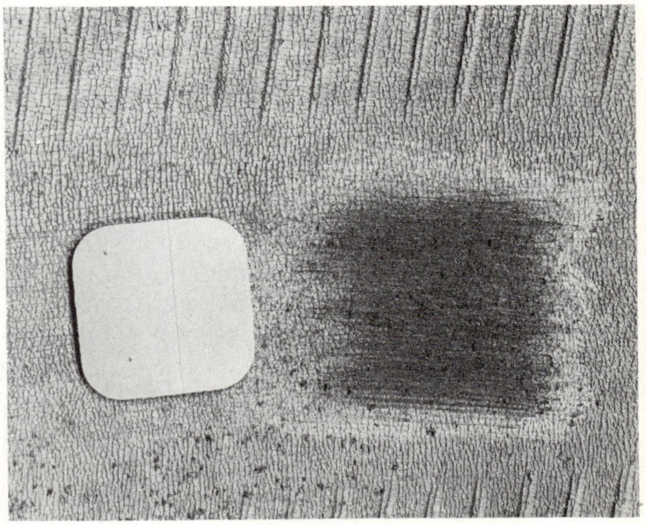

corrugated wheel and the rolling process is called *stitching* the patch.

If the leak was between the bead and wheel rim, carefully inspect the complete tire and wheel. Clean all corrosion and dirt from the bead and rim before remounting.

Remounting. Several things should be done before remounting the tire. The tubeless tire valve stem should be examined for cracks or damage. New valve stems should be installed when mounting new tires. The life of the valve stem rubber is close to the life of the tire rubber. The old stem is pulled out of the rim. The new valve stem is coated with rubber lubricant and installed by drawing it through the wheel from the inside to the outside until it is properly seated. The bead seat area of the wheel rim flange is cleaned with a wire brush. The bead area of the tire is then coated with either a soap solution or rubber lubricant. The tire is guided onto the wheel, lower bead first, with the use of the tire-mounting equipment. One side of the bead is held in the drop center of the rim as the tire changer rolls the other side of the bead over the rim flange onto the wheel. The outer bead is rolled over the wheel rim flange in the same manner.

The tire is then rotated on the wheel until the valve stem crayon mark that you put on the tire before dismounting is in line with the valve stem. The tire is then inflated. The beads must be held against the rim flanges until the tire begins to pressurize. A bead expander may be needed. This will either be a mechanical clamp or an expanding sleeve that squeezes in on the center of the tire tread. This forces the beads against the wheel rim as shown in Figure 5-32. Without the use of the expander, air may leak past the bead as fast as the air is being put in through the valve stem. The tire will not inflate when this happens.

Tire inflation on passenger car wheels should always be done with the wheel fastened securely to the tire machine. This will hold the assembly safely as the bead snaps to seat against the rim. Always keep your fingers clear of the bead area while inflating the tire. They would be severely pinched if caught as the bead snaps into place.

Inflation air will enter the tire faster if the valve core is removed from the valve stem. The tire should be inflated to approximately 40 psi (276 kPa). This will positively set the beads on the rim. Replace the valve core as air is coming from the stem; then adjust the operating pressure. The tire and wheel assembly should be rechecked to make sure that there are no leaks.

FIGURE 5-32 Expanding sleeve on the tire to help keep the bead against the rim as the tire is inflated.

Balance the tire and wheel assembly and install it on the vehicle to complete the repair. With the tire properly marked during disassembly, tire balancing may be omitted, provided the balance weights were not moved as the tire was removed and remounted. Rebalancing is recommended. Align the marked lug and bolt holes as the wheel is installed. Tighten the wheel lug nuts in the proper sequence to the correct torque.

Retreading. Tire retreading is a means by which a new tread can be fitted on a good tire carcass. This is done by specialized retreading shops.

One of the most important parts of the retreading process is the selection of tire carcasses that are in excellent condition. They are expected to provide satisfactory service for the life of the new tread. Tire retreading is most commonly done on tires used on heavy-duty road trucks where high mileage is accumulated in a short period of time. Their tires usually have a good quality carcass with as many as 12 plies.

The old tread rubber remaining on the tire carcass is ground down to the cord material. New tread rubber is wrapped around the carcass. The carcass is then placed in a tire mold and the new tread is molded and bonded to the old carcass under heat and pressure. When properly performed, retreading provides a lower-cost tire with a good service life.

STUDY QUESTIONS

Section 5-1

1. How can you recognize tire wear caused by underinflation?
2. How can you recognize tire wear caused by overinflation?
3. How can you recognize tire wear caused by incorrect camber?
4. How can you recognize tire wear caused by too much toe-in?
5. What causes tire cupping and spotty wear?
6. What should be done after a tire is reinstalled on a wheel?
7. What is the basic cause of tire separation?
8. What causes wheel pull?
9. What causes ply steer?
10. What causes waddle?

Section 5-2

11. Describe lateral runout.
12. Describe radial runout.
13. How is tire-wheel runout corrected?
14. What allows loaded runout to be different from radial runout?
15. What must be done to correct a tire that shakes the vehicle as a result of incorrect loaded runout?

Section 5-3

16. In what ways do wheels usually fail?
17. What should be done when one lug bolt fails?
18. What is likely to happen if a lug bolt is removed with a punch?

Section 5-4

19. How does static imbalance affect the motion of the suspension?
20. How can you tell if more weight is needed when the wheel balance is rechecked after adding weights?
21. How does the location of weights on the wheel rim differ between static and dynamic wheel balance?

Section 5-5

22. Describe the procedure used to find a leak in a tire.

23. Why is plugging considered a temporary repair?
24. What conditions makes tires unrepairable?
25. Describe two different tire repair procedures.
26. How is the drop center of the rim used in demounting and remounting a tire?
27. What should be done to the wheel while the tire is off the rim?
28. What is necessary to inflate a remounted tubeless tire?
29. What safety precautions should be observed when inflating remounted tires?
30. What are the advantages and disadvantages of retreading tires?

6

Suspension Systems

The primary purpose of the suspension system is to keep the tires on the road surface and to provide a smooth ride for the driver and passengers. The suspension is designed to absorb and dampen the road surface variations before they reach the passenger compartment. This is done with springs, linkages, bushings, and shock absorbers. This chapter presents some suspension design requirements and introduces a number of suspension designs.

Suspension Requirements. The suspension system in vehicles must meet a number of design requirements. In addition to keeping the tires on the road and providing a smooth ride with little transmission of road noise, the suspension must also handle braking and acceleration forces. It must also allow for free steering of the front wheels and handle high lateral (side) forces as the vehicle is turned at speed. The final job of the suspension system is to hold the required wheel angle changes for maximum handling and minimum tire wear as the vehicle moves over various surfaces.

To summarize, a modern vehicle's suspension system must keep the tires on the road, provide ride smoothness, have good handling quality, and be designed to handle all forces imposed on it. In addition, the suspension system must be lightweight to reduce the overall weight of the vehicle, compact to allow large interiors in short-wheelbase vehicles, and allow maximum underhood room for the engine and transaxle. It must also be strong enough to withstand abnormal forces without damage, such as hitting a pothole when traveling at high speeds.

The suspension system provides ride smoothness by being soft (compliant) and having the proper amount of damping. The system provides handling quality by being relatively firm, minimizes body roll while providing maximum tire contact with the road surface at all times, and distributes weight transfer in proper proportions to each outside wheel during cornering. The suspension is made strong and durable by selecting the right materials, properly designing each part, and positioning the parts to distribute high forces without damage. Today's suspension systems provide a compromise between ride and handling. This compromise is based on whether the vehicle is a luxury- or a sports-type vehicle. In recent years, the trend in domestic automobiles is toward firmer ride with improved handling.

Suspension Types. Many suspension designs have been used over the years. During the early days of the automobile, rugged solid-axle front suspension designs were used. These provided the reliability necessary for traveling over rough roads at low speeds. Today, solid-axle front suspensions are used on heavy trucks for the same reasons. As roads became smoother, vehicle speeds increased and the demand for improved handling increased. This demand led to lower ride height with improved ride quality, leading, in turn, to the need for independent front suspension. For years, automobiles had independent front suspension with dependent (or solid-axle) rear suspension. Today, four-wheel independent suspension is becoming the more popular design.

Independent front suspension primarily used the *long-short arm* (unequal length) design that was shown in Figure 1-10. This design allowed for independent action of the front wheels, lower ride heights, improved handling, and good tire life. Independent front suspension designs on some vehicles took the form of either a leading or trailing arm design. This design allowed for vertical movement of the wheel without any change in the angle between it and the vertical plane. Today, an increased number of vehicles use a *strut-type* of design that was shown in Figure 1-11. The strut design provides good ride, good handling, and increased underhood clearance for transverse engine mountings.

Dependent rear suspension, which was primarily used for simplicity with front-engine, rear-wheel drive (see Figure 1-13), is being replaced by independent rear suspension. Most independent rear suspensions are used with front-wheel-drive vehicles. However, a few rear-wheel-drive passenger vehicles use independent rear suspension. The independent rear suspension can take the form of a strut-type or a trailing or semitrailing arm suspension (see Figure 1-14). All independent rear suspensions tend to have improved ride quality over dependent suspensions. Suspension designs are discussed in more detail in the following sections.

6-1 SPRING REQUIREMENTS

Some type of spring is needed to support the vehicle weight and load. At the same time, the springs must have flexibility to minimize the shock transmitted to the passengers and vehicle load from a rough road surface. While the vehicle is in motion, the deflections of the suspension system must be absorbed to maintain tire-to-road contact and to provide passenger comfort. Part of the roughness of the road surface is dampened within the tire as it flexes. Large deflections are transmitted to and are absorbed by the vehicle springs. Vehicle springs are designed to provide a compromise between handling characteristics and riding comfort. *Spring rate*, as illustrated in Figure 6-1, is the load required to deflect a spring 1 inch. A vehicle designed with a low spring rate will allow more suspension movement, giving a soft ride for passenger comfort. That same vehicle designed with a higher spring rate would have improved handling characteristics but it would give a harsh ride with less passenger comfort.

Small subcompact vehicles, vans, pickups, and station wagons place an even greater demand on vehicle springs. With less empty vehicle weight, a lower spring rate is needed to support the weight at each wheel. A low spring rate provides satisfactory ride and handling characteristics during normal day-to-day use. This same vehicle may be used to transport four more persons and cargo for an occasional trip, thus carrying an additional weight of 1000 lbs (454kg). This extra weight would lower the ride height of the vehicle to the point where the suspension would hit bottom as the vehicle rolls over a small bump. If the designer put high-rate springs on the vehicle to carry an additional load, it would produce a harsh ride in normal day-to-day use under lightly loaded conditions. What is required is a spring or mechanism that can sense vehicle load and provide additional support for the added load. This would provide the necessary support, handling, and ride quality under all operating conditions. Air suspension, air-assist suspension, mechanical assist, and hydraulic assist have been used or are being studied to provide the required variable rates in future vehicles.

6-2 SPRING TYPES

Spring types that are currently being used in vehicles include leaf springs, torsion bars, coil springs, and air springs. These were shown in Figure 1-8. Leaf springs may be separated into two types: multiple-leaf and single- or monoleaf design. The leaf spring has a long history of proven performance, from the days of the horse-drawn buggy to today. This spring design can provide all the necessary restraint for the wheel assembly during braking, acceleration, cornering, and suspension flexing that results from road surface irregularities. The other types of springs only support weight. They use arms, struts, and links to hold the angle of the suspension.

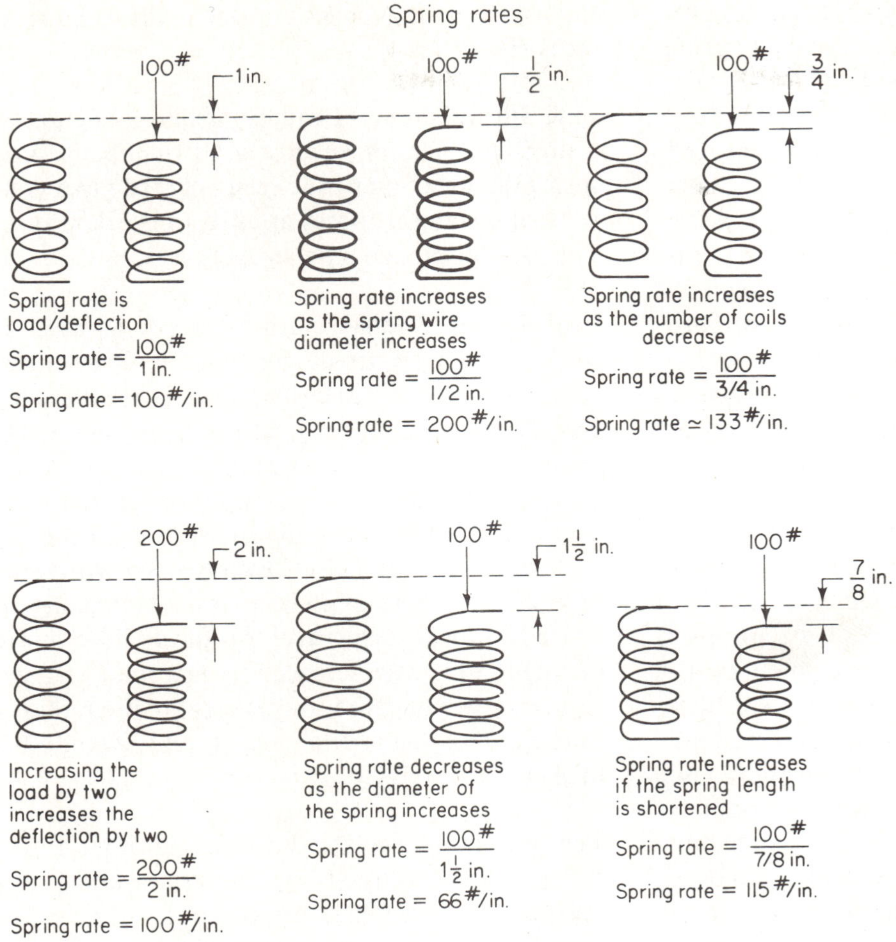

FIGURE 6-1 Effect of coil spring design on spring rates.

Multiple-Leaf Springs. The multiple-leaf spring has a number of slightly arched sections of different lengths of steel. Figure 6-2 shows a typical leaf spring. The spring sections, or leaves, are fastened together by a through-bolt placed near the center of the spring. This bolt is called the *center bolt*. The use of multiple leaves provides the necessary suspension control and compliance with high fatigue strength. As the spring flexes, the individual leaves slide over one another. This sliding can be a source of noise and it also produces friction. These problems are reduced by interleaves of zinc or plastic placed between the spring leaves. Friction, as the leaves slide, produces ride harshness as the spring flexes but it also dampens the oscillations produced by large spring deflections.

The multiple-leaf spring incorporates a number of thin leaf sections. This design minimizes the stress buildup in the leaf that occurs as the spring flexes. Additional leaves increase the strength as well as the weight of the spring.

FIGURE 6-2 Typical rear-leaf spring.

Monoleaf Springs. The monoleaf spring design provides the same restraints as the multiple-leaf spring in controlling movement. It does not have the noise and static friction of the multiple-leaf spring. Steel monoleaf springs are made of chrome carbon steel. They vary in thickness from a maximum at the center to a minimum at the spring eyes on each end. After steel monoleaf springs are rolled to form the correct taper, they are shot-peened to provide maximum fatigue resistance. Fatigue results from the continued flexing of the spring. As the thickness of the spring section is increased, the resistance to fatigue is decreased. This inverse relationship results from the change in length that must occur on the upper and lower leaf surface as the spring flexes. It is proportional to the distance from the centerline of the leaf to its surface, as illustrated in Figure 6-3.

In 1981, General Motors introduced a composite fiber reinforced epoxy single-leaf spring on the Corvette. It weighs about half as much as the steel spring it replaces. The spring is 70 percent fiber and 30 percent epoxy. The fiber is wound from end to end. This gives the spring high strength and resiliency. The center of the spring is thick and narrow to give the spring its strength. The ends are thin and wide to give the spring flexibility. This composite spring isolates noise and vibration better than steel to improve vehicle ride. By 1987, this type of spring was mounted transversely on several models of General Motors vehicles. The spring is clamped to the vehicle body through insulators. Metal around the ends of the springs slides on control arm contact surfaces (Figure 6-4).

Leaf Spring Shape. Leaf springs are designed with a positive, flat, or reverse arch (camber). A leaf spring designed to operate relatively flat gives greater control of lateral forces applied to the suspension. This will minimize side sway.

Leaf-Spring Failure. Spring failure of steel leaf springs often occurs at the center bolt because of the reduced spring section width at the bolt hole. Failure may also occur near the spring mount as a result of localized stress buildup caused by overtightening the spring pads. Overloading the vehicle is another major cause of leaf spring failure. Premature failure may also be caused by worn shock absorbers which result in little or no dampening. The suspension will then travel too far and the number of oscillations will increase for each road surface irregularity encountered. As springs are used, they will fatigue. This leads first to sag and eventually to broken leaves.

Torsion Bars. Torsion bars are another type of spring used to support vehicle weight and provide the necessary suspension movement. Torsion bars are made of heat-treated alloy spring steel. They support the vehicle load by twisting. One end of the torsion bar is fastened to the vehicle frame while the opposite end is connected to the lower control arm.

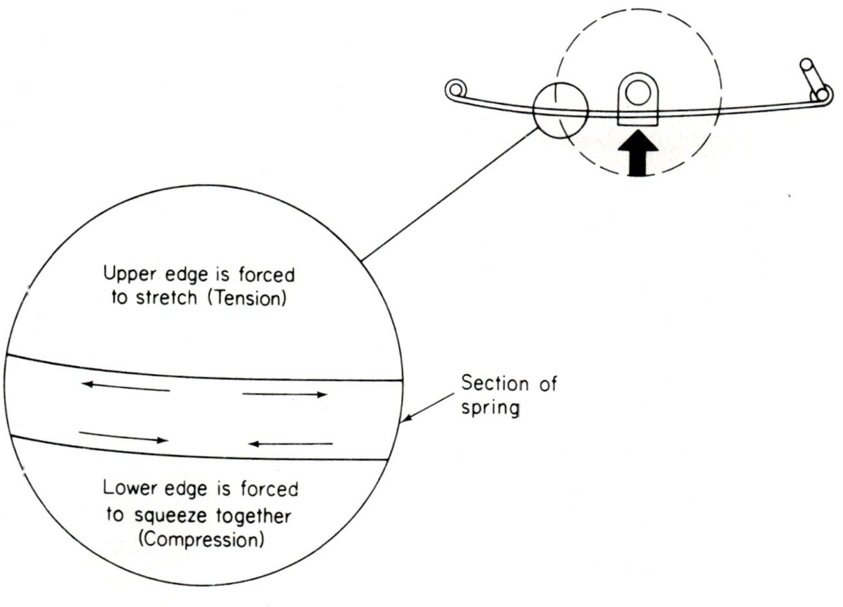

FIGURE 6-3 Stress within a spring leaf.

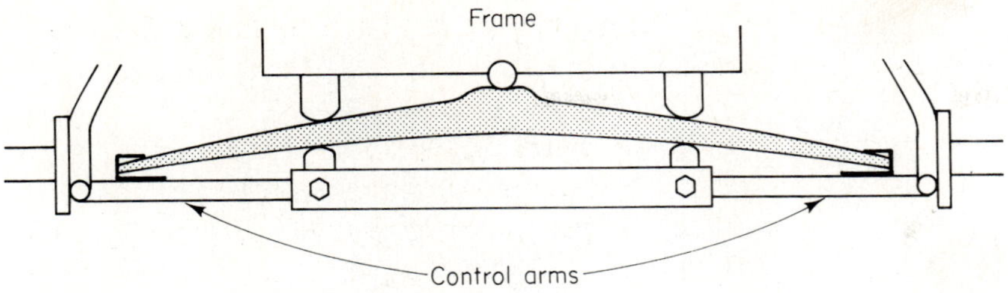

FIGURE 6-4 Typical mounting of a composite fiber monoleaf spring.

One type is shown in Figure 6-5. The bar is usually formed with a round cross-section having ends formed into a hexagonal shapes to fit into a socket in the lower control arm (moving end) and the vehicle frame (stationary end). In other cases, one end of the arm is bent about 90 degrees. Torsion bars are most commonly found on the front suspension, with coil or leaf springs supporting the rear suspension.

The independent front suspension is readily adaptable to the use of torsion bars. The torsion bar uses very little space under the vehicle. Using a torsion bar with a conventional mounting is difficult on dependent rear suspensions.

During manufacturing, torsion bars are prestressed to provide fatigue strength. Because of directional prestressing, the torsion bars are directional. The torsion bar is marked right or left to identify the side of the vehicle on which it is to be used. Surface imperfections on the bar will produce an area of high localized stress which may result in premature failure. This type of failure can be prevented by using care during removal and installation of the torsion bar so that you do not make a mark that will cause stress concentration.

FIGURE 6-5 Logitudinal front torsion bar on a Chrysler automobile.

Torsion bars are mounted longitudinally (parallel to the centerline of the vehicle) or transversely (perpendicular to the vehicle centerline). With the bars mounted longitudinally, the front suspension noise is transmitted along the bar to its fixed end. The longitudinal bar often has its fixed-end socket in the cross-member that is located beneath the front seat of the vehicle. Noise and vibration are transmitted to the cross-member. It is difficult to isolate the noise from the passenger compartment. Mounting the torsion bars in a transverse location, as shown in Figure 6-6, allows easier isolation at the front of the vehicle. The fixed end of the transverse torsion bar is attached to the front cross-member. There are large soft rubber insulators between the cross-member and the vehicle body. This isolates the noise from the passenger compartment. The use of torsion bars gives good underhood space because coil spring towers are not needed.

Coil Springs. Coil springs are the most common type of vehicle spring. They support the vehicle load while allowing the wheels to deflect as they hit bumps and holes in the road surface. Coil springs, like monoleaf and torsion bar springs, do not have inherent noise problems nor do they develop the static friction that causes ride harshness as multiple-leaf springs do. Coil springs have load-carrying, lower spring seats on the suspension and reinforced upper spring seats in the vehicle frame or underbody as illustrated in Figure 6-7. This type of mounting requires reinforcement not required with other types of springs. A coil spring is also used on a MacPherson strut suspension. One spring seat is on each part of the strut (Figure 6-8).

Each short section of a coil spring develops its force by twisting. If you could examine a short section of the coil while the spring was compressed, you could see that it had a twist like the twist in the short

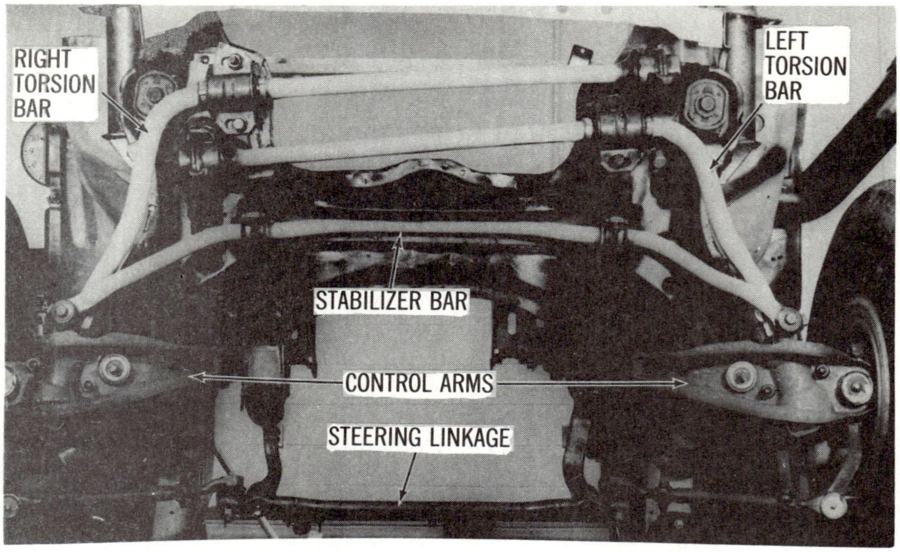

FIGURE 6-6 Bottom view of a transverse and bent torsion bar on a front suspension of a Chrysler automobile. Front is toward the top of this view.

FIGURE 6-7 Typical front suspension with the coil spring mounted between the frame and lower control arm.

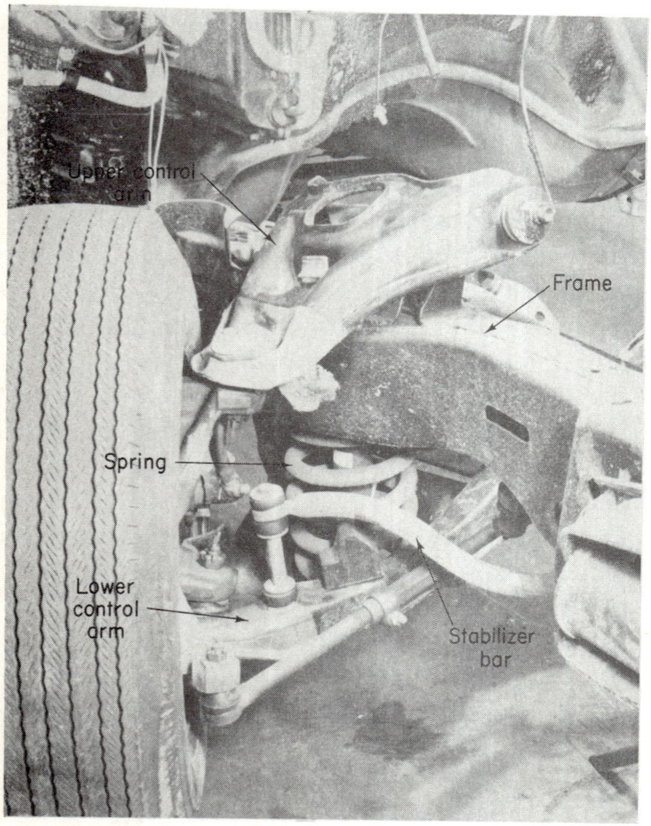

FIGURE 6-8 Coil spring on a MacPherson strut. There is a spring seat on both the upper and lower parts of the strut.

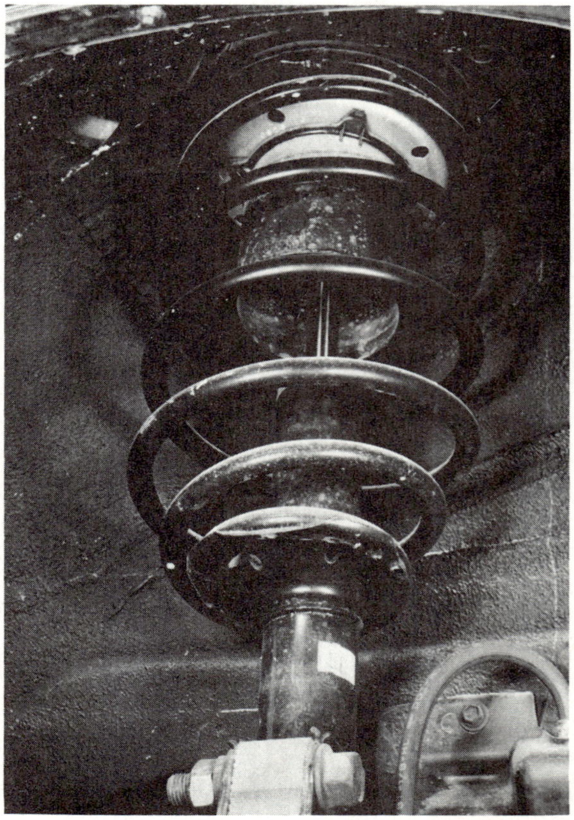

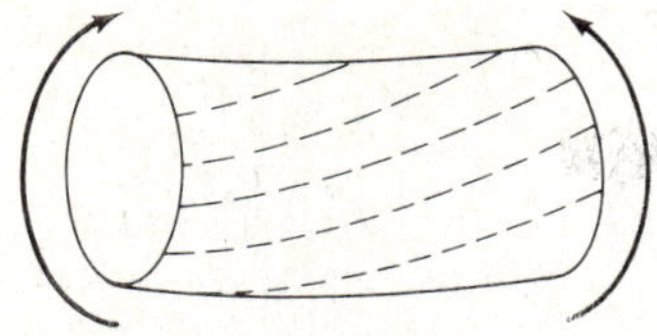

FIGURE 6-9 Twist in a short section of a coil or torsion bar spring.

section of the torsion bar. This twist is illustrated in Figure 6-9. You can see that a coil spring is actually a torsion bar wound in a coil to make a compact spring that will absorb vertical loads.

Coil springs can be designed with a variable spring rate. This means that they become stiffer as they are forced to deflect as illustrated in Figure 6-10. The spring in Figure 6-8 is also a variable-rate spring. The variable-rate design provides ride smoothness while operating under light loads and, at the same time, it has the ability to carry heavy loads. One type of variable-rate spring can be identified by the closely wound coils in one part of the spring. Another type of variable-rate spring is taper wound. As the vehicle is loaded with more weight, the closely wound coils come in contact with each other and cease to function. The more widely spaced coils will then provide the spring deflection. Because the widely spaced part of the coil has a higher spring rate, the spring

FIGURE 6-10 Types of variable-rate coil springs.

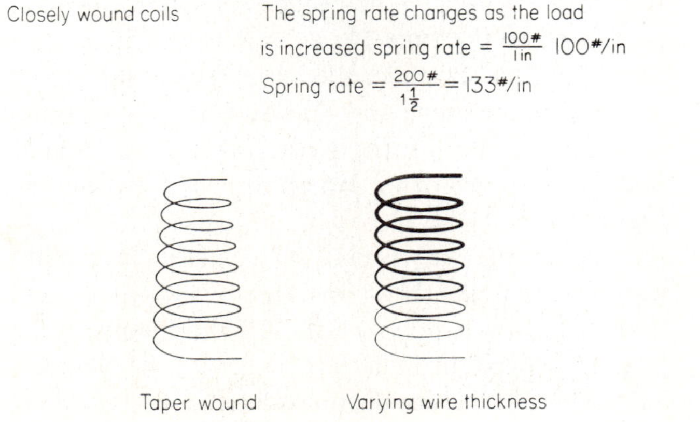

has the ability to carry a heavy load. This design has been used in the rear of light trucks that are required to carry varying loads. Variable-rate coil springs are more costly to make because the manufacturing requires two separate coil spring winding processes. Coil springs only have the capacity to absorb loads applied axially (vertically). They provide no control of the lateral (sidewise) or longitudinal (front and rear) forces that develop in the suspension system. For this reason, a complex linkage system must be used when coil springs are used to control the torque loads that occur during acceleration and braking and the side loads that occur during cornering.

Air Springs. As vehicles were downsized and the wheelbase became shorter, the ride became more choppy. This was noticed most by customers accustomed to the smooth, soft ride of luxury automobiles. Computer-controlled air springs were developed and introduced to give a soft ride while the vehicle was lightly loaded and still give support when the vehicle was loaded.

Air springs (see Figure 1-8) were introduced on Continental and Mark VII automobiles in 1984. The air spring was fitted in place of the coil spring used on earlier models of these vehicles. An electric-motor-driven air compressor is used to supply the high-pressure air. A computer, mounted in the trunk, receives signals from the sensors, processes these signals, and controls solenoids that either add air to or remove air from the springs to hold the normal vehicle trim height.

Air springs are used with MacPherson struts, front and rear. Air springs are neoprene synthetic rubber bags with two nylon cord plies. They have nylon reinforced caps at each end. The bag is vented and filled through a solenoid valve on top of each air bag. The valve has a leak rate of 1/20, the leak rate of a tire valve. Three Hall effect height sensors, two in front and one in the rear, let a microprocessor know the suspension height. The system also senses when the ignition is on, when a door is open, and when the brake pedal is depressed. The microprocessor is in the left trim panel in the trunk. The air pump is on the left front fender panel. The air pump has a reservoir-dryer to keep ice from forming at low temperatures. Air escaping from the air bags goes out through the dryer. As it escapes, it picks up the moisture the dryer absorbed when the air bags were filled.

The air spring gives the vehicle maximum riding comfort during all types of vehicle operation. It holds the trim height with all vehicle loads. To do this, the springs must have variable rates, as air springs do. This keeps the headlights aimed correctly and the mirrors adjusted normally. The suspension geometry remains constant with all vehicle loads so the vehicle has constant handling characteristics. Air springs have a longer travel than coil springs. Jounce and rebound of air springs are controlled with gas-charged shock absorbers.

A number of import vehicles have an active form of air suspension created by adding air chambers to the shock absorbers. Coil springs support most of the vehicle weight when the vehicle is empty. The air chambers in the shock absorbers help the springs to support vehicle weight when the vehicle is loaded to maintain the normal trim height. Air pressure in the chambers is controlled by a computer in a similar way to the air spring just described.

6-3 SHOCK ABSORBERS

Shock absorbers are direct double-acting suspension dampers. Most vehicles use four shock absorbers, one near each wheel. Additional shock absorbers have been used to produce special ride qualities. The shock absorbers are called direct-acting because of their direct connection between the vehicle body or frame and the suspension member near the wheel. The double action describes the direction of motion control, both jounce (compression) and rebound (extension).

Shock absorbers are placed near each wheel to help dampen uncontrolled up-and-down wheel motion as well as to control the rolling motion of the body that results from rises and dips in the road surface. The standard shock absorber differs from an air-lift or spring-assist shock absorber because it does not support vehicle weight. The shock absorber's only job is to dampen body oscillations and wheel movements that could affect control of the vehicle. It helps to keep the tires on the road, improve ride quality, and lengthen the life of suspension parts.

Air- and spring-assist shock absorbers aid the vehicle springs when the vehicle is carrying heavy loads. These assist-type shock absorbers are easily identified by a coil spring mounted around the conventional shock absorber on a spring-assist, or a synthetic rubber, air chamber between the upper and lower parts of the air-assist shock absorber. A spring-assist shock absorber is illustrated in Figure 6-11. Figure 6-12 shows an air-assist shock absorber.

FIGURE 6-11 Typical spring-assist front shock absorber.

Shock Absorber Operation. The shock absorbers use hydraulics to convert the kinetic energy of the vehicle body motion (in an up-and-down direction) to thermal energy (energy of heat). You should recall that energy cannot be created or destroyed, but it can be converted to other forms. The shock absorber dampens the motion by forcing hydraulic fluid through small openings between chambers within the shock absorber. For all practical purposes, the hydraulic oil contained within the shock absorber is incompressible. If the incompressible fluid is forced through a small-sized opening called an *orifice*, it needs time for the fluid to flow from one side of the orifice to the other. The length of time, compared to the time for an uncontrolled suspension to deflect, dampens the initial deflections. Fluid is moved in the shock absorber by a piston connected to a rod, which, in turn, is connected to the vehicle body. Shock absorber operation is illustrated in Figure 6-13.

Each time the wheel of the vehicle is forced upward (jounce), the shock absorber is compressed and the piston moves downward. The fluid beneath the piston is forced through the orifice into the chamber above the piston. The piston rod takes space so some

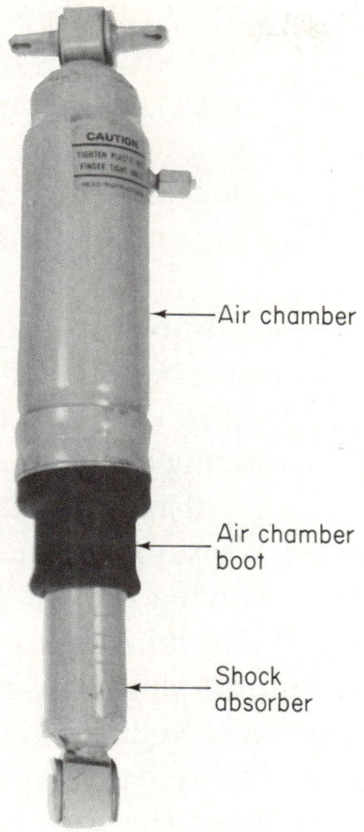

FIGURE 6-12 Typical air-assist shock absorber.

of the fluid goes through a separate valve into the fluid reservoir.

As the suspension returns to its normal position, the piston moves upward, causing the fluid above it to return to the lower chamber through a different orifice. It also draws some fluid from a reservoir where fluid was displaced when the wheel went into jounce.

The shock absorber requires a fluid reservoir to ensure that enough fluid is available as the shock absorber travels through its range. Reserve fluid will prevent air from being drawn in on either side of the piston. When air is drawn in, it is called *aeration*. Air is a gas and it is compressible. When air is mixed with the fluid, the mixture will produce erratic shock absorber operation. Aeration normally occurs during rapid cycling. It is minimized in standard shock absorbers through the use of properly placed baffles. The baffling on standard shock absorbers can usually be seen as spiral grooves around the shock absorber body. A small air chamber at the top of the fluid reservoir column allows for ex-

FIGURE 6-13 Operation of a typical shock absorber.

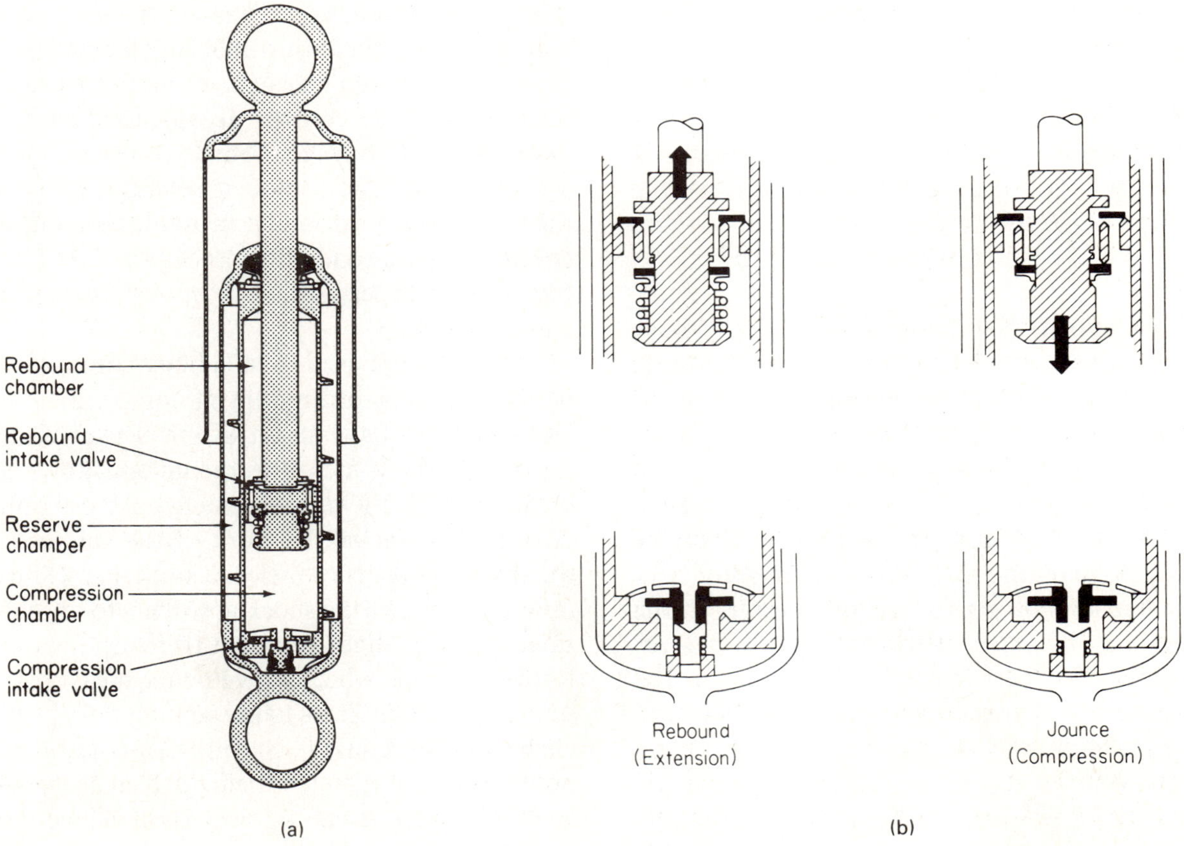

pansion of the fluid as it warms and for contraction of the fluid as it cools. The fluid reservoirs are of different sizes in shock absorbers made by different manufacturers, even when pistons have the same diameter. A large fluid reservoir will improve the life expectancy of the shock absorber and it will operate better under heavy-duty operation because it has the ability to reject more heat. The heavy-duty shock absorbers may have a larger diameter piston than standard shock absorbers. The fluid in a shock absorber develops more heat when the vehicle is driven over rough roads for an extended period of time. The shock absorber may reach a temperature that is too hot to touch. Because of the heat loads, the fluid used in shock absorbers must be able to withstand high temperatures.

Gas-Charged Shock Absorbers. Many shock absorbers are gas charged. An inert gas, freon or nitrogen, is used to pressurize the air chamber in standard shock absorbers. The pressure is not great enough to support the vehicle load. The gas-charged chamber keeps all air out of the shock absorber fluid. This will prevent fluid aeration even when a shock absorber is cycled rapidly as the vehicle is driven over rough roads. The inert gas tries to expand with increasing temperature. This increases the pressure in the gas chamber. Thus, if the temperature rises during high demand, it puts more pressure on the hydraulic oil. The higher pressure eliminates any possibility of fluid aeration. The gas-charged shock absorber provides the required control as the temperature of the shock absorber rises due to rapid cycling and this results in improved control.

Damping Control. Shock absorbers usually have different damping control in jounce versus rebound. Shock absorbers designed to improve handling approach an equal amount of control in both jounce and rebound, such as 40-60 (40 percent jounce and 60 percent rebound) or 50-50. Shock absorbers used on drag racing vehicles generally are 90-10 on the front and 50-50 on the rear. The 10 percent of control on extension allows for weight transfer to the rear wheels while accelerating and the 90 percent retards front-end drop during the gear change. The 50-50 rear shock absorbers control rear-wheel motion for maximum traction. Public demand for more responsive vehicles has led to the use of passenger car shock absorbers approaching the 50-50 range. Compact vehicles, due to their light weight and soft springing, require a stiffer (more control in both jounce and rebound) shock absorber. This will allow the frequency of the vehicle suspension to stay in its natural 60-80 cycles/minute range. This frequency range provides the best ride characteristics. Damping rates within the shock absorbers are controlled by the size of the piston and the size of the orifices.

Selective-Control Shock Absorbers. Adjustable shock absorbers have been used for many years. The amount of dampening on these shock absorbers can be adjusted by rotating the piston rod to line up with soft, medium, or firm control marks before the shock absorber is installed. The dampening can be changed by removing the shock absorber and rotating the rod. Rotating the rod changes the control valve within the shock absorber to change the rate of dampening. Reduced dampening gives the vehicle a good, soft ride on smooth roads. Increased dampening gives the vehicle good high-speed handling while accelerating, cornering, and braking. It also gives the vehicle more control on rough roads.

In the mid-1980s, remote adjustable shock absorbers were introduced. These shock absorbers use a small electric motor or rotary solenoid in the top of the shock absorber to change the control valve, somewhat like the adjustable shock absorber. The driver of these vehicles uses a switch to select soft, medium, or firm control. It was a natural further development to have the vehicle computer take over the switching. The computer selects a soft ride for the best possible ride and acceptable handling at low speeds on good roads. It selects a firm ride for best handling and acceptable ride at high speeds and on rough roads.

Sensors are used to let a body computer know when to change the amount of dampening control. Some vehicles have an ultrasonic sensor under the front bumper. It measures the up-and-down movement as the front of the vehicle goes over bumps. Slow movement of the front of the vehicle causes the shock absorbers to select a soft ride, while rapid movement causes the shock absorbers to select a firm ride. It can change from soft to firm very quickly. Rapid steering wheel movements measured by a steering sensor indicate to the computer that the vehicle needs firm control. Speed sensors tell the computer when the vehicle is being driven at high speed so the dampening can be firmed up. The computer firms dampening when the brake pressure increases above a set pressure, about 400 psi. This reduces brake dive. Computers use the throttle position sensor to firm the dampening when the throttle is opened about 90 percent.

Computer-controlled shock absorbers have been combined with air chambers to control ride height. In some systems, like the 1986 Lincoln with air springs, both the shock absorbers and springs are controlled by the computer. Other vehicles, like the Mitsubishi ECS system, have air chambers on the shock absorber to assist the coil spring. An air actuator on the strut opens and closes a valve between the two air chambers as well as changing the shock dampening. The computer automatically controls the air suspension to lower the ride height when the vehicle is operating at high speed.

When the computer adjusts the suspension to meet driving and handling demands that the sensors anticipate, the system is said to be *adaptive*. If it is designed to quickly and automatically respond to dynamic forces, it is called an *active suspension*.

Lotus has carried this one step further on some of their test vehicles. Their active suspension system has four dual-acting hydraulic cylinders, called *actuators*, one at each wheel. Each actuator is controlled independently. Each has a sensor to tell the computer the changing length of the cylinder and how fast the cylinder length is changing. Each has a load sensor to keep track of how much weight each wheel is supporting. The vehicle has roll and acceleration sensors to tell the computer the body motions. The computer uses input from all of these sensors to actively adjust the actuator to absorb road shock and to dampen rebound by controlling the pressure and oil flow in the hydraulic actuator cylinder. In doing this, the actuator takes the place of both spring and shock absorber. The vehicle body has very little roll in a turn because the suspension firms up the cylinders on the outside of the turn. Parts of this system, or an updated version of it, may be found on future luxury and performance vehicles.

Mounting Shock Absorbers. Shock absorbers are mounted at different angles to achieve the desired ride control. The shock absorbers may be angled at the top to dampen lateral motion of the body. They may be angled forward at the top to dampen road shock that forces the wheels rearward. The shock absorbers are mounted vertically when maximum damping of the up-and-down motion of the vehicle is necessary.

Care should be taken when adapting shock absorbers to different vehicles. The shock absorber must be installed correctly. If a standard shock absorber is mounted upside down by mistake, it will provide no damping action after a few cycles. Occasionally, this condition is found on home-built vehicles and even on racing vehicles.

6-4 SUSPENSION CONTROL DEVICES

Suspension-control devices are needed on all but leaf springs to control forces developed during braking, acceleration, and cornering. A spring has almost no natural dampening of body oscillations after the vehicle goes over road bumps or dips. For these reasons, suspension-control devices are installed on vehicles.

The mass of the vehicle maneuvering around turns produces side loads on the vehicle suspension. These side loads must be transferred to the tires to maintain control. During a turn, a side load force is developed as the sprung weight of the vehicle tends to remain moving in a straight line that is tangent to the turning arc, as illustrated in Figure 6-14. This force is called *inertia,* the force that tends to keep a body in motion moving in a straight line. To change the direction of a moving body, a force must be developed in the required direction. The suspension transfers this force from the tire footprint to the vehicle. In the solid-axle front suspension system design using leaf springs, the springs restrain the body movement during cornering. Leaf springs provide the required restraint to transmit the inertia side loads of the body to the suspension, wheels, and

FIGURE 6-14 Effect of side loads on the tires and body as a vehicle moves through a turn.

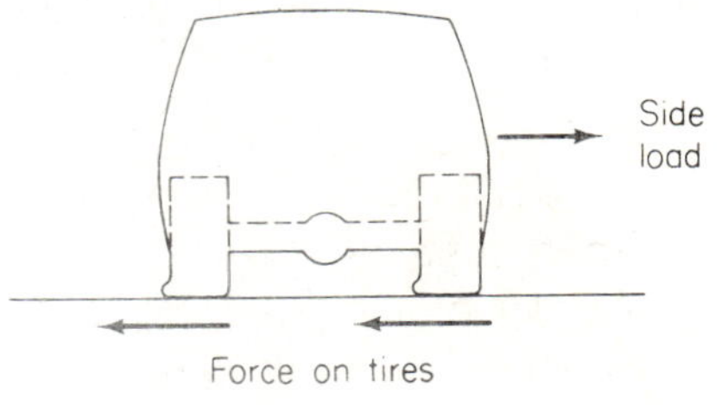

tires. A vehicle equipped with torsion bar springing in the front will transfer the inertia side load forces to the control arms or A-frames and then to the road surface through the wheel and tire. The control arms provide the restraint for side load forces.

Coil springs use many different linkages to transmit side load forces to the road surface. In front suspensions using upper and lower control arms, the side loads are restrained by the control arms. Strut-type suspensions use the lower control arm and the strut to control the forces. With twin I-beam suspension, the front axles (I-beams) transfer these loads to the wheels. Rear suspension, depending on manufacturer, may transfer the side loads through angular linkages, A-frames, a Panhard rod (commonly called a *track bar*), strut and control arm, or any combination of these.

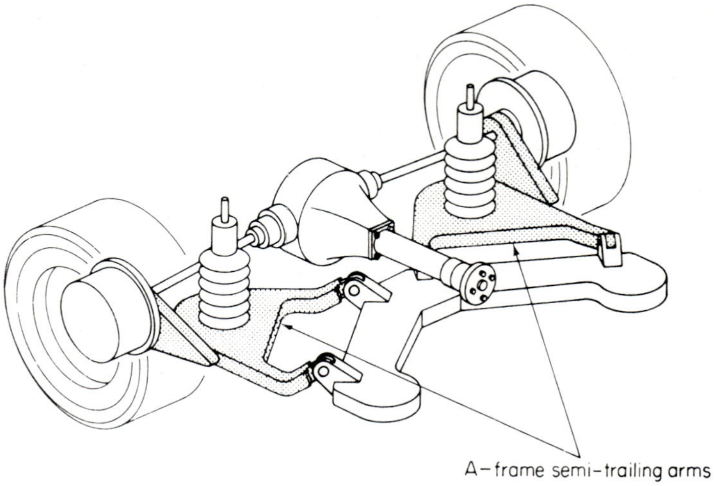

FIGURE 6-16 Independent rear suspension using A-frame semitrailing control arms to retain side loads and torque loads.

Angular Linkages. The most common suspension control on a dependent rear drive is angular linkages connecting the vehicle body and frame to the rear axle housing as shown in Figure 6-15. By angling the links toward the centerline of the vehicle, the inertia side loads from turning and the torque loads from both braking and acceleration can be transmitted to the frame. When an independent rear suspension such as that shown in Figure 6-16 is used, the A-frame control arms restrain the side loads. A variation of the conventional A-frame design may be used with dependent-type rear suspension on front-wheel drives where the axle is behind the pivot. This was illustrated in Figure 1-16. This H-shaped frame tends to twist in the middle as the wheels jounce different amounts. The center section acts as a sway bar. When the pivot point trails the fixed-pivot axis, the link is called a *trailing arm.* In these suspension systems, the body side loads are transferred to the suspension, wheels, and tires through the arms.

Track Bars. Some vehicles use parallel linkages for torque control and a track bar to transfer the side loads to the tires and road surface. A *track bar* or

FIGURE 6-15 Typical suspension control linkage on a dependent rear-wheel drive.

Panhard rod is a link connecting the vehicle frame on one side to the rear axle housing on the opposite side. This can be seen in Figure 6-17. The track bar transfers all the side loads to the axle housing. This design frees the trailing arm so it will only have to control braking and acceleration torque. Control of the side loads is through the track bar. Small vehicles sometimes use a short torque tube (a tube connected to the nose of the differential) approximately 2.5 feet (0.76 m) long. The front end of the torque tube is rubber-mounted. It also uses two trailing arms and a track bar or Panhard rod to provide full control. The track bar controls side loads while the trailing arms and torque tube transfer the torque from acceleration and braking to the vehicle frame.

6-5 ADDITIONAL SUSPENSION DEVICES

During braking, the vehicle body attempts to maintain its forward motion while the vehicle wheels and tires are attempting to stop. The resulting braking forces (Figure 6-18) develop a torque or twisting force that is transmitted to the suspension system. If these forces, as well as the inertia force of the vehicle, were not restrained, the vehicle body would continue moving, even though the vehicle brakes were attempting to stop its motion by slowing the forward motion of the suspension.

During acceleration, the differential housing attempts to rotate in a direction opposite to the direction of rotation of the tires (Figure 6-19). The forces on the suspension of the driving wheels during acceleration must restrain the axle housing to prevent rotation. Vehicles with independent suspension at the driving wheels transmit the torque of acceleration directly to the differential mounted to the vehicle frame or the vehicle power plant which is mounted on the vehicle frame. This system is shown in Figure 6-20.

The leaf springs provide enough strength to control the braking and acceleration forces if the leaf springs are placed from front to rear (longitudinal direction) on each end of the axle assembly, as shown in Figure 6-21. Transverse leaf springs require ad-

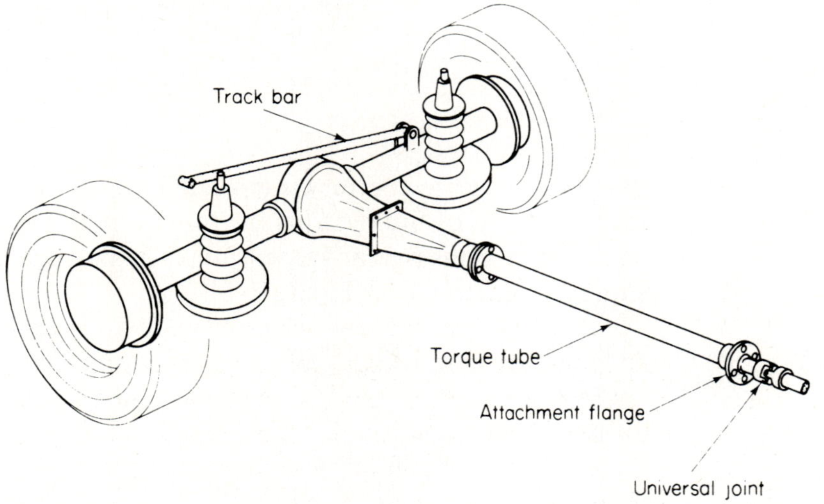

FIGURE 6-17 Rear dependent suspension using a track bar to retain side loads. This suspension uses a torque tube to retain braking and acceleration loads.

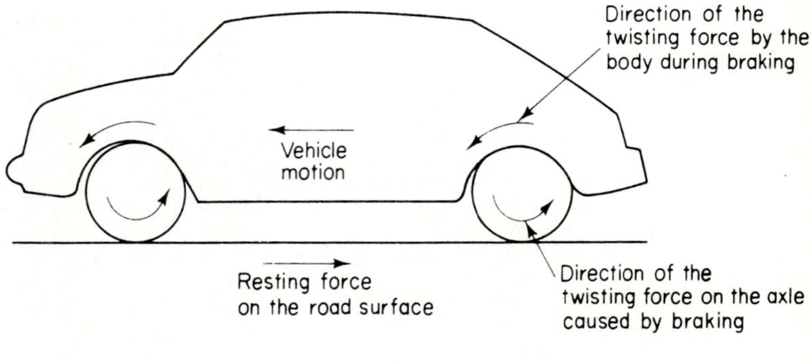

FIGURE 6-18 Torque loads during braking.

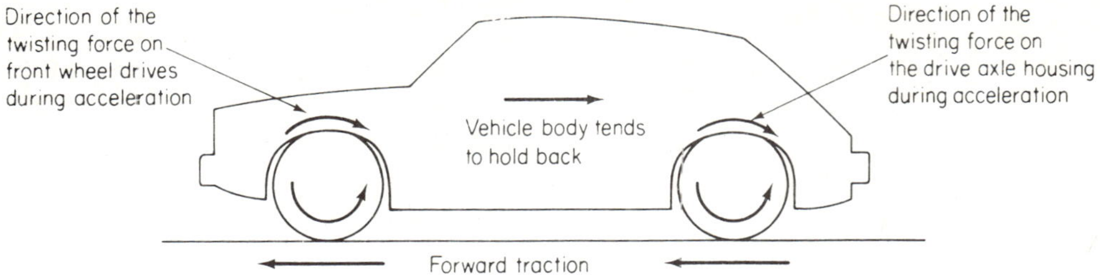

FIGURE 6-19 Torque loads while accelerating.

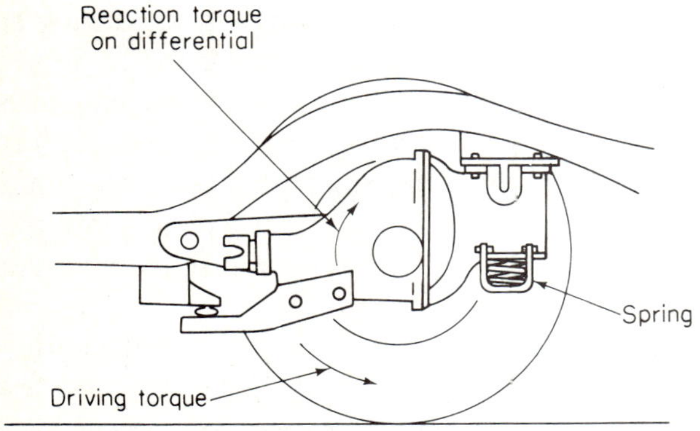

FIGURE 6-20 Frame-mounted differential transmits acceleration torque loads directly to the vehicle frame. Half-shafts are used with this type of drive line.

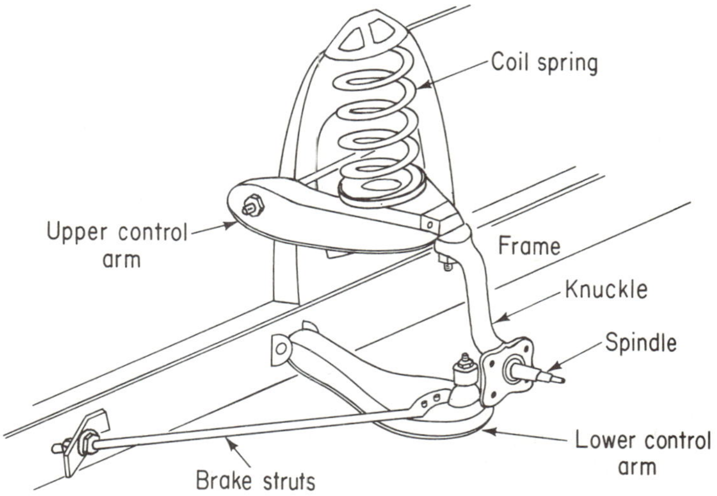

FIGURE 6-22 Two-piece lower control arm with a coil spring above the upper control arm. This type of suspension is used on some intermediate-sized domestic automobiles with rear-wheel drive.

FIGURE 6-21 Torque and brake loads controlled with a leaf spring.

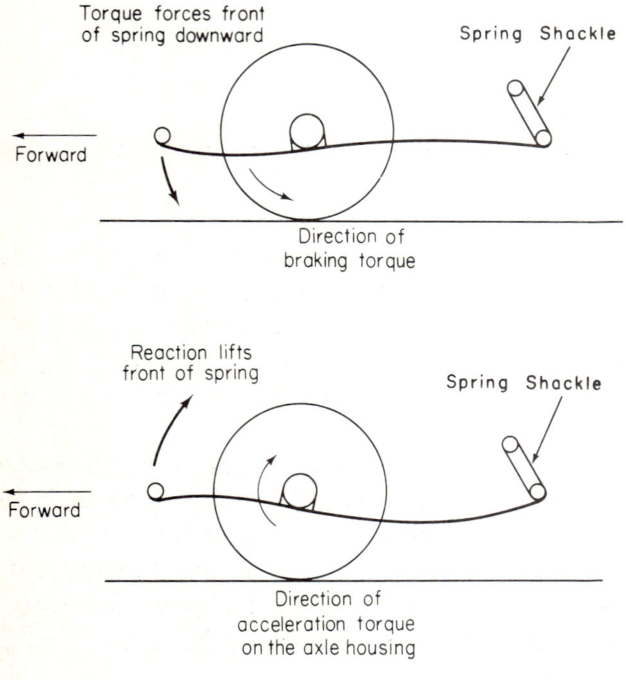

ditional linkages to control the forces. These linkages take the form of brake struts in front suspensions and trailing arms in rear suspensions. This suspension was illustrated in Figure 1-14.

Coil springs by themselves provide no lateral or longitudinal restraint. Coil springs on long-short control arm type front suspensions use the control arms to locate and restrain both upper and lower wheel pivots. An example was shown in Figure 1-10. Their design uses widespread fixed pivot points that enable the control arms to transmit the braking forces. Some suspensions have the coil springs mounted between the upper control arm and the body or spring tower, as shown in Figure 6-22. This suspension uses a smaller, lower control arm with a brake strut to absorb braking forces.

Brake struts are usually made as a heavy rod or they are a stamped steel strut, mounted one on each side of the vehicle. One end of the strut is attached to the frame and the other end is connected

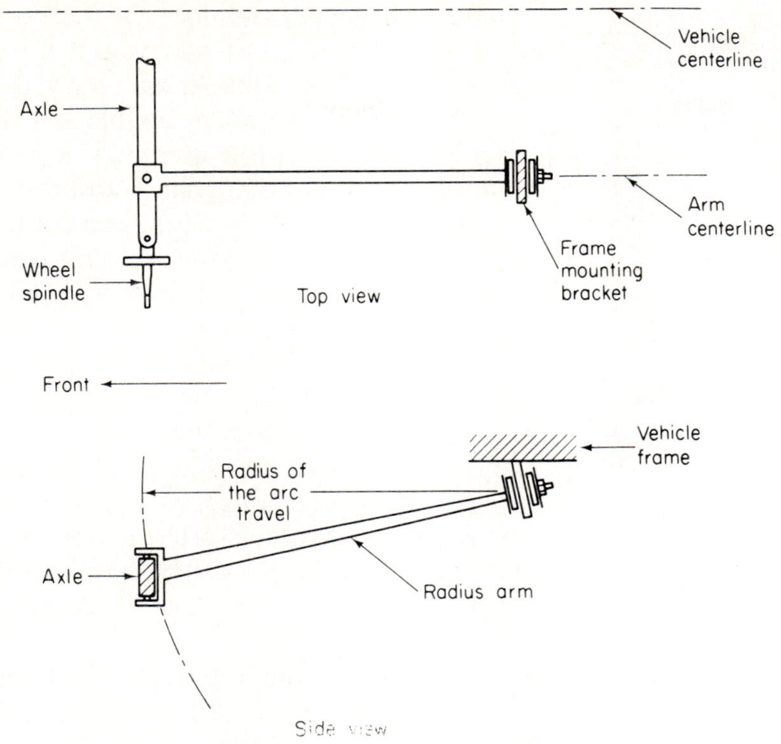

FIGURE 6-23 Radius arm used with dependent and twin I-beam front suspensions.

to the axle housing or the lower control arm near the wheel pivot. The strut can extend either forward or rearward. During braking, the brake strut transmits the rearward force on the lower control arm to the vehicle body. It also controls the rearward forces on the suspension caused by the road surface.

A radius arm is used on some front suspensions for the same purpose as the brake strut. The name *radius arm* implies one more important function, that of controlling the arc or radius through which the suspension travels, as viewed from the side. This can be seen in Figure 6-23. The centerline of the radius arm is parallel to the vehicle centerline. Rear suspension *trailing arms* are canted or angled from the centerline of the vehicle to absorb the forces from acceleration and braking. Canted trailing arms are illustrated in Figure 6-24.

Torque tubes are also used to absorb braking and accelerating forces. The *torque tube* is a large tube that connects the nose of the differential to either the vehicle underbody or the transmission extension housing, as illustrated in Figure 6-19. The torque tube absorbs rearward motion and torque forces during braking, as well as forward movement and torque forces during acceleration. A Hotchkiss drive, which has one universal joint on the differential companion flange and another at the rear of the

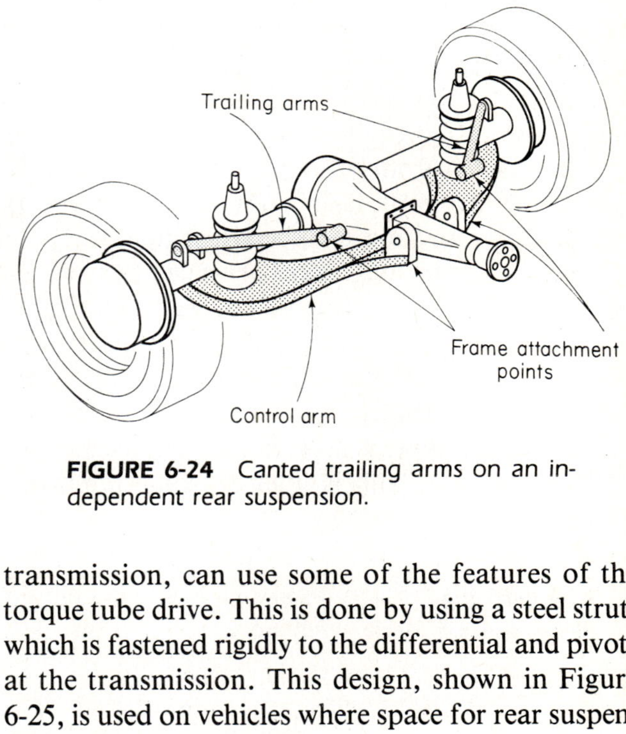

FIGURE 6-24 Canted trailing arms on an independent rear suspension.

transmission, can use some of the features of the torque tube drive. This is done by using a steel strut, which is fastened rigidly to the differential and pivots at the transmission. This design, shown in Figure 6-25, is used on vehicles where space for rear suspension linkages is limited.

Longitudinally mounted torsion bars (front to rear, as shown in Figure 6-5) use the same method for absorbing forces on the suspension arms as coil springs. When the torsion bars are mounted transversely (side to side, as shown in Figure 6-6), the torsion bars may be used to absorb forces. This is

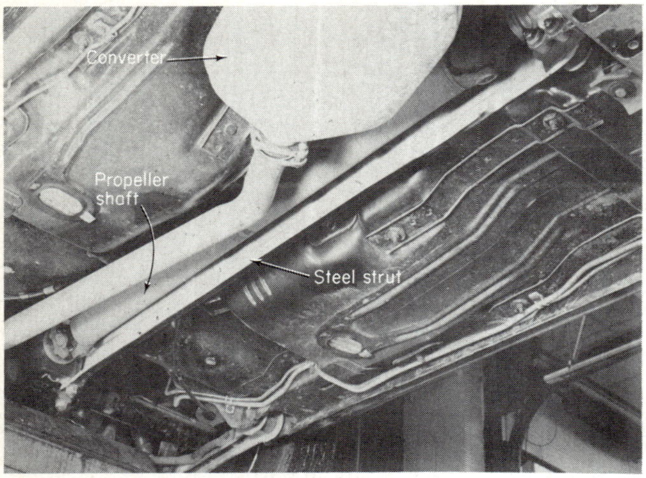

FIGURE 6-25 Steel strut next to the drive shaft controls breaking and acceleration torque.

accomplished by positioning the rotating end of the torsion bar near the wheel pivot.

Control arms on the rear suspension, used with a dependent axle, have one more important function: they control the differential nose angle. Most modern rear-drive vehicles use the Hotchkiss drive dependent suspension as shown in Figure 6-15. The two universal joints must operate near the same angles. These angles will remain nearly equal as the vehicle travels through jounce and rebound only when the control arm length, pivot point location, and operating angle are correctly engineered.

Trailing arms and radius arms can be used on either front or rear suspensions. A leading arm suspension shown in Figure 6-26 is another design alternative where the movable pivot leads the fixed pivot point. The leading arm suspension, as well as the trailing arm type, is adaptable to transverse torsion bar springing and is used on some imported vehicles.

FIGURE 6-26 Leading arms hold the wheel ahead of the frame mounting.

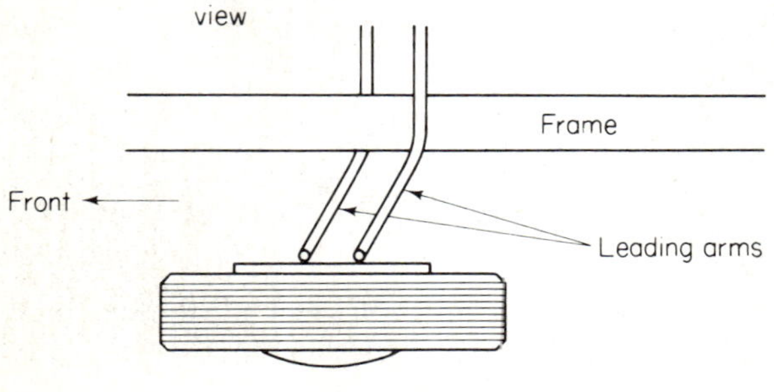

MacPherson Strut Suspension. On the MacPherson strut front suspension, a lower control arm provides lateral support. The stabilizer bar used with the control arm may be separate or double as a strut bar. MacPherson strut suspension has a coil spring mounted over a telescoping shock absorber. The upper end of the shock absorber is connected to the vehicle underbody. A lower control arm and a stabilizer bar complete the assembly. This suspension was shown in Figure 1-11. A modified design places the coil spring between the control arm and frame, inboard of the strut (Figure 6-27). Light weight, simplicity, and a minimum space requirement are the major benefits of the MacPherson suspension design. During braking, the stabilizer bar acts as a brake strut to absorb rearward forces while the upper end of the shock column absorbs the brake torque.

Stabilizer Bar. A stabilizer bar (antiroll bar or antisway bar) usually has a solid cross-section. It is connected from the lower suspension arm on one side of the vehicle to the lower suspension arm on the other side. The bar may be connected to the suspen-

FIGURE 6-27 Modified MacPherson strut.

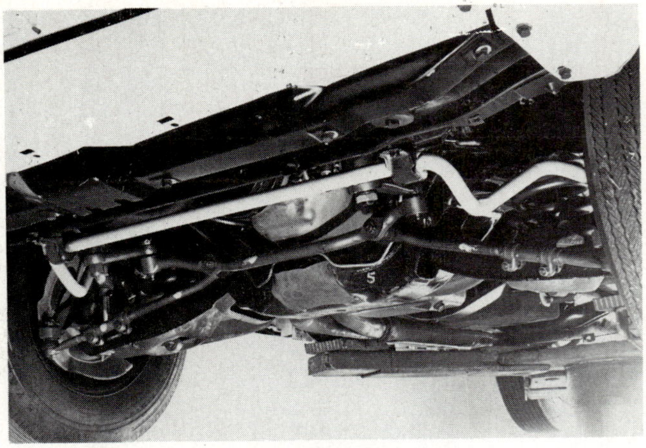

FIGURE 6-28 Typical front stabilizer bar installation.

sion arms by short links, allowing the bar to pivot on mounts fastened to the vehicle frame. You can see these links in Figure 6-28.

Connecting the stabilizer bar between the suspension control arms produces many benefits. When jounce affects both left and right wheels equally, the stabilizer bar has no effect on the vehicle. As one tire hits a rise in the road surface, the upward force on the tire is transferred across the stabilizer bar, producing an upward force on the opposite suspension. This is illustrated in Figure 6-29. This transfers part of the jounce load to the opposite suspension. The action twists the stabilizer bar much like a torsion bar. This twisting action increases the effective spring rate of the suspension in jounce and reduces the effective spring rate of the suspension in rebound.

The stabilizer bar can be considered a spring that resists the independent action of the suspension. It transfers some of the load from one suspension to the other to minimize vehicle body roll. As a vehicle enters a turn, the vehicle body leans toward the outside of the turn and this forces the outside suspension into jounce (compression) while the suspension on the inside goes into rebound. The stabilizer bar opposes both of these motions by increasing the effective spring rate of the outside wheel and decreasing the effective spring rate of the inside wheel. As a result, the stabilizer bar reduces the degree of body roll—hence the name *antiroll*. The stabilizer bar dampens body movements from wind gusts and rolling road surfaces by applying forces on the suspension that oppose vehicle input forces (one wheel going into jounce while the other goes into rebound).

Stabilizer bars were first used on vehicles designed for handling, such as road race cars. They have been incorporated into other vehicle designs, first on the front suspensions of passenger vehicles and then on the rear suspensions (Figure 6-30). Many vehicles now have both front and rear stabilizer bars to provide the handling and ride quality demanded by the operator. A reduced vehicle spring rate can be used when sturdy stabilizer bars are used. This provides the vehicle with soft ride while, at the same time, it has the handling characteristics of firm springing. The diameter and shape of stabilizer bars vary according to the manufacturers' design objectives. Due to the heavy weight bias on front-engine vehicles, the front stabilizer bar usually has a heavier

FIGURE 6-29 Action of a stabilizer bar.

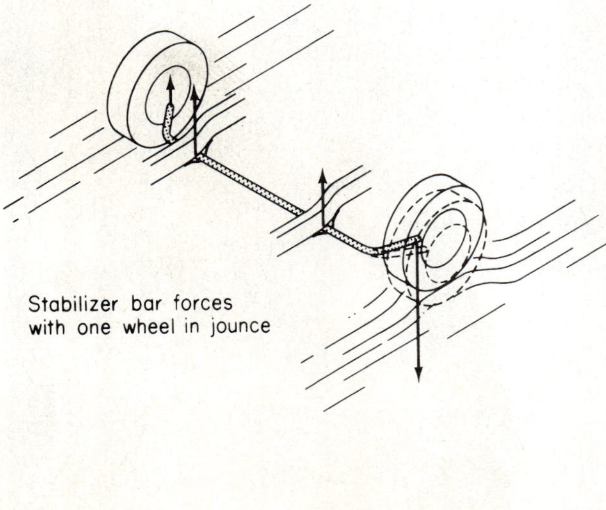

FIGURE 6-30 Typical rear stabilizer bar installation. This stabilizer bar connects between the lower trailing arms. It is not connected the body or frame.

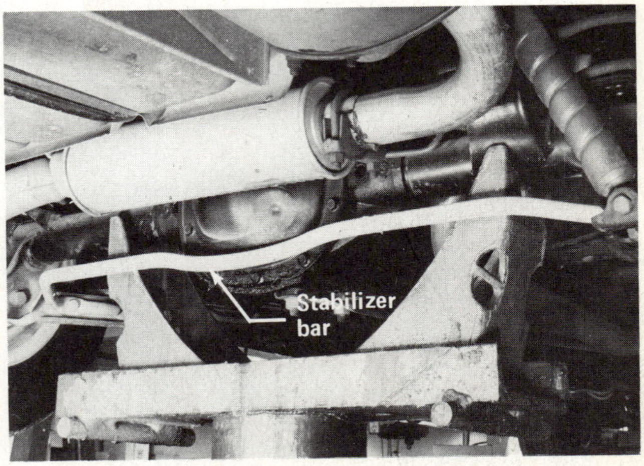

cross-section than the rear stabilizer bar. If stabilizer bars have cross-sections that are too large, the vehicle will be difficult to control on rough roads. Again, a compromise must be made between rough-road and high-speed, smooth-road handling.

Bushings. Suspension control devices such as control arms, stabilizer bars, A-frames, track bars, brake struts, radius arms, and vehicle springs are isolated from the vehicle body and/or frame. Isolation of the suspension components from the frame or underbody reduces the transmission of noise from the tires and road surface into the driving compartment. Bushings of a rubber or elastomer material are placed in the pivots of control arms, spring eyes, and idler arms or the steering linkage. These bushings usually have an inner and outer liner of steel with the rubber bonded to the liners as illustrated in Figure 6-31. Rotation can take place between the liners due to the flexing of the rubber. Bushings of this design allow relative movement (compliance) of the suspension, help absorb some of the road shock, and prevent the transmission of noise into the vehicle body. Due to the bonding of the liners to the rubber or insulating material, lubricant should *never* be used on them. Lubricants would soften the rubber-type elastomer material and separate the bonding from the metal bushing. Elastomer bushings will tolerate a slight amount of misalignment. They, therefore, reduce the need for closely controlled tolerances in the manufacturing of the suspension parts. This, in turn, reduces the manufacturing cost.

Bushings are also used between suspension and cross-members, at the ends of the brake struts, and at stabilizer bar supports. These bushing may or may not have steel liners, although their function is similar to those requiring steel liners. In some cases, wide tolerances are satisfactory, which reduces the cost of manufacturing. Examples of these wide tolerances are the typical stabilizer bar mounting brackets shown in Figure 6-32, where carefully maintained tolerances are not required.

Other types of insulators may be found between the coil springs and the spring seats, between the leaf springs and their mounting pads, and between cross-members that carry all of the suspension parts for

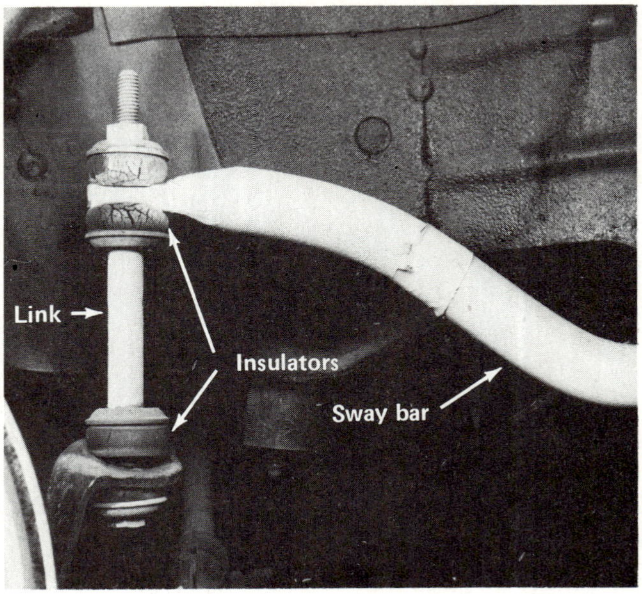

FIGURE 6-32 Stabilizer bar insulator bushings. The bushings shown here have aging cracks.

FIGURE 6-31 Steel bonded to rubber to form typical suspension bushings.

FIGURE 6-33 Bushing designed with voids that allow more suspension compliance than a solid bushing.

the front and rear of the vehicle. The main purpose of this type of insulator is to prevent the transmission of irritating vibrations and noise into the passenger compartment. Radial ply tires have increased the problem of noise isolation due to their low-speed harshness. With properly tuned suspensions, radial tires are used with the same passenger comfort as obtained with other types of tires. Large-diameter spring eye bushings with voids allow more suspension compliance for extra damping and vibration control. An example of this is pictured in Figure 6-33. Torsion bars are isolated at front cross-members. In some cases, weight is added to the suspension to move the vibration frequency out of the range of annoying vibration. An example of this type of weight is shown in Figure 6-34. These are some of the ways the chassis engineer reduces or changes the frequency of the noise to keep vibration from reaching the vehicle occupants.

6-6 STEERING PIVOTS

A vehicle must have some type of wheel pivot to allow for both front suspension movement and front-wheel steering. Vehicles using both independent twin I-beam and dependent I-beam front suspensions use a *kingpin* fitted through the steering knuckle and into the end of the axle as shown in Figure 6-35; this allows the steering knuckle to turn. The kingpin has a thrust bearing fitted between the axle and the lower steering knuckle eye to support the vehicle wheel through the bearing. Kingpins were used on independent suspension systems up to 1954. The kingpin is well suited for use in types of axles where the vertical wheel angle does not change in relation to the axle on I-beam-type suspensions and on straight axles.

FIGURE 6-34 Weight added to the suspension to change the vibration frequency.

FIGURE 6-35 Typical dependent I-beam front axle assembly with a kingpin pivot.

Ball Joint Wheel Pivot. Independent front suspensions of the long-short arm design (upper and lower control arms) and MacPherson struts use spherical ball joint pivot points to allow for vertical suspension movement while turning the wheel. The ball and socket allows the suspension to move freely. Depending on its function in the particular suspension system, a ball joint may be a load-carrying joint or a non-load-carrying joint. Both types are shown in Figure 6-36.

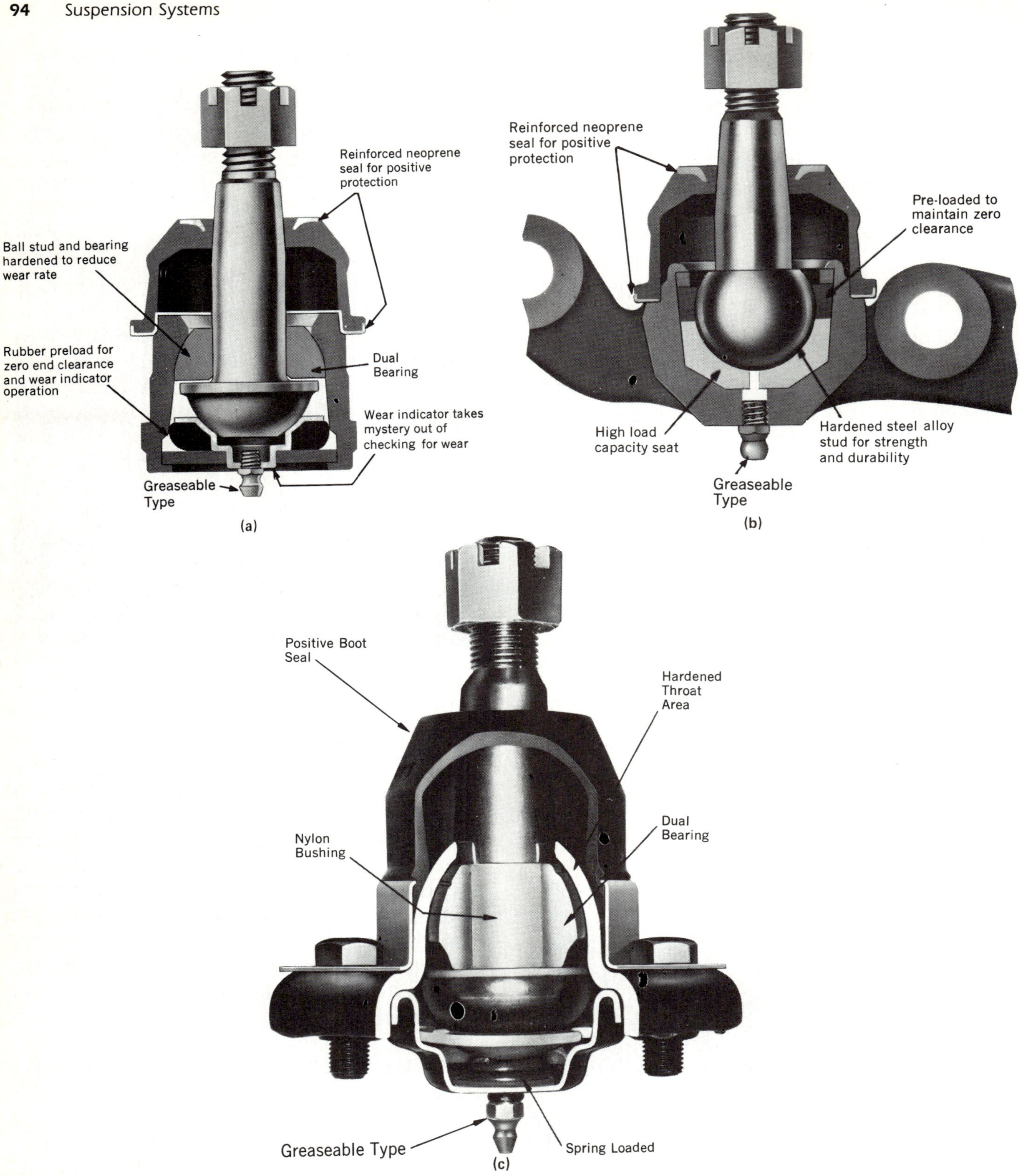

FIGURE 6-36 Typical ball joints: (a) tension ball joint, (b) compression ball joint, and (c) preload ball joint. (Courtesy of TRW Incorporated)

Vehicle-Loaded Ball Joint. The suspension spring forces one of the control arms on the suspension downward. The opposite end of the spring forces the vehicle frame or body upward. The ball joint on this load-supporting arm is called a *load-carrying* or *vehicle-loaded ball joint*. The load may be in tension, attempting to pull the ball from the socket, or in compression, attempting to force the ball into the socket.

Friction Ball Joint. One ball joint on each suspension acts as a pivot point for the suspension system and controls the direction of one end of the steering knuckle. It does not support vehicle weight and so it is called a *friction* or *follower ball joint*. This unloaded ball joint has a way of taking up wear that will normally occur. Suspension looseness would occur without this feature. A preload spring or elastomer disc is used to maintain zero clearance in the friction-loaded ball joint. A vehicle-loaded ball joint will maintain zero clearance as long as vehicle weight is on the ball joint. Some vehicle-loaded ball joints have a preloaded spring or disc to remove clearances that may occur as the weight is removed from the ball joint during rebound. Endplay in the loaded ball joint may be a source of the additional suspension noise and wear due to impact loading.

6-7 LEVEL CONTROL SYSTEMS

Air-assist shock absorbers are intended for use as leveling devices when the vehicle is carrying loads. Conventional shock absorbers do not have the ability to level the vehicle. Air-assist shock absorbers are offered as standard or optional equipment on many passenger vehicles. When required, they may also be fitted to vehicles not originally equipped with them.

Level-control air shocks are not intended to be used for replacement of worn or damaged springs; nor are they intended to be used to support vehicle overloads. They are used to keep the ride height near normal to maintain the suspension ride clearance. This is called *trim height*. It prevents the suspension from bottoming against the stops as the vehicle goes over bumps. Maintaining the proper rear suspension height prevents contact of the underside of the rear of the vehicle while driving over sharply inclined surfaces. When the vehicle stays at the correct ride height, the suspension geometry angles will operate near normal. This gives the vehicle better handling than it would have if the rear end sagged. It also helps to keep the correct headlamp aim on a loaded vehicle. The effect of rear suspension height is discussed in Section 7-3.

Leveling-Type Shock Absorbers. Air-assist shock absorbers designed to level the vehicle are very similar to conventional dual-acting shock absorbers. They are mounted in place of standard shock absorbers. The internal valving and operation are identical to the conventional shock absorbers discussed in Section 6-3. The difference is that these shock absorbers have an air chamber connected between the upper and the lower parts of the assembly. The air chamber is fitted to the upper housing. It takes the place of, or is part of, the protection shield on conventional shock absorbers. An air shock is pictured in Figure 6-12. A fitting provides a place for high-pressure air to be put into the air chamber. The high-pressure air forces the upper and lower shock absorber sections apart. This helps to level the body by increasing the distance between the centerline of the axle and the body. Since air is compressible, it has no real effect on the overall dampening action of the shock absorber. The air only supports some of the load, similar to an additional spring. Failure of the air circuit or the assist part of the air shock will cause the system to operate as a standard shock absorber without leveling control. Air pressure in the shock absorber may vary from as low as 15 psi (103 kPa) to nearly 200 psi (1380 kPa), depending on the type of shock absorber and the amount of leveling assist required. Maximum pressure will normally support up to about 250 pounds (113 kg) per shock absorber, or about 500 pounds (227 kg) for the entire rear suspension. The maximum assist is designed to support the weight of three average adults plus some luggage without appreciably lowering the rear of the vehicle.

Three types of air-level control systems are used to adjust air-assist shock absorbers: the first is manually adjusted by the operator, the second is adjusted by a switch in the vehicle, and the third is automatically adjusted.

Manual-Adjust Systems. Manual-adjust level control systems can be ordered on new vehicles. They can be retrofit on almost any type of vehicle. Manual adjust is the least expensive and the least complicated of the leveling systems. The only active part of the system is the shock absorber itself. Air lines are routed from the shock absorber to a tire-type inflation valve located in a convenient and accessible place on the vehicle. Factory-installed systems and most retrofit systems use one valve to pressurize both shock absorbers. Separate valves can be installed for each shock absorber if the vehicle has a different load on each side or if side-to-side trim balancing is needed.

Manual-adjust shock absorbers are inflated *after* the load is in the vehicle. After the load is removed, the air must be released to readjust the level of the vehicle.

Operator-Controlled Air Leveling Systems. Various air shocks are made for vehicles that do not have level controls. The shock absorbers are the same as those used on manual-adjust systems. These retrofit systems can change the air pressure to control leveling by using a switch located within easy reach of the driver.

The operator-controlled air leveling system has a pair of air-assist shock absorbers, an air compressor, a control switch, and connecting lines. By positioning the control switch, the shock absorbers can be inflated or deflated to adjust the trim height. The compressor is connected to the shock absorbers through small-diameter, high-pressure plastic tubing. When the vehicle is to be leveled to compensate for loading, the control switch is moved to the raise position. This starts a small compressor to increase the pressure in the shock absorber system. The vehicle is lowered by venting some of the air from the air shocks to reduce the pressure in the system.

Air Leveling Compressors. Two types of air leveling compressors are used: engine vacuum-driven diaphragm compressors and electric motor-driven piston compressors.

The working force in the engine vacuum-driven compressor, shown in Figure 6-37, uses engine vacuum on one side of a diaphragm and atmospheric pressure on the opposite side. The difference in vacuum and atmospheric pressure on opposite sides of the diaphragm moves the diaphragm. Diaphragm movement from side to side moves a small two-headed piston. The piston pumps up high air pressure to control the shock absorber assist.

The piston movement in one direction compresses the air ahead of the large piston in the first stage. This air is directed to the other small piston head. When the piston reaches the end of its travel, a diverter valve mechanism reverses the vacuum and

FIGURE 6-37 Section view of a vacuum-driven diaphragm compressor.

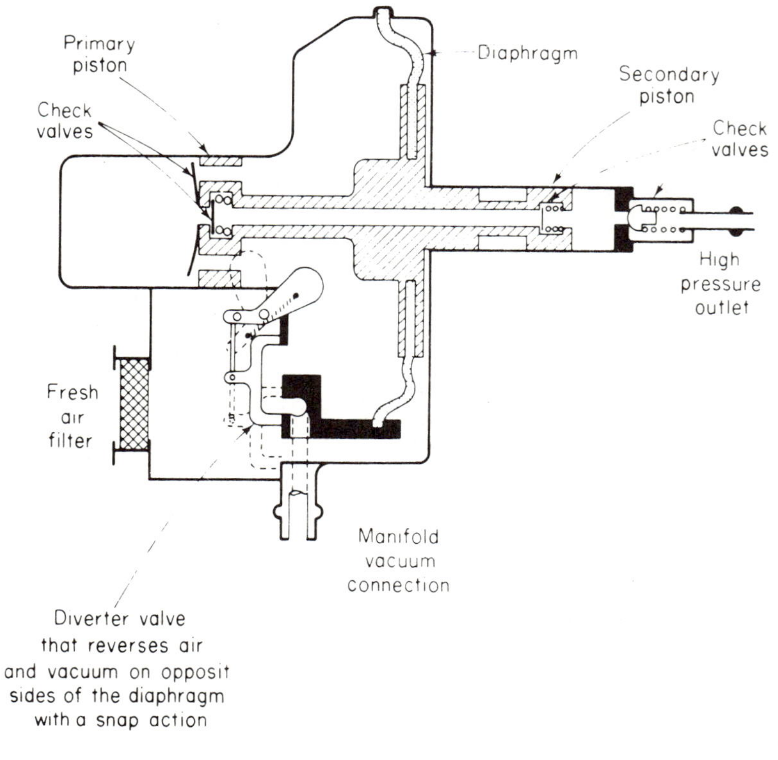

atmospheric pressure to opposite sides of the diaphragm. This drives the piston back to build up second stage higher pressure ahead of the small piston. The reversal of the diverter valve mechanism keeps the piston moving back and forth, compressing air alternately on each end of the two-headed piston. This pumping fills the reserve tank with high-pressure air.

The vacuum-operated compressor will operate only when the difference between vacuum and atmospheric pressure is great enough to overcome the air pressure on the piston head. This limits the maximum system air pressure and maximum assist from the air shocks.

The electric motor-driven piston compressor is similar to an air compressor found in a service facility for shop air, only much smaller. An electric motor drives the compressor crankshaft. This, in turn, moves a piston which draws in outside air and discharges compressed air to the system. You can see these parts in Figure 6-38. The valve assembly is built into the cylinder head. A relief valve on the system limits maximum pressure and a timer keeps the compressor from running continuously. The timer will allow the compressor to run long enough to adjust trim height. If trim height cannot be attained in the time set, the compressor will automatically shut off. This prevents continual operation of the compressor in case of a system malfunction. The trim is set again when the ignition is turned off and then turned back on.

FIGURE 6-38 Partly disassembled electric motor-driven piston compressor.

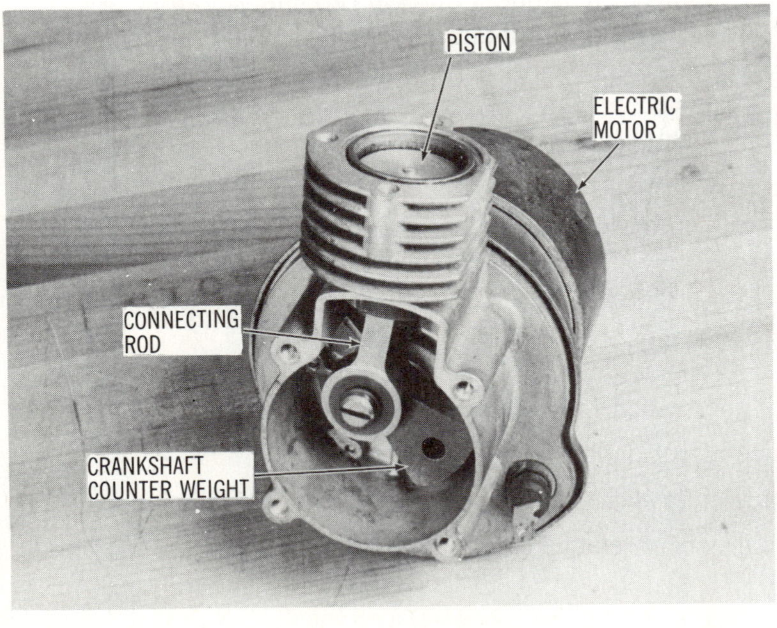

A high-pressure reserve tank is used on some systems for relatively rapid adjustment when the system is equipped with a small compressor. The reserve air will adjust the pressure in the air shocks to level the trim height. Then the small compressor will slowly rebuild the pressure in the reserve tank. The vacuum-diaphragm compressor uses a reservoir to provide some adjustment capacity when the engine is not operating. Some piston-type compressors do not use a reserve tank because the compressor has enough displacement capacity.

Automatic Leveling Control Systems. Automatic leveling control is available from the manufacturer on most full-size domestic passenger vehicles and it is standard on many. The fully automatic height-control leveling systems do not require operator control. The system has two air-adjustable shock absorbers, an air compressor, a reserve tank where needed, and a height-control sensor assembly.

The height-control sensor or leveling valve assembly automatically adjusts the pressure in the air shock absorbers. It senses the vehicle body position in relation to its trim height position and determines whether air must be added to or released from the shock absorbers. The height-control sensor is connected between the vehicle body and a suspension member. When a change in the load level is noted by the ride-height sensor, after a short delay period, air is forced into the shock absorbers until the vehicle has returned to the trim height. When the load is removed, the vehicle body rises. This causes the suspension to extend beyond its normal ride height. The height-control valve exhausts air from the shock absorbers to once again set the trim height.

An electrical switch in the sensor and a relay start the electric motor-driven compressor to raise the vehicle. To lower the vehicle, the sensor opens a solenoid to allow enough air to escape to reset the trim height. Another type of sensor is completely mechanical in operation. In this system, a valve is connected to the suspension through a linkage. The movement of the linkage and valve directs air from the reserve tank to the shock absorbers to increase height. It exhausts air from the shock absorbers to the atmosphere if the vehicle height is to be lowered. This type of system is shown in Figure 6-39.

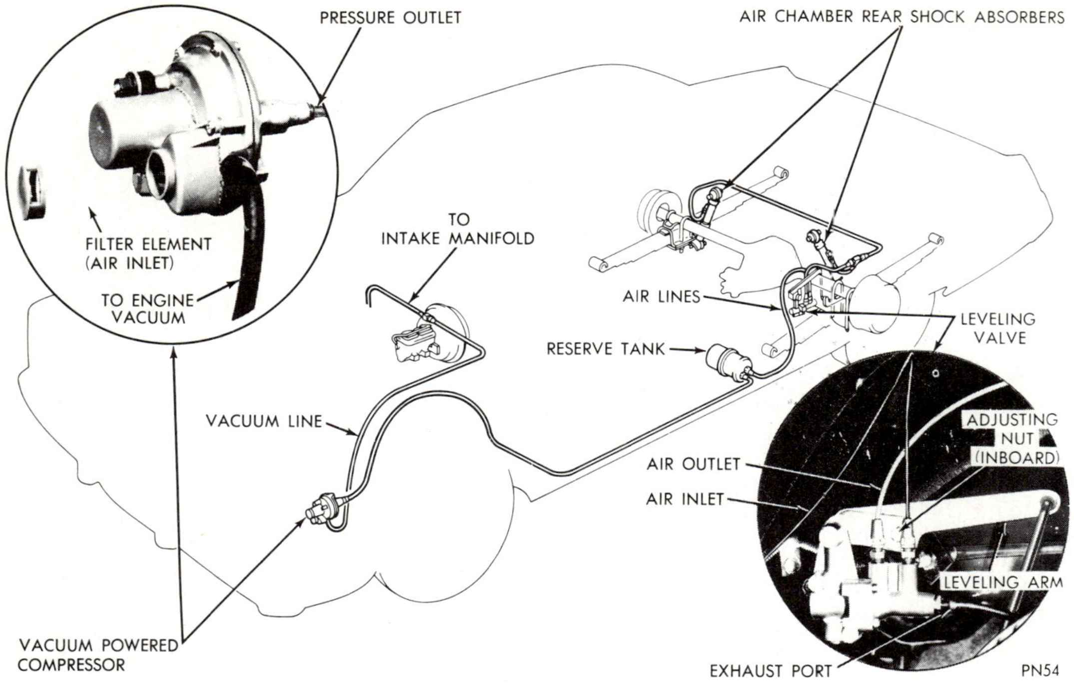

FIGURE 6-39 *Typical leveling control system using a vacuum-powered compressor. (Courtesy of Chrysler Corporation)*

Each type of sensor is equipped with a delay mechanism. The delay time is usually 10 to 15 seconds. The delay prevents the system from continually trying to adjust as the vehicle encounters bumps and dips in the road. If the position of the vehicle body remains in the new position longer than the delay time, the sensor detects this as an actual change in loading, not a momentary change. The automatic adjustment system will then begin to make adjustments to the trim height. As adjustments occur, the switch or valve gradually moves toward the neutral position. When it reaches neutral, it stops further adjustment at the trim height.

The only periodic maintenance required on the adjustment systems is air filter cleaning and/or dryer replacement. Air leakage can occur and is easily detected. Occasionally, problems result from improper positioning of the ride-height sensor. This should to be readjusted. Failure can also occur inside the shock absorbers. All air shock absorbers require a minimum air pressure of at least 15 psi (103 kPa), even when the vehicle is completely unloaded, to provide a normal service life. If this is not maintained, the rubber boot will rub and this will cause the boot to fail. Failure to maintain minimum pressure could be due to tubing failure, improperly adjusted height sensors, a compressor failure, or a failure in the valving mechanism.

STUDY QUESTIONS

Introduction

1. What is the primary purpose of the suspension system?
2. How does suspension compliance affect handling?
3. Name two popular front suspension designs.
4. What is the difference in ride quality between dependent and independent suspensions?

Section 6-1

5. What does a spring rate of 1500 lbs/in. tell you about a spring?
6. What compromise must be made between spring rate and vehicle weight?

Section 6-2

7. What does a center bolt do in a leaf spring?
8. What are the results of the leaves sliding against each other in multiple-leaf springs?
9. What are the advantages of a monoleaf spring over a multiple-leaf spring?
10. What are the common causes of leaf spring failure?
11. How do torsion bars support the load?
12. What does a variable-rate spring do?
13. What kind of loads do coil springs support?

Section 6-3

14. What is the purpose of shock absorbers?
15. Why do shock absorbers get warm in use?
16. What do air- and spring-assist shock absorbers do?
17. Describe the flow of the fluid in a shock absorber during jounce and rebound.
18. What is the advantage of gas-charged shock absorbers?
19. What part of the shock absorber controls dampening?
20. How does an adaptive suspension differ from an active suspension?
21. What care is required when mounting shock absorbers?

Section 6-4

22. Why are suspension-control devices needed?
23. What do angular linkages do for the suspension?
24. What does the track bar do for the suspension?

Section 6-5

25. How are braking and acceleration torque transferred from the wheels to the body?
26. How do braking and acceleration control differ between leaf and coil springs?
27. How do brake struts and radius arms differ?
28. What controls the side loads when a MacPherson strut is used?
29. What controls the brake torque when a MacPherson strut is used?
30. How does a stabilizer bar help control vehicle motion?
31. What are the advantages of rubber bushings with bonded steel liners?
32. Why do some rubber bushings have voids?
33. Why are weights added to some suspension parts?

Section 6-6

34. Name three types of ball joints.
35. Where is a follower ball joint used?

Section 6-7

36. List the advantages of level-control air shocks.
37. How much pressure is put into air shocks?
38. When is air put in manual-adjust air shocks?
39. How do the operator-controlled air shock systems differ from manual-adjust systems?
40. What limits the maximum air pressure developed by a vacuum-operated compressor?
41. What limits the maximum air pressure developed by an electric motor-driven compressor?
42. What is the purpose of having a timer in the circuit of an electric motor-driven compressor?
43. Why is a reserve tank used with some air shock systems?
44. Why do some air shock systems operate without a reserve tank?
45. Why do air shocks use a delay in the control system?
46. Why is it necessary to maintain a minimum air pressure in the air shocks?

7

Suspension Design

The suspension system, like all other parts of the vehicle, is designed to meet specific requirements. These requirements consider the number of passengers to be carried, the useful load, the type of ride desired, and the required handling characteristics. As is generally true, when one suspension property is improved, the quality of another property is reduced.

The first priority of suspension design is to carry the vehicle weight. It must hold the wheels and their supports in the proper position during acceleration, braking, and turning. At the same time, it must flex enough to absorb road bumps and dips so that the passengers and load, and even the vehicle itself, will not be shaken harshly. Shock absorbers must quickly stop the vehicle oscillations after jounce and rebound.

Controlling the vehicle in the desired track down the road is called *handling*. This assumes that the tires are in constant contact with the road surface. Handling becomes more critical as vehicle speed increases or when turns become more abrupt. The location and design of the suspension linkages, the pivot points, and the steering linkage greatly affect the way the vehicle handles. The handling characteristics of the vehicle are modified by the type of tires being used.

The vehicle suspension, combined with the steering linkage and tires, is designed to provide the driver with safe, positive vehicle control that is free of irritating vibrations. The design should produce minimum wear on the tires and on other parts of the suspension system.

7-1 SUSPENSION CHARACTERISTICS

The objective of suspension geometry is to provide the best possible vehicle handling control while ensuring maximum tire life and a comfortable ride. *Suspension geometry* is the arc followed by the suspension and the wheel movement as the suspension travels through jounce and rebound. An arc is always produced as one end of the suspension member is rotated about a fixed point on the other end. The sideways displacement of the wheel (the distance the wheel moves from the maximum outboard to the maximum inboard position as the suspension moves through the arc) is reduced as the length of the suspension arm is increased. You can see this in Figure 7-1.

I-Beam Axle. A single I-beam axle will produce an arc when one wheel is moved through jounce and rebound with very little sideways wheel displacement. The small displacement is produced because the pivot point is around the wheel on the opposite side of the vehicle. The radius for the arc produced is equal to the distance between the tires. The twin

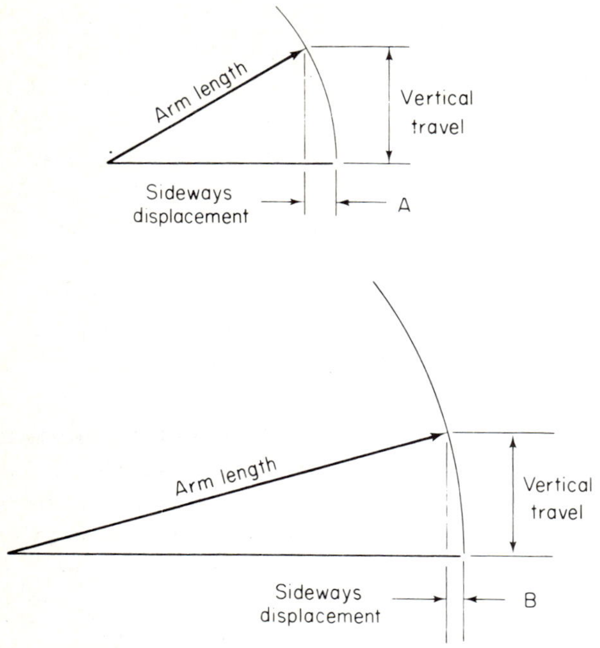

FIGURE 7-1 With equal vertical travel, a short arm length causes a greater sideways displacement (marked A) than the sideways displacement of a long arm (marked B).

FIGURE 7-2 Different camber changes and sideways displacement on I-beam and twin I-beam axles.

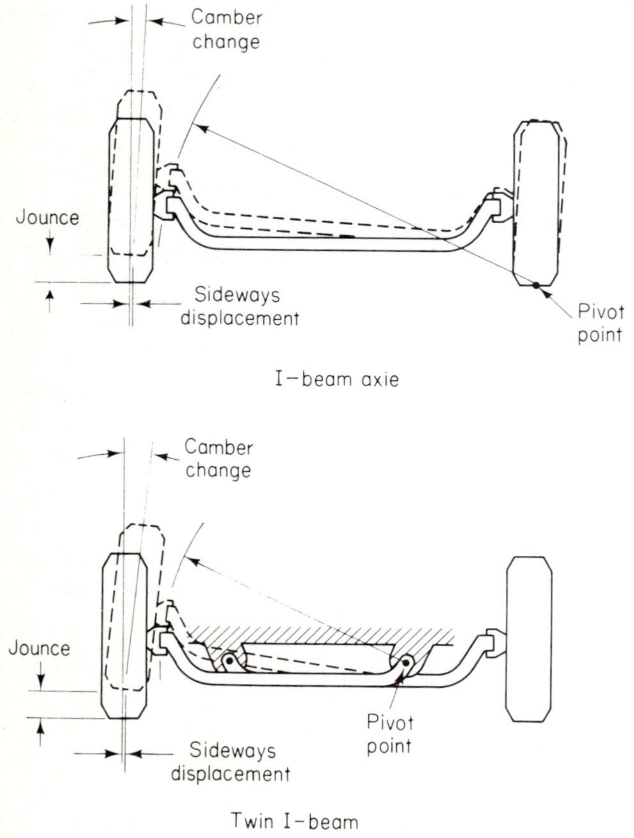

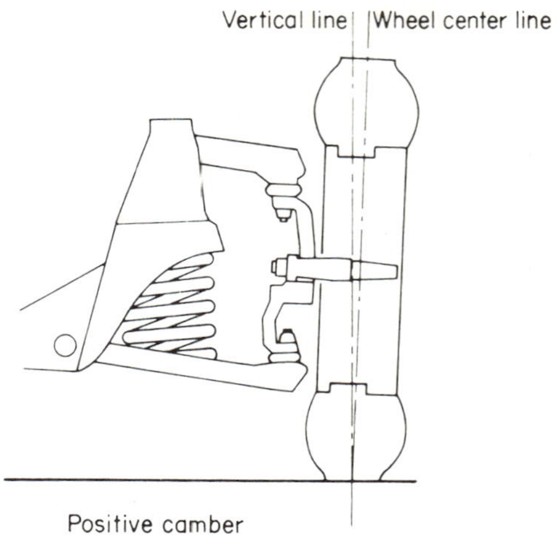

FIGURE 7-3 Front view showing positive camber when the top of the tire and wheel lean away from the vehicle.

I-beam suspension produces an arc similar to the single I-beam axle. The tire has a slightly greater sideways displacement because the axle pivot point is inboard of the opposite wheel.

You can see in Figure 7-2 that the top of the wheel tilts inward and the bottom of the wheel tilts outward as the wheel moves into jounce. This change in angle between the tire and the road surface is called *camber*. **Camber is the angle formed between a line drawn through the tire and wheel centerline and a vertical line as viewed from the front of the vehicle.** Camber is shown in Figure 7-3.

Long-Short Arm Suspension. The independent suspension using two control arms, or A-frames, is called a *long-short arm suspension*. It is so called because the lower control arm is always longer than the upper control arm. Due to the relatively short length of the lower control arm, approximately 18 inches (46 cm), compared to a single I-beam axle which may be in excess of 60 inches (152 cm), you can see in Figure 7-4 that the arc and the sideways displacement of the tire is greater for the same jounce distance. This displacement causes a greater amount of side travel and more tire wear.

Scuff Travel. The amount of tire side travel on the road surface is referred to as *scuff travel*. To reduce the amount of scuff travel in the long-short arm suspension, the angle of the arms and their lengths are critical. The upper control arm is shorter than the lower control arm to make more room for the

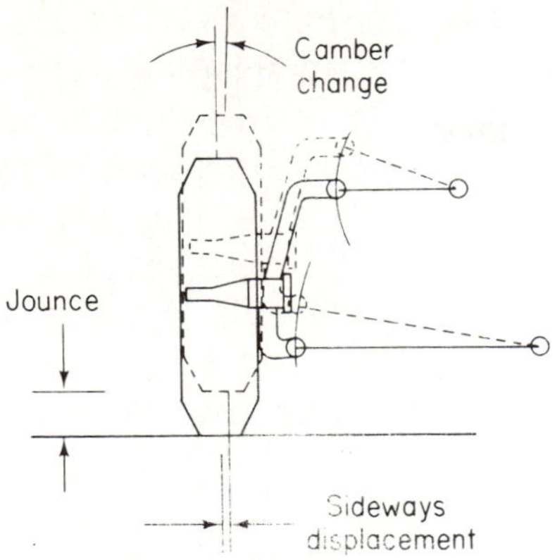

FIGURE 7-4 Effect of the same amount of jounce as shown in Figure 7-2 on camber and sideways displacement of a long-short arm suspension.

vehicle engine. Therefore, its displacement is greater as it travels through the same jounce as the lower control arm. Placement of the pivot point and the correct arm length will control the amount of scuff travel. This is illustrated in Figure 7-5. A small amount of scuff travel is useful because it produces some damping of the vehicle suspension when the wheel encounters road surface bumps and dips. If no scuff travel were present, all of the damping would take place within the shock absorber and friction would occur within the suspension joints.

FIGURE 7-5 Control arm lengths and pivot points that produce minimum skuff travel.

$\frac{L^2}{2A}$ ··· Sideways displacement of upper ball joint

$\frac{L^2}{2B}$ ··· Sideways displacement of lower ball joint

No scuff travel is present if the suspension is moved an amount L when HB = (H + C)A

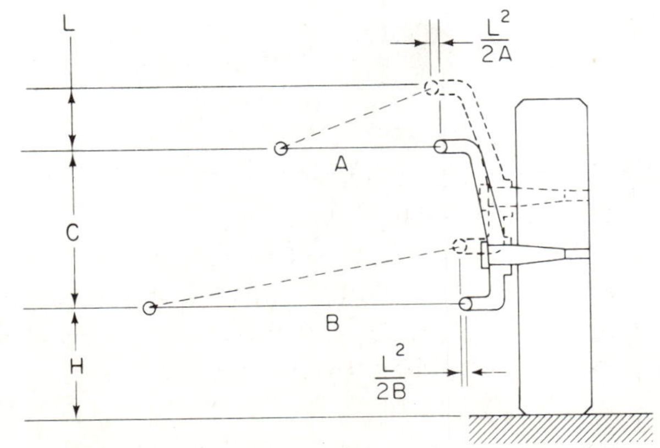

Effect of Camber Angle. The camber angle is important as the suspension travels through jounce and rebound. The desired operating characteristics are achieved through a compromise in the suspension design and its effect on scuff travel and camber change. Negative camber during jounce and positive camber on rebound is the design aim for many performance vehicles. As a vehicle with this suspension design enters a turn, centrifugal force and inertia will force the vehicle's center of gravity toward the outside of the turn. This transfers weight from the inside tires to the tires on the outside of the turn. The outside suspension will go into jounce and the inside suspension of the turn is moved into rebound. The twist or roll of the vehicle body moves the upper control arm pivot further than the lower arm pivot. Without careful design, this action will produce a tendency for positive camber on the outside wheel which, in turn, will reduce the tire-to-road contact area. If the suspension is designed so the outside suspension produces negative camber, the tire will maintain maximum road contact at all times to help vehicle handling. This is shown in Figure 7-6. The negative camber designed into the suspension allows the tire to have better support for the increased load in a turn while the suspension is in jounce.

This same effect will occur on a crowned road that tends to place the outside suspension in jounce and the inside suspension in rebound. Camber will increase on the inside wheel to correct for road crown as shown in Figure 7-7. This suspension geometry produces excellent handling characteristics, but it may also reduce tire life. As the vehicle with this geometry is loaded, the suspension goes into jounce. This produces negative camber that causes tire wear.

FIGURE 7-6 Long-short control arm suspension design to change camber as the body rolls in a turn; it helps the tires keep maximum road contact.

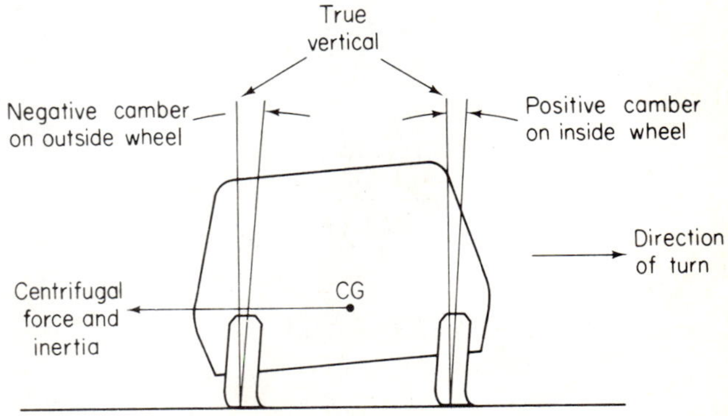

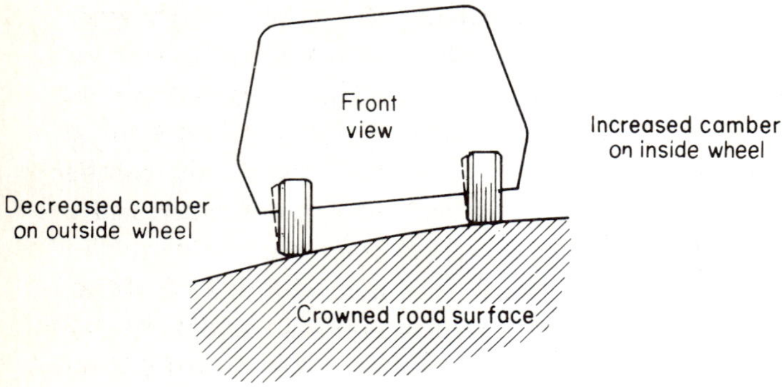

FIGURE 7-7 The camber effect of long- short control arm suspension shown on a crowned road surface.

Ride Quality. The suspension designer may choose to sacrifice some handling performance for improved ride quality. It is possible to produce a suspension with these two goals by designing a suspension having positive camber on jounce and negative camber on rebound. In a vehicle with this design, the passengers hardly feel a bump in the road surface. You can see this in Figure 7-8.

Other suspension geometries are also used. These depend on the suspension designer's objectives. The MacPherson strut suspension shown in Figure 7-9 has the wheel spindle attached to the bottom end of the vertical spring-shock strut. Geometry in this type of suspension is predicted by the lower control arm pivot point, the length of the strut and arm, and the angle between the spring-shock strut and the lower control arm. A MacPherson strut suspension is usually designed to produce negative camber in jounce and positive camber during rebound.

FIGURE 7-8 Long-short control arm suspension design that causes positive camber during jounce and negative camber during rebound.

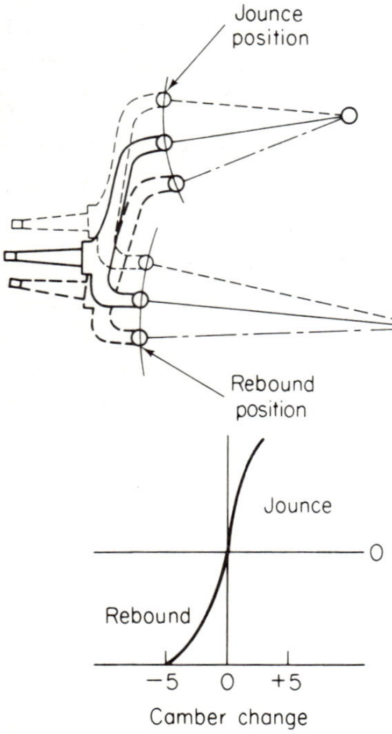

Caster Angle. As the vehicle suspension travels through jounce and rebound, the upper pivot point of the steering knuckle (upper ball joint) on most suspensions moves slightly rearward in relation to the lower pivot point (lower ball joint). This changes

FIGURE 7-9 The effect on camber during jounce and rebound of a MacPherson strut suspension.

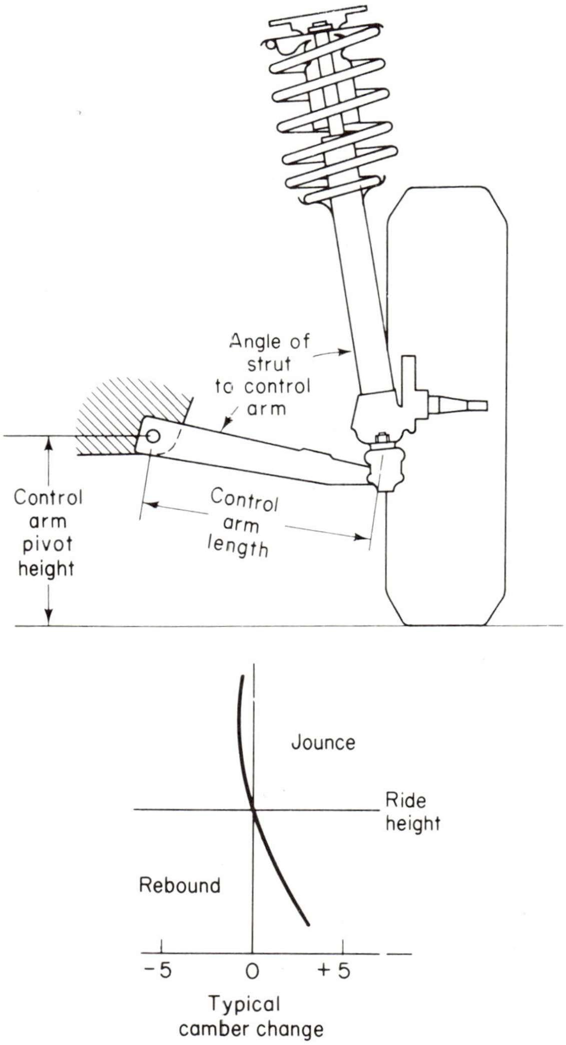

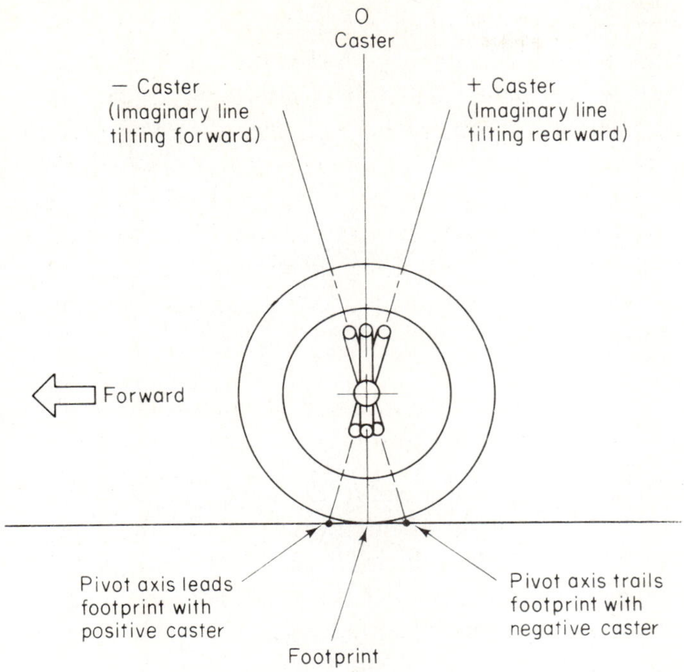

FIGURE 7-10 Caster angle of the left front suspension.

the suspension caster. **Caster is an angle between an imaginary line drawn through the steering pivots and a line perpendicular to the road surface, as viewed from the side of the vehicle.** If the top of the line tilts rearward, as shown in Figure 7-10, the vehicle is said to have positive caster. If the top of the line through the wheel pivots forward, the vehicle is said to have negative caster.

The caster change during jounce-rebound on vehicles using I-beam axles and semi-elliptical springs can be predicted by the fore and aft position of the axle on the leaf springs, the amount of arc of the springs at ride height, and the horizontal relationship of the front-to-rear spring eye height. Vehicles that use I-beam axles and leading radius arms have increased caster in jounce and reduced caster during rebound (see Figure 6–23). This is true because the arc in which radius arms travel produces a rearward tilt of the top of the axle in jounce.

During braking, the forward shift of the center of gravity tends to lower the front and raise the rear of the vehicle. This has the effect of shifting some weight from the rear wheels to the front wheels. This forward weight transfer places the front suspension in jounce and the rear suspension in rebound. The front of the vehicle goes down and the rear goes up. Because the suspension is fastened to the vehicle frame, the change in the angle of the frame to the road surface temporarily reduces the amount of caster during braking.

The effect of the vehicle frame angle to the road surface on caster is offset by the suspension design to increase caster during jounce. The caster angle then remains unchanged during braking and this maintains the vehicle stability. The radius arm on twin I-beam suspension is lower in the front where it is attached to the axle and higher at the rear frame attachment. The angle that the arm produces as the axle springs upward during jounce increases caster as well as reduces dive during braking. You can see this in Figure 7-11.

On vehicles with long-short arm suspension, the upper control arm pivot shaft is angled upward at the front. This angle of the upper control arm tends to move the upper ball joint rearward in relation to the lower ball joint as the suspension goes into jounce. You can see this as a dotted line in Figure

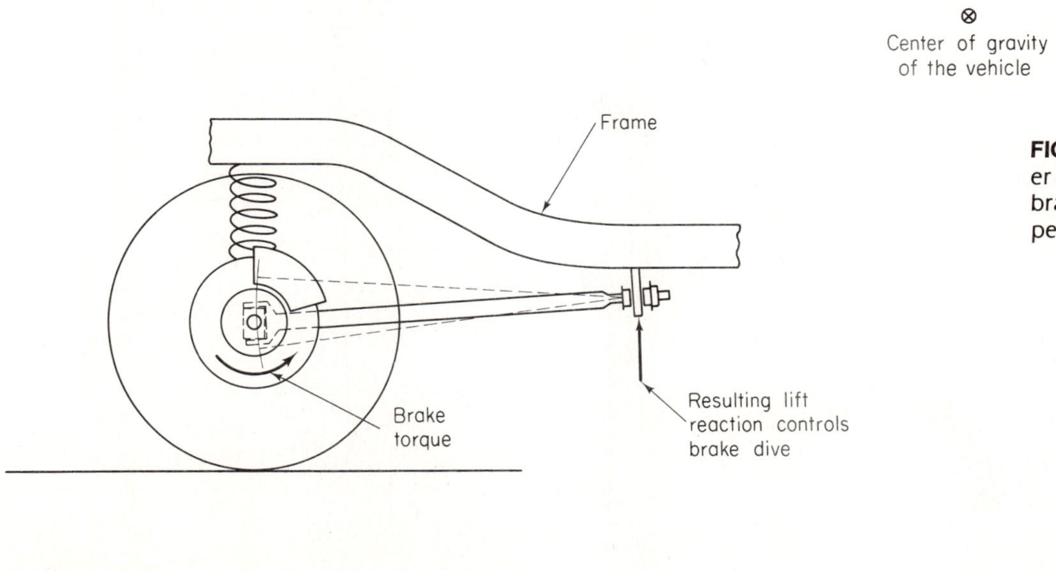

FIGURE 7-11 Control of caster caused by brake dive while braking on a twin I-beam suspension.

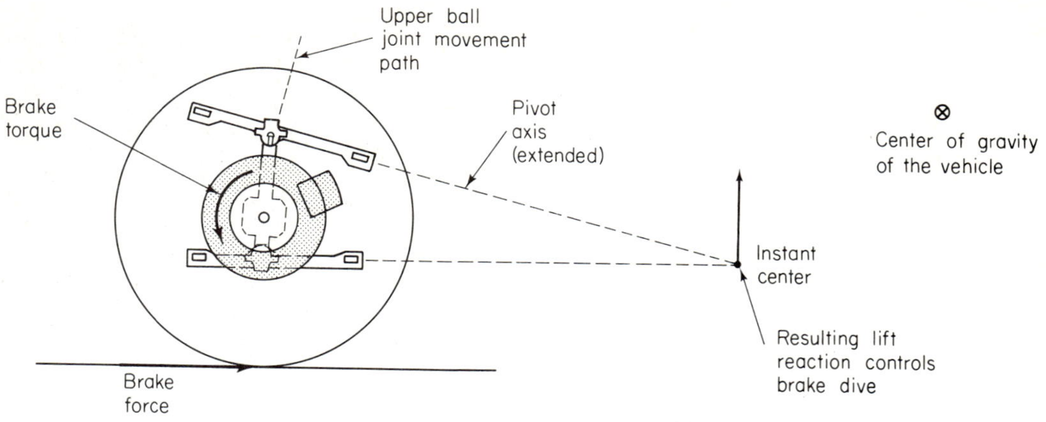

FIGURE 7-12 Control of caster during braking on a long-short control arm suspension.

7-12. This movement results in an increase in caster during jounce and a decrease in caster during rebound. The angled pivot axis produces an upward reaction on the axis to help reduce brake dive.

The caster change on MacPherson strut suspension is largely controlled by the angle that the strut makes with the vertical, as viewed from the side of the vehicle, and the arc through which the stabilizer bar swings while traveling through jounce and rebound (Figure 7-13).

A new, double-wishbone front suspension design was introduced on the 1984 Honda Prelude to better control the changes in front-wheel geometry during jounce and rebound. The most obvious change from a typical MacPherson strut system was to remove the knuckle from the shock strut. The shock strut has a forged steel arm on the lower end. Its forked curved arm goes around the half-shaft and connects to the outer end of the lower control arm. A strut leads forward from the outer end of the lower control arm to the vehicle frame to control forward and backward motion of the arm. The upper end of the shock strut is connected to the vehicle in a way similar to a typical MacPherson strut. The bottom of the steering knuckle connects to the lower control arm through a ball joint. The upper end of the steering knuckle is curved over the top of the tire to reduce the steering axis inclination to 6 degrees. The upper end of the steering knuckle is connected through a ball joint to a short upper control arm. The upper control arm is connected to the vehicle at an angle pointing forward at about 45 degrees, nearly like a leading arm. A sway bar is connected between the rear side of both lower control arms. This suspension system design has made it possible to lower the hood line and still not take any

FIGURE 7-13 Control of brake dive on a MacPherson strut suspension.

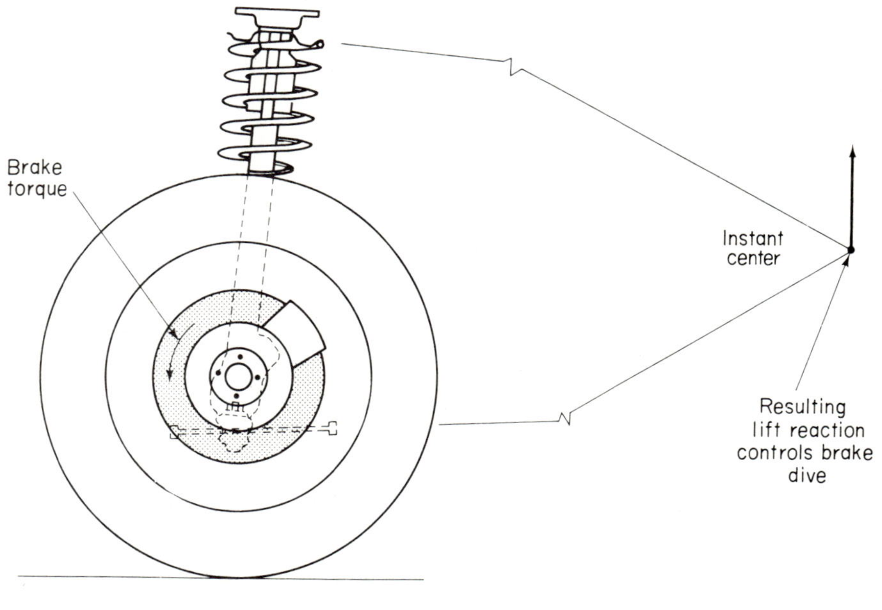

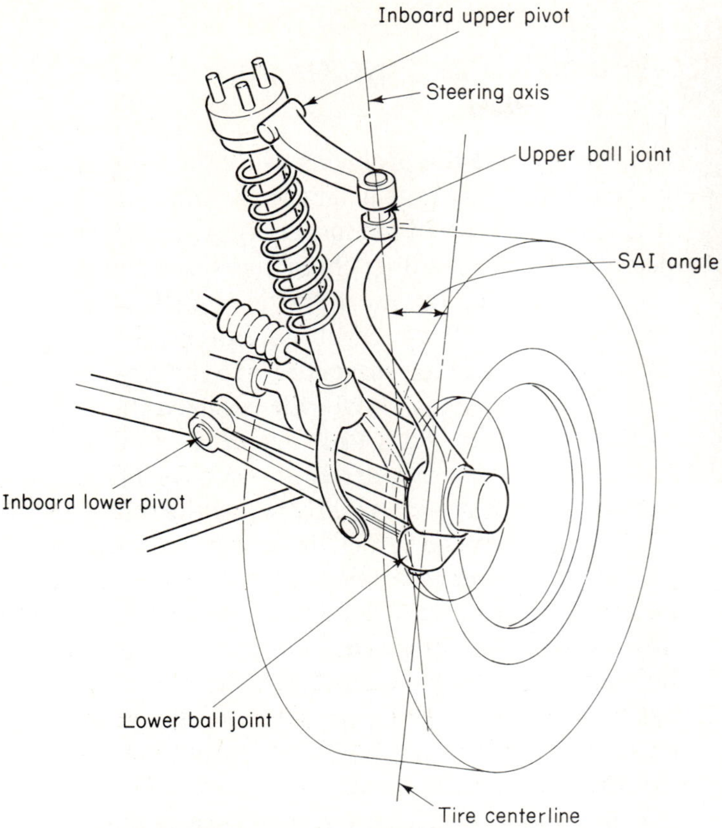

FIGURE 7-14 Honda's double wishbone front suspension.

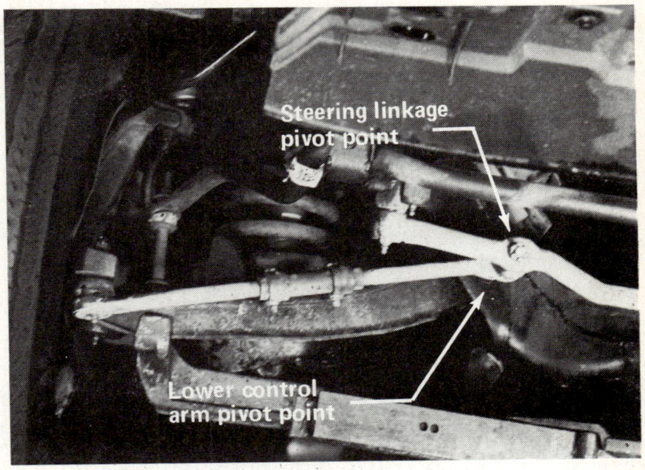

FIGURE 7-15 The pivot points of the steering linkage align with the outer control arm pivot points.

more underhood space than a MacPherson strut. You can see this suspension in Figure 7–14.

This Honda front suspension design, at the cost of complexity and expense, results in the wheel staying upright with zero camber at the road surface. It stops bump steer and improves high-speed stability, even on uneven road surfaces. The near-zero caster produces very fast steering response. The design increases caster during jounce like the long-short arm suspension to maintain effective zero caster during braking.

7-2 STEERING LINKAGE CHARACTERISTICS

The camber change during jounce and rebound is caused by the different arcs through which the suspension arms travel. The camber change would turn the wheel either in or out if the steering linkage pivots were not connected at the correct places. The inner pivot for the outer steering linkage is usually placed directly in line with the inner, lower control arm pivot. The outer steering linkage connection is in the same horizontal plane as the ball joint. As a result, the outer steering linkage is the same length as the lower control arm. The arc followed by the steering linkage is nearly identical to the arc followed by the lower control arm.

The alignment of the steering linkage tie rod and control arm pivots can be seen in Figure 7-15. When the arcs of the steering linkage and control arm are nearly identical, no toe change will result when the wheels are in the straight-ahead position. **Toe is the difference in distance between the front of the two tires and the rear of the tires at the spindle height.** *Toe-in* occurs when the tires are closer together in front than they are in the rear. If the two tires are farther apart in front than they are in the rear, they have *toe-out*. Toe is shown in Figure 7-16.

FIGURE 7-16 Steering toe-in and toe-out.

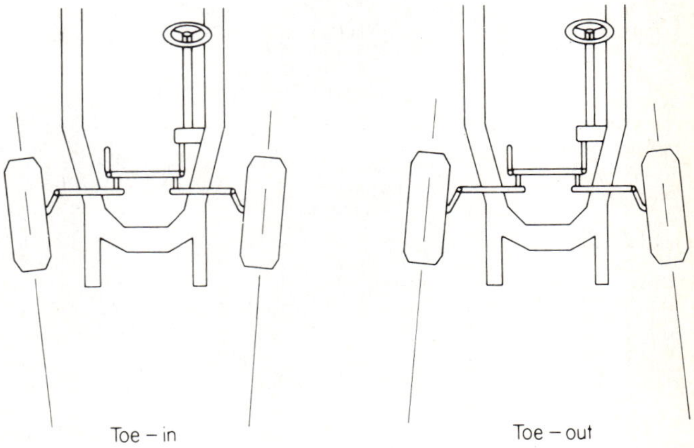

A typical steering linkage is shown in Figure 7-17. The outer link of the steering linkage is called a *tie rod.* A ball and socket, similar to the ball joint, connects each end of the tie rod to the steering arm. As the vehicle enters a turn, the steering linkage is moved to force the front wheels to turn. A vehicle entering a turn causes the suspension to react. The outside suspension will go into jounce and the inside suspension will go into rebound. A suspension designed with toe-out in jounce and toe-in during rebound generates a useful steering effect while making a turn. This design reduces the steering angle of the outside wheel as the vehicle body rolls toward the outside of the turn. The outside wheel has the most effect on steering because it is carrying the most effective weight. Steering effort is increased and the steering angle is further reduced as the body roll increases. This action is called *roll steer.* Roll steer on domestic automobiles reduces the steering angle of the outside wheel.

As a vehicle goes through rises and dips in the road surface, both suspensions may go into jounce and rebound at the same time. This will turn the wheels in and out, producing toe changes. These changes are referred to as *wheel fight.* Wheel fight produces tire scuffing, tire wear, and increased rolling resistance. In racing circles, terms such as *bump steer* are mentioned. *Bump steer* is the steering effect that results as the wheel moves through its suspension travel. On racing vehicles, bump steer is reduced to a minimum so that body roll does not change the steering angle.

FIGURE 7-17 Typical linkage placed in front of the control arms.

7-3 RIDE HEIGHT AND HANDLING

You can now see the problems caused by changing the ride height from the height set by the manufacturer. The geometry of the suspension provides completely different handling characteristics when ride height is changed. The most drastic effect is created when the vehicle is raised above ride height. The caster, camber, and toe curves begin to change rapidly when the chassis is above ride height (rebound) as shown in Figure 7-18. Ride height is established without passengers or a driver aboard. As a result, the ride height is always slightly below the specified ride height when the vehicle is driven. In a severe turn, the wheel on the outside of the turn is carrying a greater load than the wheel on the inside of the turn. The higher the vehicle is raised, the greater the load on the outside wheel. The increased load makes the outside wheel support more of the turning force. At the same time, the lightly loaded inside wheel provides less turning force. It is, therefore, more important to have correct geometry on the outside wheel than on the wheel on the inside of the turn.

While cornering, the rear axle centerline of a vehicle with a dependent rear suspension may twist in relation to the body centerline. When a vehicle with independent rear suspension is in a turn, the centerline of each wheel turns independently, causing an angular change between the wheel centerline drawn through the tread of the tire and the body centerline. This can be seen in Figure 7-19. On front suspensions, it is possible to design the suspension and steering system so the tires will turn either into or out of a turn, as required. In this way, it is possible to increase or decrease the effective steering input from the driver. The changes that take place in front and rear suspension systems are due to the geometry of the linkages and the amount of body roll.

The tires introduce another input force into the steering of the vehicle. Tires have a slip angle. *Slip angle* is the angular difference between the centerline of the wheel and the tangent to the turn the vehicle is following. You can see this in Figure 7-20. All tires have some slip angle while making a turn. The greatest slip angle occurs at the maximum turning force.

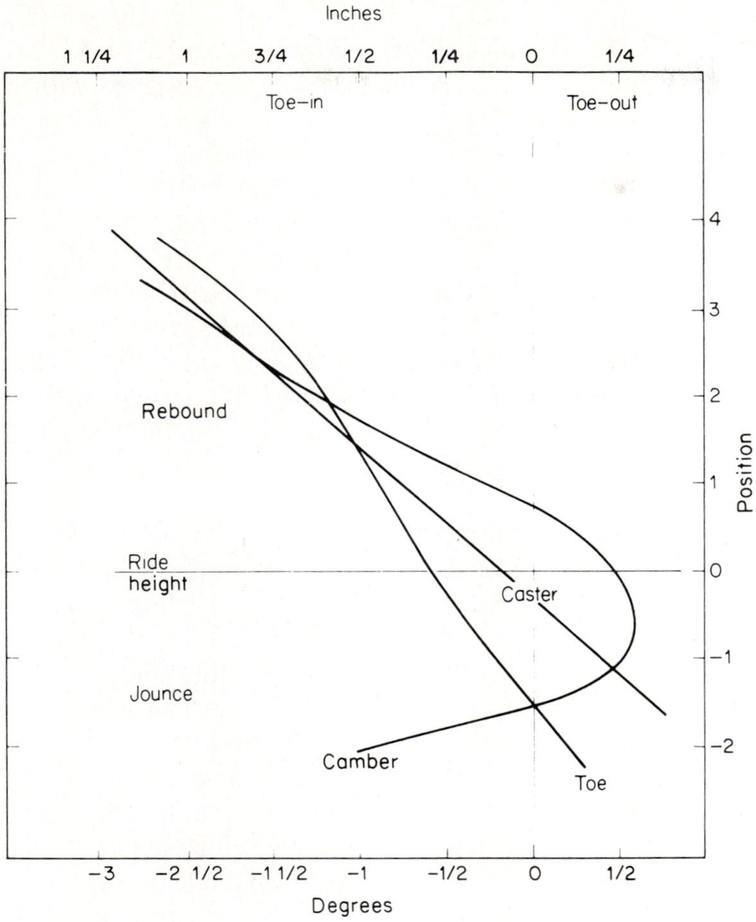

FIGURE 7-18 Graph of the change in suspension geometry of a typical long-short control arm suspension as it goes through jounce and rebound. Note that the change is greatest during rebound.

FIGURE 7-19 The angular steering effect of the rear suspension.

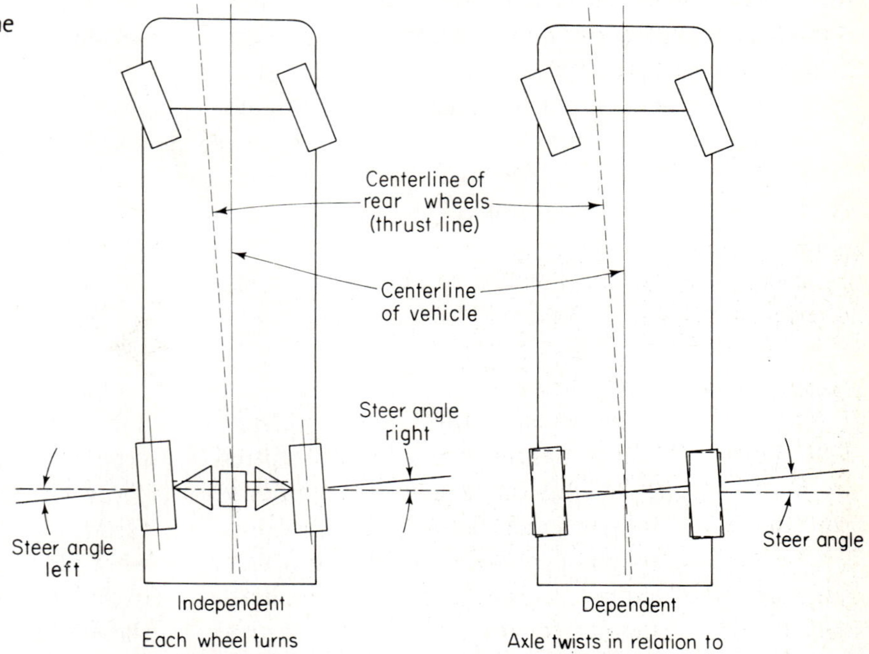

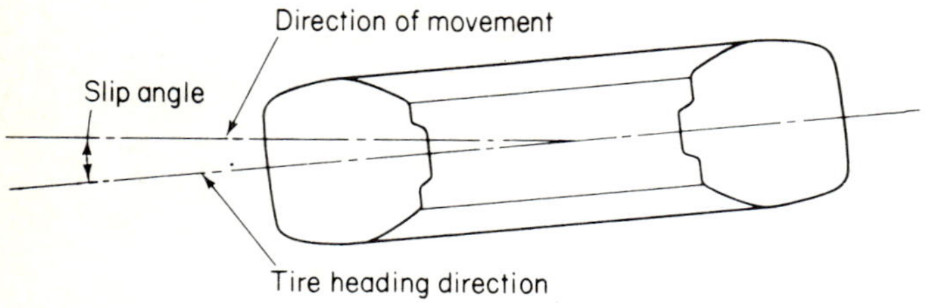

FIGURE 7-20 Slip angle.

This happens just before the tire starts to slide sideways. The slip angle depends on the rubber compound of the tire, tire construction, and the road surface, as discussed in Chapter 4. The tread rubber composition affects road adhesion (coefficient of friction) and the maximum cornering power of the tire.

7-4 VEHICLE STEER

A vehicle is said to *understeer* when the front suspension, rear suspension, and tire slip angles combine to require increased steering effort as speed is increased while making a turn. Understeer is the tendency for a vehicle to continue to increase the diameter of the turning circle as vehicles speed increases while holding a given steering wheel position.

When the front suspension, rear suspension, and tire slip angles combine to produce a vehicle that responds directly to the driver's steering input effort, the vehicle has *neutral steer*. The turning circle will remain constant as the vehicle speed increases when the vehicle has neutral steer until the speed is reached when the tires completely lose traction and slide.

Oversteer occurs when the suspension geometry and tire slip angles cause a vehicle to continue to turn sharper as speed increases while holding the steering wheel constant.

The understeer, neutral steer, or oversteer tendency of a vehicle making a turn is not affected by the driver. It depends on suspension geometry, the amount of body roll in a turn, and the slip angle of the tires. Incorrect suspension ride height affects suspension geometry. Worn or incorrect shock absorbers and springs affect the amount of body roll in a turn. Incorrectly inflated tires affect the tire slip angle. All of the above, either singularly or combined, will drastically affect the steering response of a vehicle during a turn.

Passenger cars are designed with understeer. Understeer provides the average driver with a large safety margin when entering a turn because the vehicle has a slow response. Understeer prevents the unskilled driver from overreacting in an emergency situation. Late-model vehicles with more precise handling are designed with less understeer tendency.

Tire slip angles increase with speed. Any change in slip angle will change the handling characteristics of the vehicle. The slip angles of the tires will not affect the steer angle when the front and rear tires are matched and have equal loads.

Steer Angle. *Steer angle* is the vehicle heading angle plus the rear slip angle less the front slip angle. Figure 7-21 shows that the steer angle equals the heading angle when the front and rear slip angles are equal. The problem becomes more complex when loads on the tires are not the same. The average front-engine passenger vehicle has front-weight bias. A greater percentage of the vehicle weight is on the front wheels than on the rear wheels. This is especially true of front-wheel-drive vehicles. The weight bias may produce a greater slip angle on the front tires than on the rear tires. This tends to create a steering angle that is less than the heading angle, which is understeer. The reverse happens when the vehicle is loaded with luggage and passengers. Loading may produce a rear tire slip angle that is greater than the front tire slip angle. The steer angle is now greater than the heading angle and the vehicle will oversteer. Oversteer is considered to be dangerous and the vehicle may be uncontrollable by the average driver trying to make evasive maneuvers.

You can see that a problem develops when tires that have different body construction (radial, bias belted, bias ply) are mixed on a vehicle. In addition, tires of a similar body construction but from different manufacturers often have different slip angles. This is discussed in Section 4-3. Each construction type and tread compound has a different slip angle. Tires with different amounts of wear will have different slip angles. If radial tires with low slip angles were placed on the front of a vehicle and bias ply tires with large slip angles were placed on the rear, the resulting steer angle would be much greater than the heading angle and this would produce oversteer. If bias ply tires were placed on the front and radial

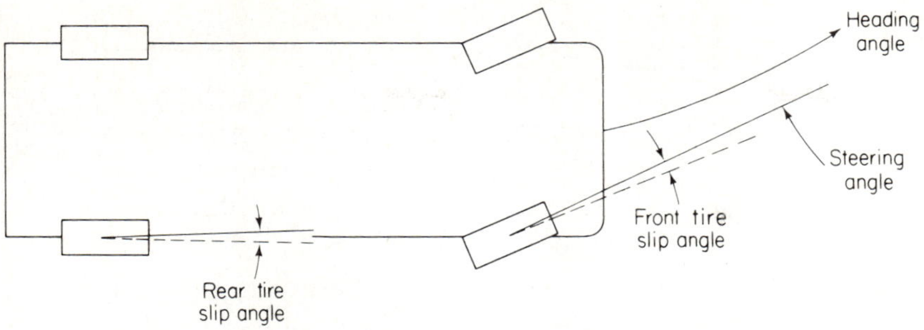

FIGURE 7-21 When front and rear tire slip angles are equal, steering angle equals heading angle.

tires on the rear, the heading angle would be much less than the steer angle and this would cause understeer. Tires of similar construction from different manufacturers may produce a slight variation in steer characteristics when placed on the same vehicle. Tire effects on steering are shown in Figure 7-22.

Suspension Design Versus Tire Types. Vehicle suspension is designed by the manufacturer for use with specific tire types. The vehicle will produce the handling characteristics the manufacturer has carefully engineered only when the specified tires are used. The manufacturer may design a vehicle suspension with a relatively large understeer and then install bias belted tires on the vehicle. This will produce a safe vehicle with the desired handling characteristics. If the owner exchanges the original tires for radial tires, the vehicle will approach a neutral steer condition. Some drivers consider neutral steer as essential to "better handling." If a vehicle is manufactured for use with radial tires, the amount of steer built into the suspension is different than it would be for bias belted tires. If the radial tires are exchanged for bias belted tires, a greater amount of understeer caused by these tires will result in "poor handling." Some vehicles have suspensions designed to produce neutral steer with bias belted tires. These vehicles may oversteer if radial tires are installed. This could produce a vehicle with poor or possibly dangerous handling characteristics.

Four-Wheel Steering. Tires, you recall, have a limited amount of traction. When some of the tire traction is used for braking, there is less traction remaining for steering before the tire starts to slide. The front tires on rear-wheel-drive vehicles use all of their traction for steering unless the vehicle is braking. When the vehicle has front-wheel drive, part of the traction is being used to pull the vehicle. There is less traction for steering and cornering capacity is reduced. At the same time, the cornering force has increased because there is greater weight on the front tires. These forces work together to increase the tendency of front-wheel-drive vehicles to understeer.

Four-wheel steering is being developed to reduce some of the understeer. The rear wheels turn a small percentage of the front wheel turning radius. At high vehicle speed with small steering wheel turning movement, the rear wheels turn slightly in the *same* direction as the front wheels. This reduces the rear wheel slip angle and the tendency to understeer. At low vehicle speed with large steering wheel turning movement, the rear wheels turn in the *opposite* direction from the front wheels. This reduces the turning radius of the vehicles. There is more discussion on four-wheel steering in Section 8-2.

7-5 RIDE QUALITY

Ride quality is an important factor in suspension design. It was noted in the discussion on suspensions that a multileaf spring has interleaf friction. This friction will not remain constant throughout the life of the vehicle as the parts wear. Because of this, the characteristics of a leaf spring will continuously change throughout its service life and this will change

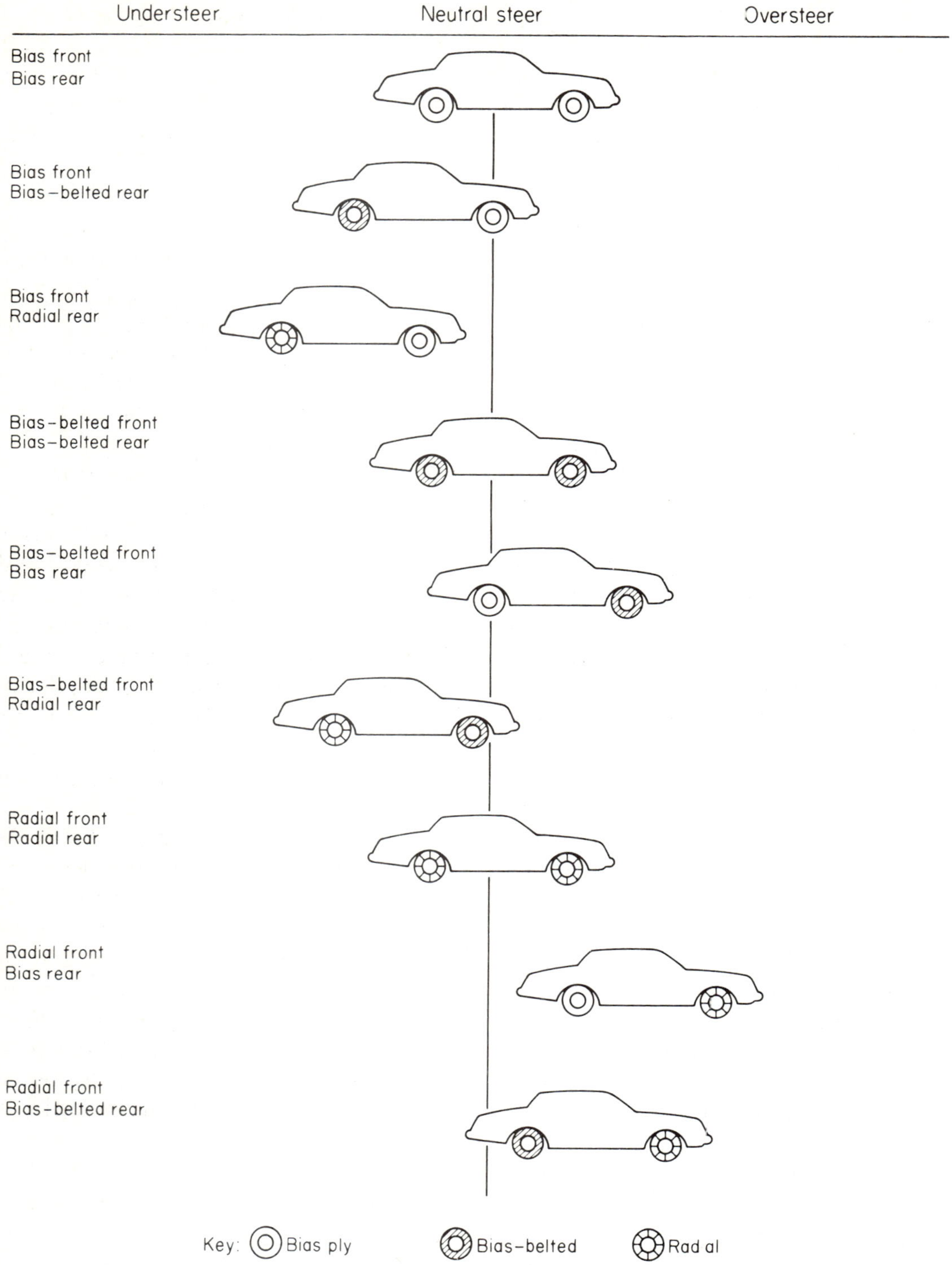

FIGURE 7-22 Steering effects on understeer, neutral steer, and oversteer caused by different types of tire construction on the front and rear of a vehicle.

the ride quality. The composite single-leaf spring eliminates this problem. The internal friction of a multileaf spring was a desirable quality before vehicles were equipped with modern shock absorbers. The internal friction of the leaf spring supplied the system with the damping required to provide an acceptable ride. Coil, single leaf, and torsion bar springs have since become popular. They have a good ride quality and they require less space. Coil and torsion bar spring suspension systems have little static friction, so they readily flex to absorb small road surface bumps and dips. Modern shock absorbers can handle all of the suspension damping required by these springs. The shock absorbers are mounted so they will minimize side sway. This improves handling and ride quality on all types of road surfaces.

Road Clearance. Hard surface roadways require less road clearance than dirt roads. This reduces the ride height and gives the vehicle a lower center of gravity. A low center of gravity produces less body roll during a turn for better passenger comfort. It gives the vehicle a lower overall height and it improves handling characteristics. Contrast this with the popular fad of "jacking" vehicles 12 to 18 inches. The independent front suspension was an important step toward reducing the height of the center of gravity. The clearance needed for solid-axle movement beneath the frame was eliminated and so the frame could be lowered.

The reduction of the body-to-road clearance required an increased tunnel height on the floor inside the vehicle to provide clearance for the drive shaft under the floor pan. This is the most undesirable result of a low vehicle having a front engine and rear-drive wheels. The reduction in the size of the wheels to further lower ride height is limited by the size of the brakes required to stop the vehicle. As the wheel size is decreased, the diameter of the wheel-mounted brakes must also be reduced. This results in lower brake efficiency. The automotive engineer must compromise the ride height while considering passenger comfort, vehicle handling, and braking requirements. Rear-engine, rear-wheel-drive and front-engine, front-wheel-drive vehicles do not have all of the same compromises for handling and comfort that are present in front-engine, rear-wheel-drive vehicles. No tunnel is needed for drive shafts in front-engine, front-wheel-drive vehicles. This results in more room for the passengers. This is important in compact vehicles. It is one of the factors leading to the increased use of front-wheel drive. Because four-wheel-drive vehicles must have a place for the drive shaft, they have more body-to-road clearance.

Engine/Drive Characteristics. Cost, durability, and wear must be considered by all design engineers. A front-engine, rear-wheel-drive vehicle is an economical design where passenger space permits.

A front-engine, front-wheel-drive vehicle provides the most passenger space in the subcompact vehicle. The drive train is expensive, however, and it is only used where it is required. Four constant-velocity universal joints are required on front-wheel-drive vehicles because they must both drive and steer. The strength of the suspension parts must be increased to provide strength for drive torque and durability. Care must be used in the design to ensure that the forces from the driving wheels are not transferred through the steering linkage to the driver. As the complexity of the suspension system increases in front-wheel-drive, the initial cost is increased as well as the wear rate of the suspension components. The cost increases still more when using four-wheel-drive.

A midship-mounted engine, forward of the rear axle, has ideal handling characteristics but it reduces the interior space for the passengers and load. As a result, it is usually found on vehicles designed for only two passengers. It also increases the interior noise level of the vehicle. A rear-mounted engine behind the rear axle usually has poorer handling characteristics than midship-engine vehicles. Engine cooling and accessibility to service the engine and drive line are additional drawbacks of both rear and midship-engine vehicles.

Space Availability. Space considerations, especially in small vehicles, dictate the type of suspension to be used as well as the type of drive train. Front suspension of the MacPherson-strut-type requires less width than a long-short arm suspension but requires greater height. As a result, long-short arm suspensions are beginning to be used in vehicles with very low hoods. The space required by the rear suspension can be reduced by using a short torque tube or a torque reaction arm as shown in Figure 6-17 and 6-25. These are used to eliminate the long upper control arms usually required for absorbing brake and acceleration torque on vehicles with rear-wheel drive.

Tire Considerations. Vehicles designed to use radial tires require special consideration by the suspension designer. Radial ply tires have a characteristic low-speed harshness that is not present in bias belted or bias ply tires. They often cause a steering wobble that makes the steering linkage oscillate. This is due to the high cornering forces developed by the lower tire slip angles. You will remember from Section 4-3 that radial ply tires have a lower radial spring rate than bias ply tires. As a result, radial ply tires produce a lower natural frequency of total unsprung weight than when bias ply tires are used (see Figure 4-11). Vehicles designed for operation with radial tires have their suspensions tuned to dampen vibration at this frequency. This may change the vibration frequency to a range that is not as noticeable to the driver and passengers. An objectionable vibration or a vibration resulting in noise may be isolated at its source. The frequency of the vibrations will usually change when more unsprung weight is added to the system. In other cases, the frequency of the vibration may be shifted to a frequency range where it is no longer noticeable. Suspensions are dampened by large spring insulators, large suspension bushings with voids in them, isolation of the suspension parts on a cross-member, and mounting on synthetic rubber insulators between the cross-member and the rest of vehicle.

If you are modifying a vehicle, the discussion and equations in Chapter 15 will help you develop a suspension to meet your design requirements.

STUDY QUESTIONS

Introduction

1. Describe vehicle handling.

Section 7-1

2. Describe suspension geometry.
3. How does the sideways displacement of the wheel and tire differ between an I-beam axle and long-short arm suspension during jounce and rebound?
4. Describe tire skuff.
5. Why is the upper arm shorter than the lower arm in long-short arm suspensions?
6. What are the advantages and disadvantages of tire skuff travel?
7. What are the advantages and disadvantages of having negative camber during jounce and positive camber during rebound?
8. What are the advantages and disadvantages of having positive camber during jounce and negative camber during rebound?
9. How does a MacPherson strut affect camber during jounce and rebound?
10. How does caster change during jounce on an I-beam axle with radius arms?
11. How does braking affect caster?

Section 7-2

12. Why are the pivot points of the lower control arm and the tie rod on the same axis?
13. What vehicle reaction results from roll steer?

14. What causes wheel fight?

15. What are the results of wheel flight?

16. What is bump steer?

Section 7-3

17. How does body roll in a turn affect steering?

18. How does tire slip angle affect steering?

Section 7-4

19. Describe understeer.

20. When does a vehicle have neutral steer?

21. What vehicle reaction results from oversteer?

22. What are the advantages and disadvantages of understeer?

23. Why is the type of tire construction important to vehicle handling when replacing tires?

Section 7-5

24. How do multiple-leaf springs affect the ride quality differently than coil springs?

25. What is the most undesireable result of lowering body-to-road clearance?

26. How does ride height affect braking?

27. How does the type of vehicle drive line affect handling?

28. What methods are used to minimize undesirable vibration and noise coming from vehicle suspensions?

8

Wheel Alignment Principles

The primary purpose of the steering system is to provide the driver with maximum directional control of the vehicle. This control can only occur if the tires stay on the road surface and the steering control of the tires on each side of the vehicle work together. Vehicle steering, wheel alignment, and suspension angles work together to achieve the maximum possible directional control of the vehicle.

A vehicle is designed to have a neutral sense. A *neutral sense* is the tendency of the vehicle to go straight ahead with no driver input and to return the wheels to a straight-ahead position after a turn is made. The steering system is designed to require minimum corrective input effort from the driver. In addition, it should transmit as little road shock as possible to the steering wheel over irregular road surfaces. It is most important that maximum safety be designed into each steering system part.

Wheel alignment is the setting of all interrelated alignment suspension angles for maximum vehicle control, easy steering, good handling, free and true running wheels, and maximum tire mileage. Periodic wheel realignment is necessary because high mileage, poor driving habits, and rough roads will, in time, cause changes in wheel alignment angles. These changes are due to the wear of the suspension and steering parts or to a relaxing of suspension bushings and the vehicle frame structure.

Dynamic loads of a moving vehicle cause slight deflections of the suspension and steering parts. This deflection is called *compliance*. Wheel alignment is checked in the service shop with alignment equipment while the vehicle is stationary. As a result, the alignment specifications for a stationary or static vehicle take into account the change in alignment that occurs while the vehicle is in operation. Adjusting the alignment to static specifications results in the desired operating angles for a free-rolling, stable vehicle.

8-1 SUSPENSION AND STEERING ANGLES

This section describes the interrelated suspension angles. As an automotive service technician, you must know and understand the alignment angles (caster, camber, toe, steering axis inclination, toe-out on turns) and be able to relate these angles to the suspension parts and handling problems.

Camber. *Camber* is the angle of the centerline of the wheel from vertical when the wheel is viewed from the front of the vehicle. This is illustrated in Figure 8-1. Camber is said to be positive when the top of the centerline of the wheel is angled or tipped outward away from the vehicle. When the top of the centerline of the wheel is tipped inward toward the vehicle, the camber is called negative. The camber angle is measured in positive or negative degrees.

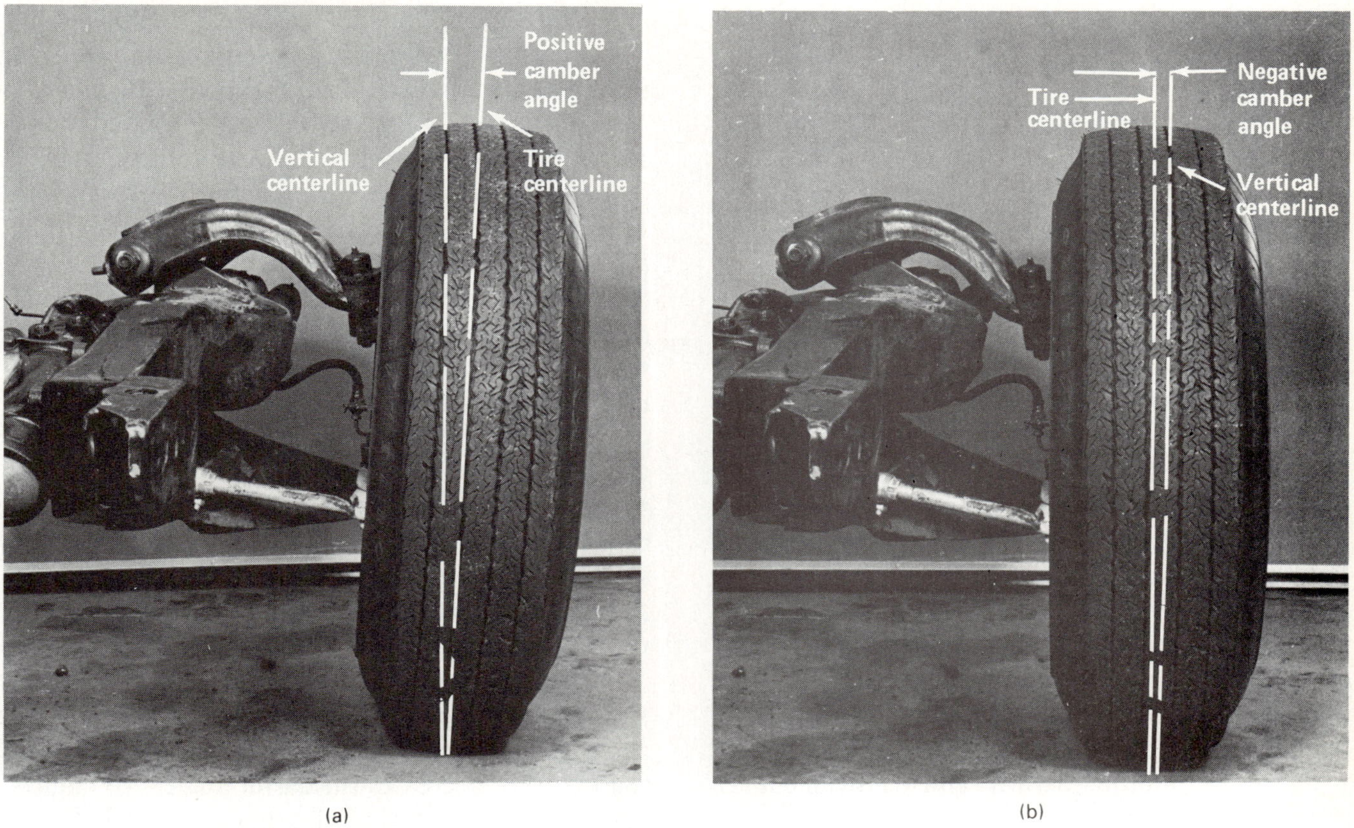

FIGURE 8-1 Positive camber (a) and negative camber (b).

Camber provides a vehicle with a directional sense. The vehicle tends to lead toward the wheel with the most positive camber. It is the normal tendency of the wheel to roll around the center of a circle when the top of the wheel is inclined toward the center of that circle as shown in Figure 8-2. Wheels with positive camber tend to turn outward. If both wheels have equal camber, either positive or negative, the turning tendencies of the wheels oppose each other so the effective steering lead will be neutral or nondirectional. Radial tires cause less lead due to camber settings than bias ply tires.

Most vehicles use a static camber setting from zero to slightly positive. The slightly positive static setting will prevent camber from becoming negative when dynamic forces cause the suspension to deflect as the vehicle is moving or when it is carrying a load. While under dynamic loading, either positive or negative camber will cause wear to occur on one side of the tire tread, thus shortening tire life (see Figure 5-4). Camber settings are more critical for uniform tire tread wear when using wide tires. As the width of the tire increases, correct camber becomes more important for even tire wear. You can see that camber is a tire-wear angle. The tire must have equal weight distribution across the entire tire tread.

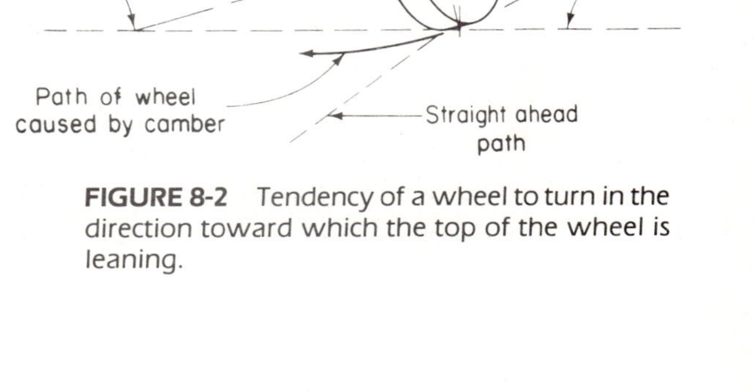

FIGURE 8-2 Tendency of a wheel to turn in the direction toward which the top of the wheel is leaning.

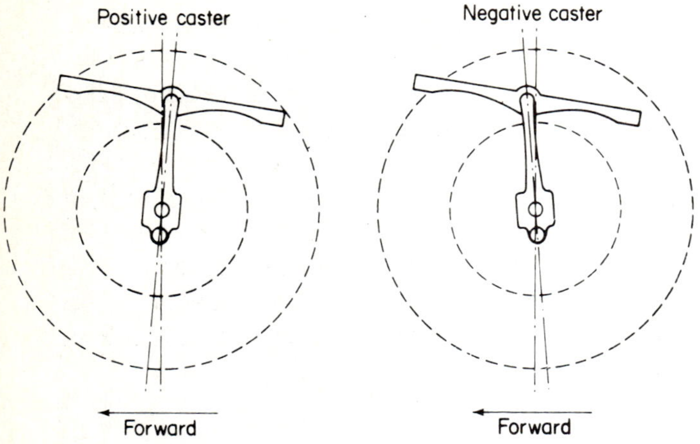

FIGURE 8-3 Positive and negative caster.

Caster. *Caster* is the angle formed between a line drawn through the steering axis and a vertical line through the spindle, as viewed from the side of the vehicle (Figure 8-3). The *steering axis* is an imaginary line drawn through the upper and lower ball joint centers or, in the case of a MacPherson strut, the upper pivot and the lower ball joint. The weight of the vehicle effectively acts through a vertical line. The steering and the forward vehicle movement act through the steering axis. The steering axis and caster angle can be seen in Figure 8-4. Caster is positive if the top of the steering axis is inclined rearward and negative if the top of the axis is inclined forward. Caster is an angle measured in positive or negative degrees.

The effects of positive caster are seen by noting where the projected steering axis strikes the road surface within the tire footprint, as shown in Figure 8-5. The vehicle weight is effectively concentrated in the center of the tire footprint. If the steering axis strikes the road surface in the front part of the footprint, as it does in positive caster, it should be clear that the tire will be pulled by the steering axis. The tire drag on the vertical line at the center of the footprint will always try to follow that axis contact point within the footprint. The forces directed within the tire footprint area during a turn are greater behind the steering axis at the footprint. This will produce a force that will try to return the wheels to the straight-ahead position to aid directional stability on nondriving front wheels. At the same time, a greater steering effort is needed to turn the tire from a straight-ahead path than would be required if caster were zero.

Weight transfer during braking will cause the front of the vehicle to lower and the rear of the vehicle to rise. This change in the angle of the body-to-road surface decreases the caster angle. An angled upper control arm pivot point prevents dive and con-

FIGURE 8-4 Caster is positive when the steering axis leads the center point of vertical weight within the footprint.

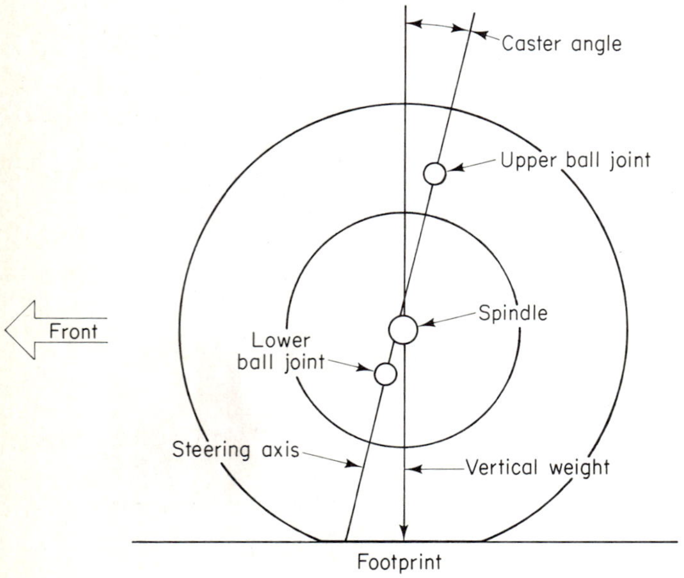

FIGURE 8-5 Directional effect of caster on vehicle steering caused by the point where the projected steering axis strikes the road in the footprint.

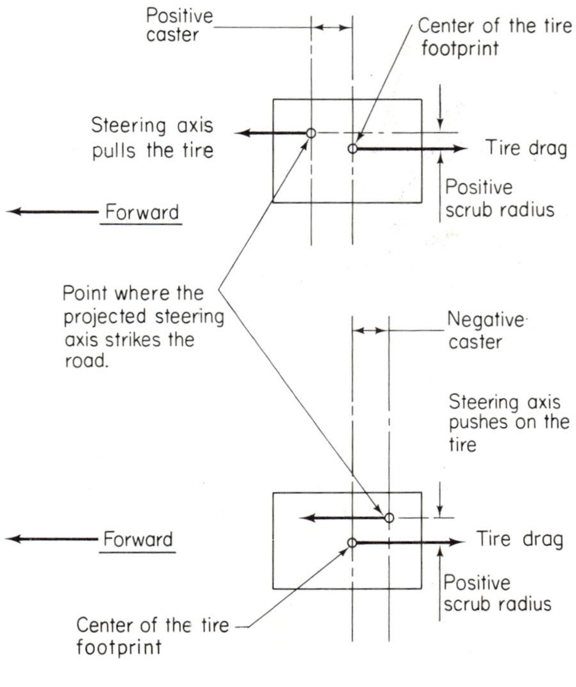

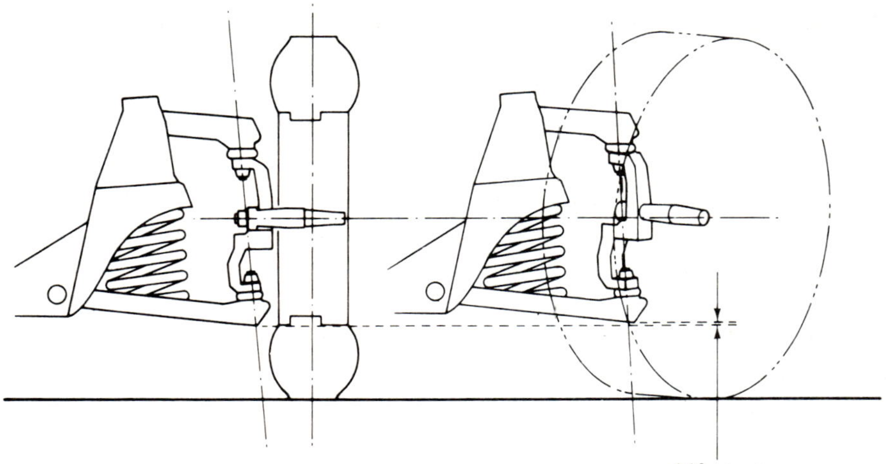

FIGURE 8-7 Directional effect of SAI on vehicle steering.

SAI is a stability angle. On a suspension with SAI, the outer end of the spindle is at the highest point in relation to the body when the wheels are pointed straight ahead. In this straight-ahead position, the body will be closest to the road surface. Vehicle weight, acting on a spindle with SAI, returns the wheels to the straight-ahead position. The effect of SAI is shown in Figure 8-7. After a turn, SAI aids in straightening the wheels. SAI also aids in directional stability by resisting road irregularities that attempt to turn the wheels away from the straight-ahead position. SAI produces many of the same benefits that improve steering stability as positive caster. SAI, therefore, reduces the need for high positive caster. The effect of SAI on directional stability is usually greater than the effect of caster. Poor steering returnability and instability caused by negative caster are overcome, within limits, by using SAI to provide steering stability. One drawback with SAI is that it causes the wheel to have positive camber in a turn and this may reduce the maximum cornering power of the tire and wheel. Vehicles with power steering require a greater amount of steering wheel returning force than those with manual steering. SAI is often used with positive caster on a vehicle equipped with power steering to provide the returnability needed. The power steering system supplies the additional power required to turn the wheels from the straight-ahead position for directional control. SAI with negative caster may be used with manual steering.

The difference in the angles between SAI and camber remain constant as the camber changes. This is shown in Figure 8-8. The spindle is fastened securely and is a part of the knuckle. This makes SAI a nonadjustable angle. SAI will only change if the knuckle has been bent. A bent knuckle is never straightened; it must be replaced.

SAI causes the steering axis to be inclined outward at the bottom. As a result, the projected steering axis strikes the road surface at a point close to the centerline of the tire footprint. You saw this in Figure 8-6. The distance between these two points is called the *scrub radius*.

FIGURE 8- 8 The difference between SAI and camber remains constant as camber changes. This is called the Included angle.

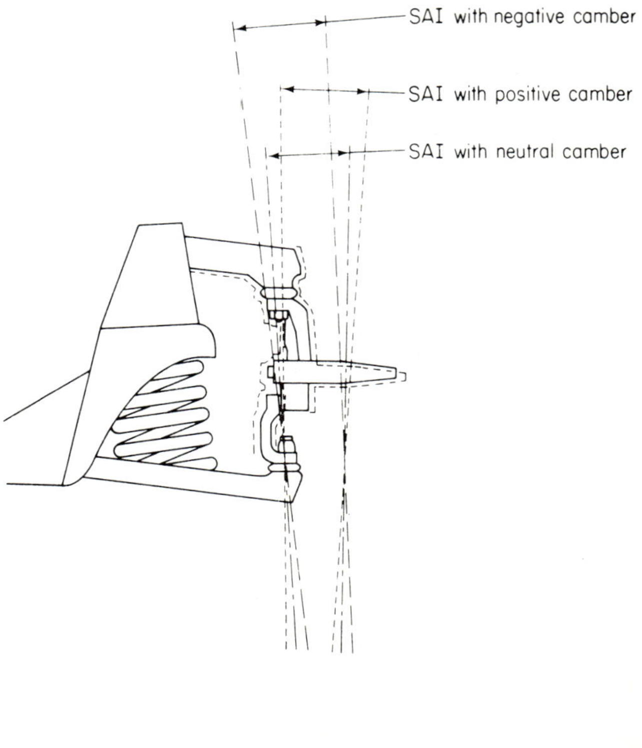

Scrub Radius. *Scrub radius* is the term used to describe the distance between the projected steering axis and the tread centerline at the road surface. Scrub radius is negative when the centerline of the tires lies inside the projected steering axis. This was shown in Figure 8-6. It is positive when the centerline of the tire is outside the projected steering axis. The scrub radius is a distance measurement and it is, therefore, measured in inches or millimeters.

The scrub radius is so named because the tire scrubs on the road surface as the wheel direction is changed. The wheel and tire will follow an arc when the wheels are turned rather than pivoting on a single point. The arc in which the tire swings depends on the radius of the circle produced as the tire footprint center point moves around the projected steering axis, as illustrated in Figure 8-9. This radius is called the *scrub radius*.

The size of the scrub radius depends on the steering axis inclination, the wheel offset, and the distance the spindle centerline is above the road surface. The vehicle manufacturer designs the required amount of scrub radius into the suspension by carefully blending SAI, wheel offset, and the spindle height. Even though the spindle height has an effect on the scrub radius, little can be done to change this height because the tire height is limited by the clearance space under the fender and body.

Aftermarket tires and wheels, those not supplied by the vehicle manufacturer, can drastically affect the scrub radius that the manufacturer has so carefully designed into the vehicle. Using wheels with increased offset in place of original-equipment wheels can double the scrub radius. An example of this is shown in Figure 8-10. Tires with lower section height will also increase the scrub radius a small amount. Tires with camber wear will also position the effective centerline and footprint of the tires at a different place and this will affect the scrub radius.

FIGURE 8-9 Directional effect of scrub radius on steering.

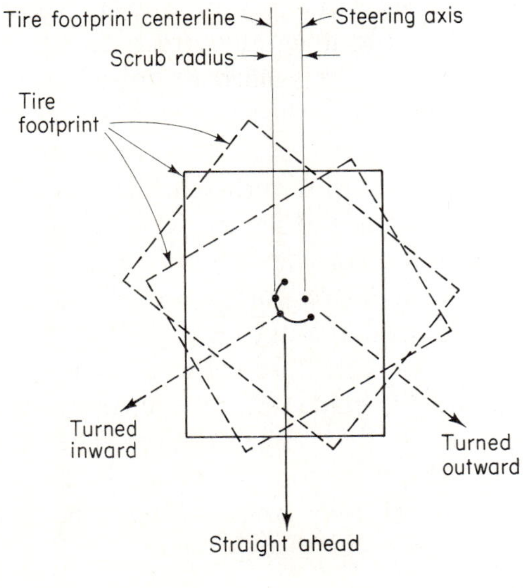

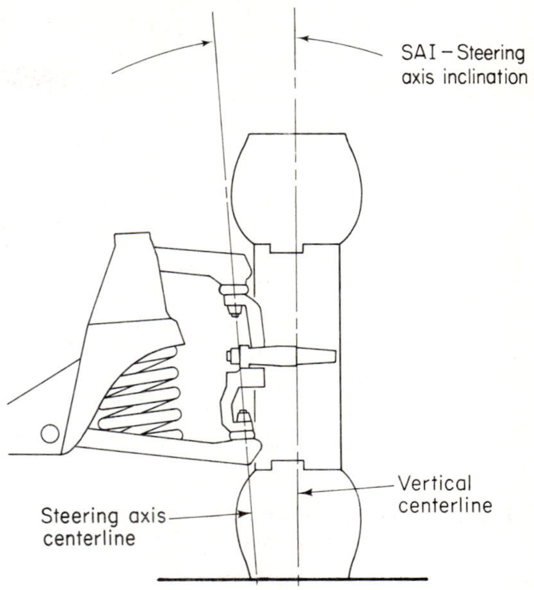

FIGURE 8-10 Effect of wheel offset on the scrub radius.

Since all handling comes from the tire footprint, the scrub radius provides the necessary feedback to give road feel to the driver. All rolling resistance forces on a nondriving front wheel are working on the tire and attempting to force it rearward. This force is concentrated on the center of the tire footprint. With the tire centerline outside the steering axis, the rearward force acts to turn both front tires outward. A slight amount of scrub radius on each wheel is necessary to maintain constant tension or compression of the steering linkage to steady the linkage and prevent steering wobble or shimmy. As the scrub radius increases, greater torque is applied to the tire to turn the wheel outward; this places higher loads on the steering linkage and suspension parts. The result will be increased linkage wear and the probability of premature failure. The effect of unequal braking on each wheel is exaggerated with a large scrub radius; this may make the vehicle dangerous to drive.

Some manufacturers are placing the tire centerline inside the projected steering axis (negative scrub radius). This was illustrated in Figure 8-6. This is most effective when using diagonal-split braking. It minimizes the wheel pull when braking if one-half of the brake system fails. It is also used on some front-wheel-drive vehicles. Negative scrub radius is helpful when a tire fails at highway speeds.
Tire forces normally cause the failed tire to immediately turn outward. There is less steering pull with less scrub radius.

The amount of scrub radius required depends on vehicle weight, steering wheel feedback, type of braking, the type of road for which the vehicle is designed, and front- or rear-wheel drive. With front-wheel drive, traction effects are also fed back to the driver as the front wheels pull the vehicle. The correct scrub radius is even more critical on front-wheel-drive vehicles.

Toe. The wheels have zero toe when the tire centerlines are running parallel to the vehicle centerline, as illustrated in Figure 8-11. Toe-in is measured in fractions of an inch, millimeters, or, on some vehicles, in degrees. As a linear measurement, toe is the distance between the front of the tires to the distance between the rear of the tires on the same axle. As an angular measurement, it is the angle formed between the vehicle centerline and each tire. Slight toe-in is preferred over toe-out on the front wheels of most vehicles because steering is aligned while the vehicle is stationary. While under dynamic loads with the vehicle moving, linkage parts flex and cause a change in alignment angles. The static toe setting will differ from the desired running toe. For maximum tire life and least rolling resistance, the running toe must be zero. The usual tendency is for the tires to turn outward on nondriving front wheels while under dynamic load, so most rear-wheel-drive vehicles are designed with a static toe-in setting. The toe will become zero as the linkage is dynamically loaded when the vehicle is in motion. Front-wheel-drive vehicles have toe specified at very near zero.

FIGURE 8-11 When A is less than B, the wheel has toe-in. When B is less than A, the wheel has toe- out.

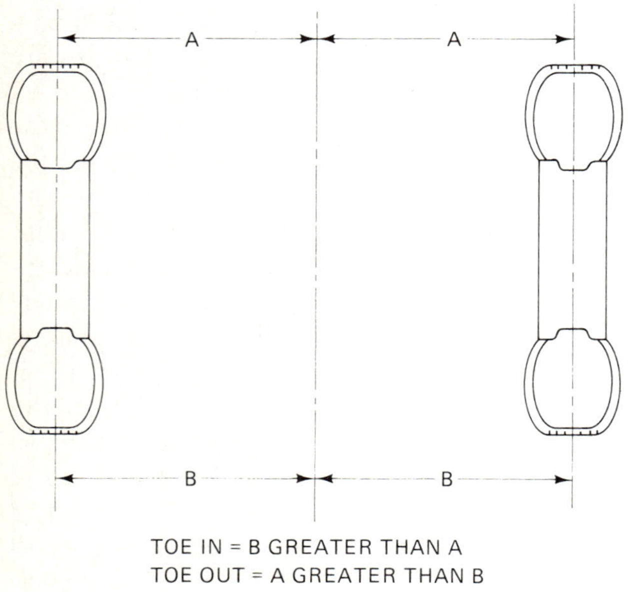

Incorrect toe is one of the factors that cause tire wear. Although a running toe-in will produce as much tire wear as running toe-out, a slight running toe-in is preferred. One-quarter-inch, out-of-correct toe will reduce tire mileage by 25 percent. A toe-out condition under dynamic loading will tend to cause the vehicle to wander. The reason for a static toe-in setting, then, is to maintain running toe between a slight toe-in and zero.

Toe-Out on Turns. When the vehicle is in a turn, the outside wheels are on a different, larger turning radius than the inside wheels. The tire centerline is always tangent to the turn circle, unless it is skidding. The outside front wheel must, therefore, be turned at less angle than the inside wheels as illustrated in Figure 8-12. This will keep both wheels tangent to their respective turning circles and prevent tire scuffing while making a turn. A steering linkage designed to give the correct steering angle on each wheel in a turn is called an *Ackerman layout*. The *steering angle* of a wheel is the angle of the wheel centerline relative to the centerline of the body. You can see this as angles A and B in Figure 8-12. The greater steering angle of the inside wheel will create a greater distance between the extreme front of the tires compared to the rear of the tires. This increased distance will cause toe-out on turns.

Toe-out on turns, sometimes referred to as *turning radius*, is made to meet the turning radius requirements of the vehicle design. The turning radius specification for a vehicle is based on wheelbase, wheel track, angle of the turn, and tire

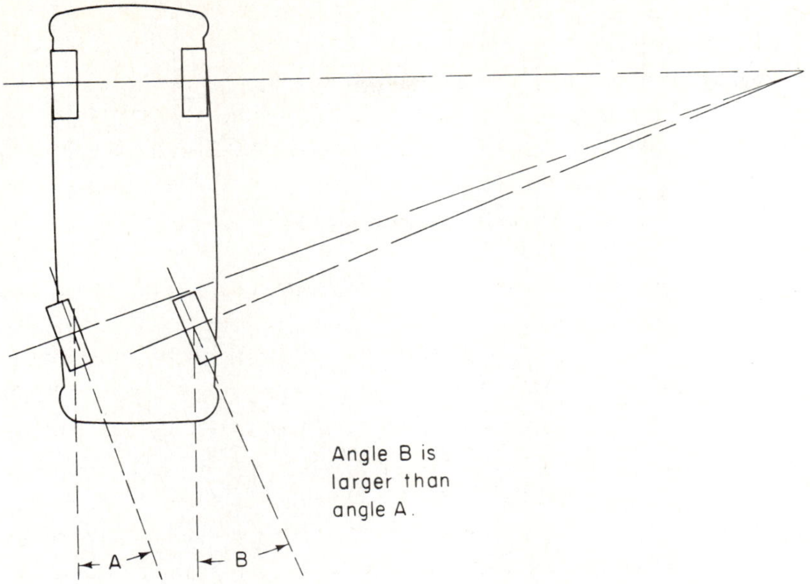

FIGURE 8-12 Toe-out on turns is needed because the rear axle is fixed and each front wheel knuckle turns on a separate axis.

slip angle. As the angle of the turn is increased, the toe-out on turns must also increase. Tire slip angle decreases the effective steering angle, which in turn increases the effective diameter of the turn (understeer). You will remember that *slip angle* is the angle between the tire centerline and the tangent to the effective turning circle. This was shown in Figure 7-20. As the effective steering angle is reduced, the amount of toe-out on turns is also reduced.

The correct amount of toe-out on turns is produced through the proper angling of the steering arms. Linkage behind the suspension is shown in Figure 8-13. Linkage ahead of the suspension is shown in Figure 8-14. The position of the steering arms in their respective turning arcs during a turn will cause the toe-out to be proportional to the steering angle. The greater the steering angle, the greater the toe-out on turns.

The turning radius or toe-out on turns of the vehicle is a nonadjustable angle. It is not specified for some vehicles. It is always checked during wheel alignment. If the turning radius is out of specification, a steering arm is usually damaged. Damaged steering arms are replaced. A damaged idler arm or Pitman arm would also create an out-of- specification turning radius. Incorrect steering linkage parts always cause an out-of-specification turning radius.

FIGURE 8-13 Angle of the steering arms causes toe-out on turns when the steering linkage is behind the suspension.

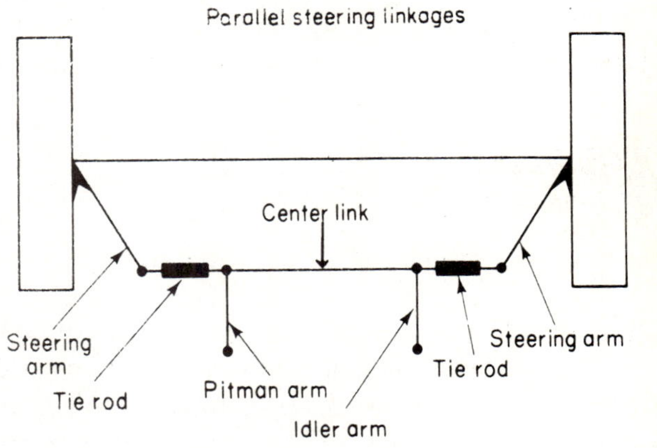

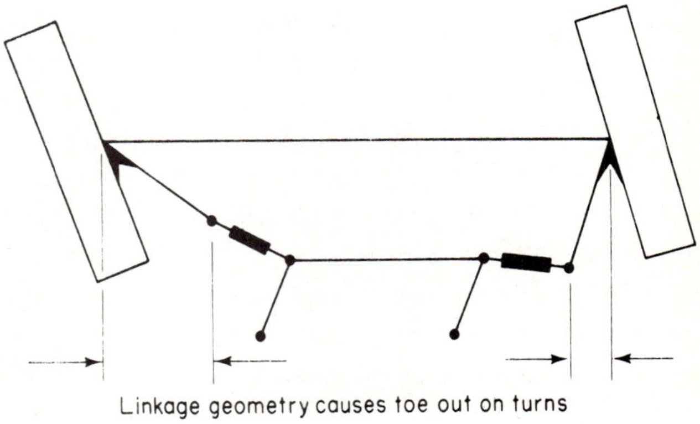

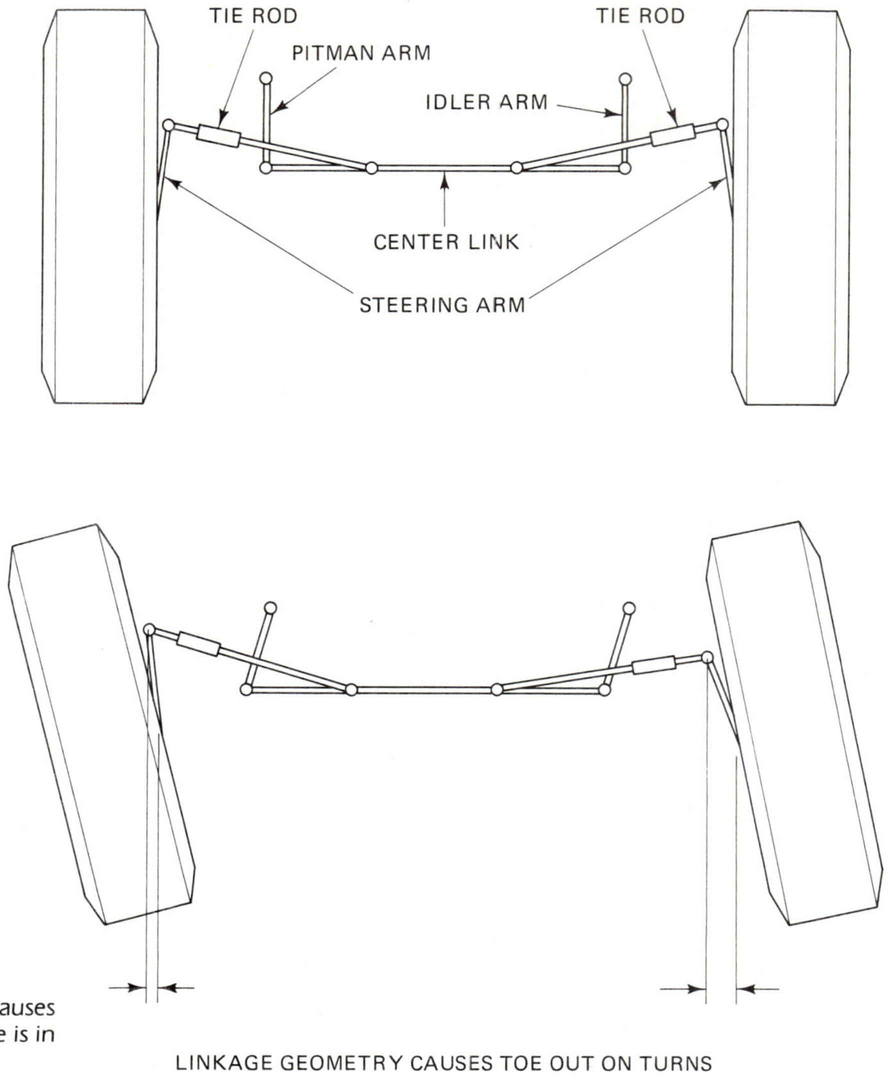

FIGURE 8-14 Angle of the steering arms causes toe-out on turns when the steering linkage is in front of the suspension.

Track. In the past, wheel alignment was limited to the front wheels. The heavy rear axle housing used with dependent rear suspension is fastened to a heavy frame so there is very little chance of misalignment. Most modern vehicles use a unitized body. The rear suspension is fastened to the body sheet metal. Any misalignment in the sheet metal between the front and rear wheels will cause misalignment of the rear wheels. This will keep the rear wheels from tracking directly behind the front wheels. When vehicle track is correct, the rear wheels will roll on the tracks made by the front wheels as a vehicle rolls straight. You can see this as the vehicle travels on a wet road surface. When vehicle track is not correct, the rear wheels will make tracks offset to the tracks made by the front wheels. This is caused by misalignment between the effective angle of the rear axle, called the *thrust line*, and the vehicle centerline. The vehicle centerline runs down the middle of the chassis. Track must be correct before the front wheels can be properly aligned.

The *geometric centerline* is the line that runs through the midpoint between the front wheels and the rear wheels. If the rear wheels are offset to either side, as shown in Figure 8-15, the thrust line does not fall on the vehicle centerline. This is one common cause of incorrect vehicle track.

The drive or thrust from rear-drive wheels is effective through a line perpendicular at the midpoint between the rear wheels. This thrust line should

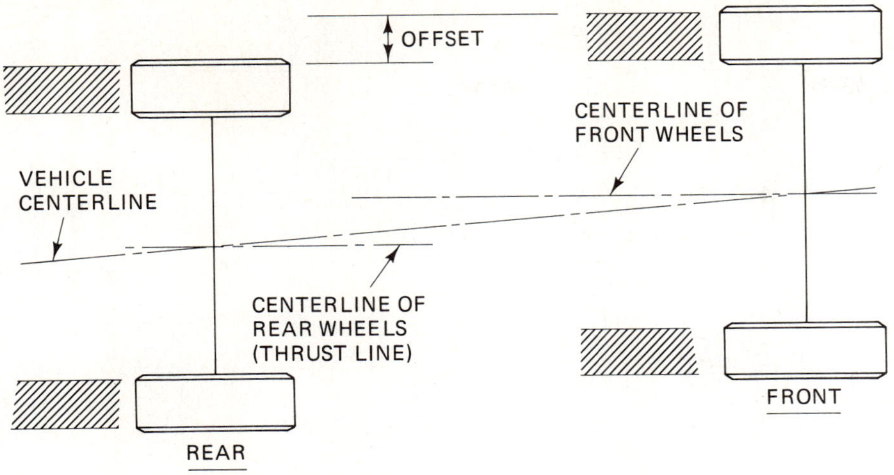

FIGURE 8-15 Track misalignment when the thrust line does not fall on the vehicle centerline.

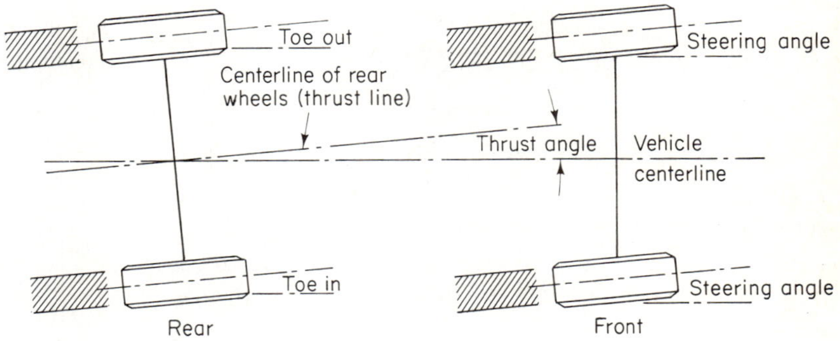

FIGURE 8-16 Track misaligned when one rear wheel has toe-out and the other has toe-in.

be on the vehicle centerline if the vehicle is to track correctly. If one rear wheel is ahead of the other rear wheel, the thrust line will be at an angle, as shown in Figure 8-16. This will cause incorrect vehicle track. The term *thrust line* for the rear wheels is also used for front-wheel-drive vehicles, even though the rear wheels do not produce thrust to drive the vehicle. The thrust line angle results in the same vehicle response on both rear-drive and front-drive vehicles.

You can see that the rear wheels control the directional sense of the vehicle. The vehicle is being turned if either the geometric centerline or thrust line is not on the vehicle centerline. As a result, the vehicle can only be driven straight by keeping the front wheels slightly turned. This upsets the geometry built into the front suspension for correct vehicle handling.

Each rear wheel of vehicles with independent suspension can operate at a different angle. As a wheel with toe-in rolls, it tends to move inward toward the vehicle centerline. As a wheel with toe-out rolls, it tends to move outward away from the centerline. One rear wheel in Figure 8-16 has toe-in and the other has toe-out. You will notice in Figure 8-15 and 8-16 that the thrust line is perpendicular to the midpoint of the dependent rear suspension. The wheels are held parallel by the dependent rear suspension. Toe must be adjusted to make the rear wheels parallel on independent rear suspensions. If the wheels are not parallel, the thrust line will split the angle between the wheels, as shown in Figure 8-17.

Toe-in or toe-out will cause the tires to scuff and wear rapidly. Rear wheel camber will cause the same kind of tire wear as camber wear on front tires. The toe and camber of each rear wheel must be checked when setting the alignment on vehicles with independent rear suspensions. Track should be checked on dependent suspensions as well as on independent suspensions to locate any worn suspension bushings that allow the rear suspension to deflect enough to change the track.

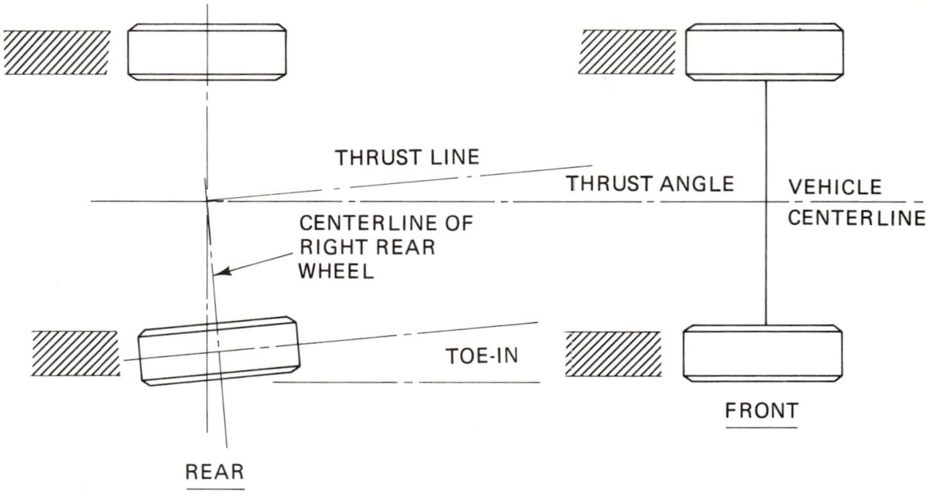

FIGURE 8-17 Track is misaligned when one rear wheel has toe-in.

8-2 FOUR-WHEEL STEERING

Four-wheel steering improves vehicle maneuverability, high-speed control, and stability. It is most useful on front-wheel-drive vehicles. The heavy front ends cause these vehicles to understeer. Their steering angle is limited by the angle through which the universal joints can operate. The steering angle is also limited by the amount of additional steering traction the front tires can provide at the same time they are providing driving traction. During acceleration, there is less tire traction remaining for steering and this increases the tendency to understeer. Four-wheel steering helps to reduce understeer, moving toward a more predictable neutral steer. This improves safety as the vehicle is operated near its handling limit.

The tires on the nonsteering rear wheels of a typical two-wheel-steering vehicle do not develop lateral acceleration to help steer until the vehicle is already in the turn. All four tires begin active steering the moment the driver turns the steering wheel when the vehicle has four-wheel steering. This improves the performance capabilities of the vehicle.

Vehicle manufacturer's have taken two approaches to four-wheel steering. Some use passive systems with suspensions designed to make use of the dynamic loads on the vehicle to change the camber and toe of the rear wheels. Suspension geometry increases negative camber of the outside wheel for added grip on slight turns to reduce tail slide. It causes a small amount of toe-in to help stability. The other systems are active. They use mechanical or hydraulic power to positively steer the rear suspension.

The first active four-wheel steering systems turned the rear wheels a small amount in the *same* direction as the front wheels. This is called *parallel* or *same-phase steering* (Figure 8-18). It reduces the chance of a skid when making a moderate to high-speed turn. It also makes the vehicle respond much faster to steering wheel movements. It is interesting to note that a small change in the steering angle of the rear wheels has a large effect on the cornering characteristics of the vehicle. Later designs added low-speed, rear-wheel steer in the *opposite* direction from the front-wheel steer. This is called *circle, contraphase,* or *antiphase steering* (Figure 8-19). It is used to reduce the turning circle. The 1988 Honda Prelude reduced the turning circle from 34.8 feet with two-wheel steering to 31.4 feet with the four-wheel steering option.

FIGURE 8-18 Same-phase four-wheel steering.

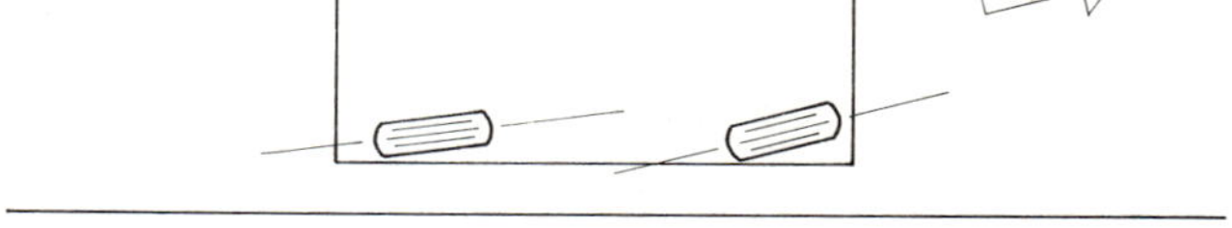

Changing lanes

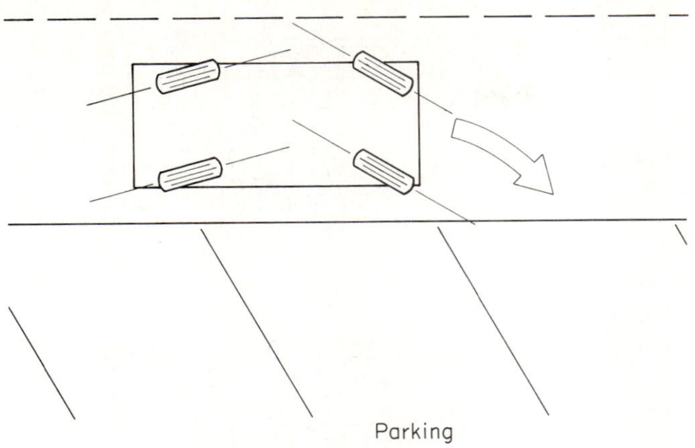

FIGURE 8-19 Anti-phase four-wheel steering.

Passive Four-Wheel Steering. The rear suspensions we usually see on two-wheel-steering vehicles are designed to give a wide margin of safety when a vehicle is driven near its handling limit. Unfortunately, this does not make the vehicle handle well under all driving conditions. An ideal rear suspension should help the vehicle understeer during high-speed cornering and it should cause toe-in while braking for stability.

Mazda introduced passive rear-wheel steering on the 1986 RX-7. It has a semitrailing, independent rear suspension. The bushings in the hub carrier are designed to control compliance or movement of the suspension. The outside wheel will toe-out slightly as the vehicle goes into a turn. This results in quick response to the steering wheel movement. It holds the toe-out until the lateral acceleration G-force reaches 0.5. At this point, roll steer changes the toe-out to a slight toe-in to increase understeer and improve stability. It also increases the negative camber of the outside wheel. This action, working with the rear tire slip angles, gives good vehicle control. It improves maneuverability and stability without reducing either characteristic.

Active Four-Wheel Steering. In 1986, Nissan introduced active rear-wheel steering on the Skyline model sold in Japan. This vehicle has no rear-wheel steer at speeds below 30 mph. Above this speed, the rear wheels steer in the same direction as the front wheels. A hydraulic cylinder on each side of the rear suspension moves a rear suspension subframe up to 0.5 degree by deflecting the suspension bushings. This is illustrated in Figure 8-20. A hydraulic pump, like a power steering pump, with its own reservoir, produces hydraulic pressure. Fluid pressure is sent to a control valve on the rack and pinion power steering housing. It is similar to a power steering control valve. When the control valve senses lateral acceleration (sideways movement), the valve moves to pressurize one end of the hydraulic cylinder to move the rear suspension. Pressure goes to the cylinder through an electromagnetic valve. The electromagnetic valve is controlled by a vehicle speed sensor working through an electronic controller. Pressure moves the rear suspension in the same direction the front wheels are steered. The newest version of this suspension moves the rear wheels as much as 7 degrees. This suspension design aids cornering performance as well as corrective steer for side wind gusts. It coordinates rear steer with rear tire slip angles to produce the desired vehicle response.

Mazda introduced computer-controlled, low-speed, rear-wheel steering on the 1988 Capella model sold in Japan. Below 22 mph, the rear suspension steers up to 5 degrees in the opposite direction of the front-wheel steer (antiphase). Above 22 mph, it switches to the same-phase steering like the Nissan described above. It is sensitive to the front-wheel steer angle. A shaft from a steering-angle transmission on the power steering rack housing runs to the

FIGURE 8-20 Simplified sketch of a steering rear suspension subframe.

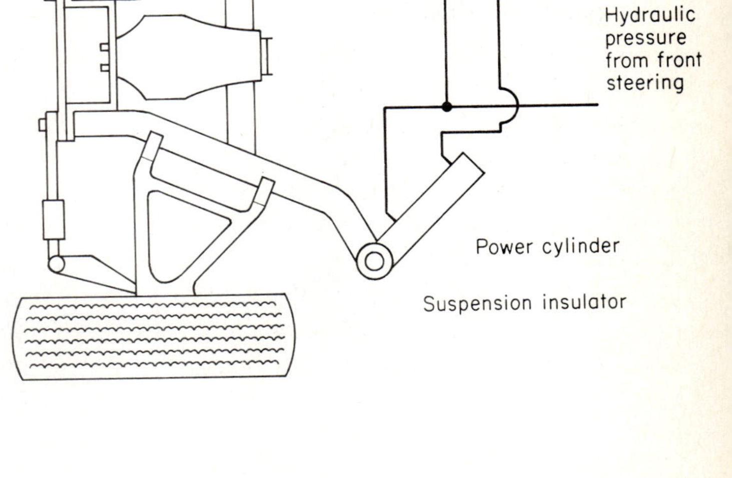

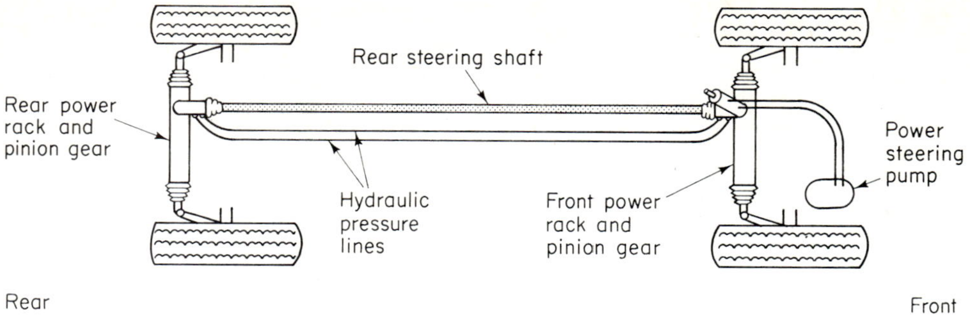

FIGURE 8-21 Simplified sketch of a computer-controlled four-wheel steering system.

phase-control unit at the rear. The phase-control unit controls the direction and amount of steering needed. The phase-control unit is operated by a steering angle transmission shaft, two vehicle speed sensors, a stepper motor, a control valve, and a power cylinder with an output rod. You can see a layout of this in Figure 8-21.

Mitsubishi four-wheel steering only steers in the same direction as the front wheels turn. The system is operated with hydraulic power. The hydraulic pump, with its own reservoir, is driven by the drive line so there is only pressure as the vehicle moves. This pump pressure is sensitive to vehicle speed until it reaches maximum pressure. The system also uses the power steering pump as a signal to indicate the amount and direction of the front wheel movement. Using both pressures, the rear power cylinders move tie rods (like front-wheel steering) to control rear-wheel steering.

Honda introduced an active four-wheel steering option on the 1988 Prelude. It is a mechanical system that turns the rear wheels 1.5 degrees in the same direction as the front wheels when the front wheels are turned 10 degrees. This takes about 115 degrees of steering wheel movement. Rear-wheel steering in the same direction as the front wheels provides good handling at medium to high speeds. As the steering wheel is turned 230 degrees, rear steering goes to zero or neutral. As the steering wheel is turned even more, the rear wheels begin to steer in the opposite direction. This reduces the turning circle when operating at low speed. They will reach 5.3 degrees and the front wheels will be turned 30 degrees when the steering wheel is against the lock at 450 degrees. This change in turning is shown by the graph in Figure 8-22.

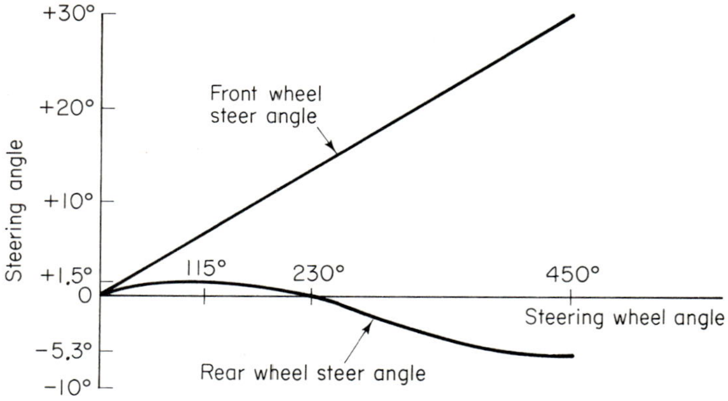

FIGURE 8-22 Angular movement of an active four-wheel steering system.

Honda engineers call their four-wheel steering ‘‘double-crank’’ steering. A long steering shaft runs from the front rack and pinion steering gear back to a double-crank gear box at the rear (Figure 8-23). The double-crank gear box is in neutral when the front wheels are straight ahead. It moves the rear suspension any time the steering wheel is turned either right or left. The worm shaft rotates an eccentric, like a crank. The eccentric carries a planetary gear meshed with a fixed internal ring gear for gear reduction. The planetary also has an eccentric or crank that changes the rotating motion to a sliding motion to move a stroke rod. The stroke rod moves the rear tie rods to steer the rear wheels.

FIGURE 8-23 Double crank mechanical steering to give the same phase steering at low steering wheel angles and anti-phase steering at large steering wheel angles.

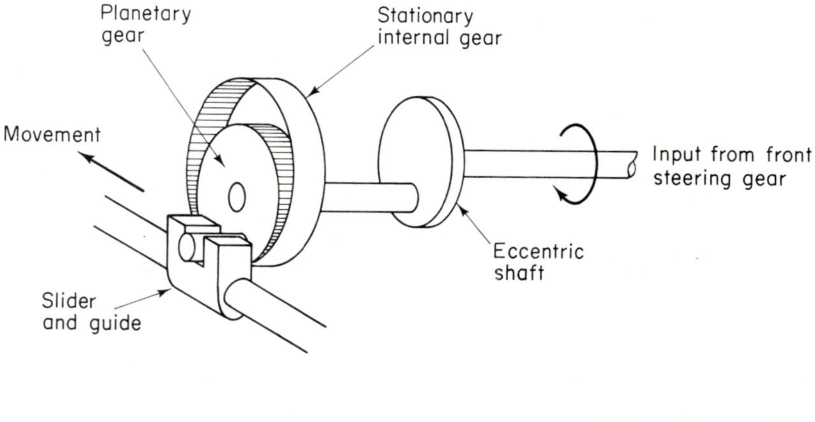

STUDY QUESTIONS

Introduction

1. What is the primary purpose of the steering system?
2. How does neutral sense of a vehicle differ from neutral steer?
3. What is wheel alignment?
4. Why is periodic wheel alignment necessary?
5. What is suspension compliance?

Section 8-1

6. Define camber.
7. What steering reaction does positive camber produce?
8. How does camber affect tire wear?
9. Define caster.
10. How is the suspension designed to minimize caster change during braking?
11. What are the advantages of positive caster?
12. What are the advantages of negative caster?
13. What prevents instability of negative caster?
14. What effect does SAI have on steering?
15. What disadvantage does SAI have on steering?
16. What effect does scrub radius have on steering?
17. What steering handling problems are caused by too much wheel offset?
18. What problem is caused by too much toe?
19. Why is a slight toe-in preferred over a slight toe-out?
20. Why is it necessary to have toe-out on turns?
21. What causes the wheels to toe-out on turns?
22. How does tire slip angle affect toe-out on turns?
23. What causes four tracks behind a four-wheel vehicle moving on a wet surface?
24. How can you find the thrust line when an independent rear suspension has toe-in?

Section 8-2

25. What is the advantage of four-wheel steering?
26. What does same-phase steering do? When is it used?
27. What does antiphase steering do? When is it used?
28. How does active four-wheel steering differ from passive four-wheel steering?
29. What is the advantage of driving the hydraulic pump used on four-wheel steering with the drive line rather than with the engine?
30. How does mechanical four-wheel steering differ from hydraulic and computer-controlled four-wheel steering?

9

Wheel Alignment

Aligning wheels requires the measurement and adjustment of suspension and steering angles. The measured angles are compared to the manufacturer's specifications for that vehicle. The alignment angles are adjusted when they do not fall within specifications. Wheel alignment is checked whenever the vehicle has abnormal handling or tire wear, as a preventive maintenance measure, or after parts that affect suspension and steering geometry have been replaced.

Prior to measuring the alignment angles, listen carefully to the customer's complaint and then do a thorough inspection of the condition of the vehicle suspension and linkages. This preliminary inspection is necessary to ensure that everything affecting alignment angles is in a normal condition.

9-1 PRELIMINARY INSPECTION

Preliminary inspection must include a check to ensure that the vehicle is at curb weight. *Curb weight* is established when a vehicle has all liquid levels filled, including a full fuel tank, one spare tire, tires properly inflated, no occupants, and no luggage or cargo. If the fuel tank is not full, 6 1/2 pounds (3 kg) properly positioned over the fuel tank may be substituted for each gallon of fuel required to fill the tank.

After establishing curb weight, the tires are inspected. Tire wear patterns are shown in Figure 9-1. Wear patterns were also shown in Figures 5-1 through 5-5. The tire size must correspond to tire manufacturer's specifications. Tread wear should be equal. The wheels should be of the size and offset specified. Most importantly, they must be matched side-to-side in axle sets. Tire pressures are adjusted to recommended pressure and inspected for any conditions that might result in tire failure. The tires and wheels are inspected for excessive radial and axial runout, as discussed in Section 5-2, when the vehicle is raised to inspect the suspension and linkages.

Since alignment angles change with suspension height, the vehicle must have proper ride height for correct wheel alignment and satisfactory handling characteristics. The ride height is measured and the distance is compared with specifications. Typical measurement locations are shown in Figure 9-2. Some older vehicles with torsion bar springs have height adjusters. When required, the height is adjusted at this point in the inspection. When the suspension height is not correct on vehicles with coil and leaf springs, suspension parts must be changed to set ride height before aligning the wheels. Slight changes may be done with shims or booster springs. The best correction method is to replace the weak springs.

The vehicle is raised on a hoist or placed on safety stands to inspect the suspension and steering

CONDITION	RAPID WEAR AT SHOULDERS	RAPID WEAR AT CENTER	CRACKED TREADS	WEAR ON ONE SIDE	FEATHERED EDGE	BALD SPOTS
CAUSE	UNDER INFLATION	OVER INFLATION	UNDER-INFLATION OR EXCESSIVE SPEED	EXCESSIVE CAMBER	INCORRECT TOE	WHEEL UNBALANCED
CORRECTION	ADJUST PRESSURE TO SPECIFICATIONS WHEN TIRES ARE COOL			ADJUST CAMBER TO SPECIFICATIONS	ADJUST FOR TOE-IN 1/8 INCH	DYNAMIC OR STATIC BALANCE WHEELS

FIGURE 9-1 Tire wear patterns indicating suspension problems. (Courtesy of Chrysler Corporation)

linkage. If the vehicle has air suspension, the system must be turned off before the vehicle is lifted. Lincoln vehicles have an air suspension switch on the microprocessor located in the left trunk panel. Steering linkage inspection should include an inspection of the tie rods, steering gear mounting, idler arm, and drag link. This includes tie rods on rear suspensions. They should be checked for pivot wear, binding, and looseness. The boots on the ends of the rack and pinion steering gear must be in good condition. Pinch the tie rod end of the boot to feel the inner ends of the rod for looseness as you try to turn the wheel inward and outward. This is a good time to inspect the universal joint boots on the half-shafts. If they are not in good condition, they should be replaced. Boot failure will lead to rapid joint failure,

FIGURE 9-2 Typical suspension height measurement locations.

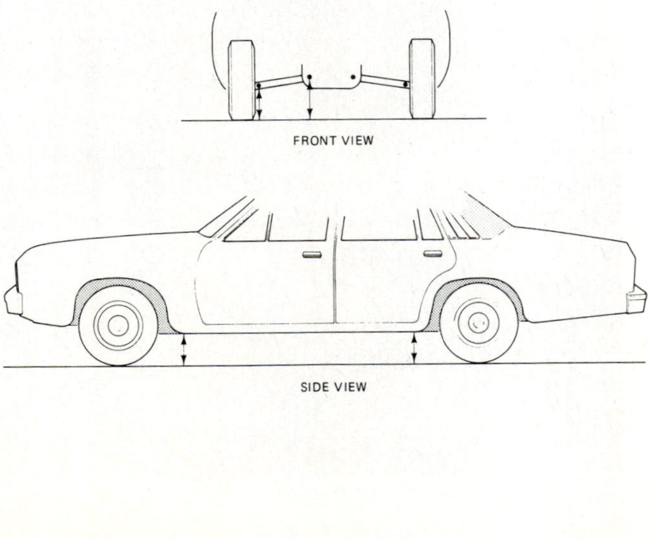

and the joints are very expensive. If the vehicle has power steering, the fluid level and belt tension should be checked and corrected, as needed. Although these items do not affect alignment, they must be corrected to ensure proper steering when alignment is completed.

Modern vehicles are made of much lighter-gauge metal than some of the older vehicles. Even though the quality of the metal is better, it is more likely to bend. As a result, the vehicle subframe should be checked for any damage that could change the location of the suspension pivot points. The pivot point of the lower control arm on the frame can be bent if a vehicle slides into a curb or hits parking blocks. Sometimes, the entire forward subframe is shifted by an angle impact on the bumper. This changes the track of the vehicle. Fresh rust, flaked paint, or lost undercoating are signs of underbody damage.

Ball joints, shock absorbers, and struts are inspected and repaired as needed. Wheel bearings must be adjusted to specifications. Wheel bearing adjustment is critical, because wheel alignment equipment is either fastened to the wheel or the wheel hub. The readings of alignment equipment will be affected by improper bearing adjustment.

All pivot points for both front and rear suspensions must be inspected. This includes control arm bushings, shock absorber bushings, and any strut bushings. If the vehicle is equipped with MacPherson strut suspension, the spring and its seat must be free to rotate on the upper bearing. This should be checked for free movement.

9-2 ALIGNMENT EQUIPMENT

After all the preliminary checks are made, the alignment angles are measured. There are several types of wheel alignment equipment. Each manufacturer of wheel alignment equipment has its own method of measuring the wheel alignment angles. Common methods for measuring the camber, caster, and SAI include the use of a bubble gauge (similar to a carpenter's level) with an adjustment and a scale for readout, light beams, and digital readouts that automatically display numerical readings. Toe can be measured with a mechanical scale that compares the distance between the front and back of the tire, a light beam, or a digital readout. Turning angle is indicated by a degree scale on the radius plates on which the front wheels are placed while checking the alignment. The radius plates allow the wheels to be turned freely inward or outward and determine the angular degree of the turn for each wheel.

Alignment equipment may be either portable or permanently mounted. Portable equipment of the type shown in Figure 9-3 allows a work area to be used for other service as well as wheel alignment.

FIGURE 9-3 Typical portable wheel alignment equipment.

FIGURE 9-4 Vehicle alignment on a rack.

Facilities with limited space and a limited volume of alignment work will generally use this type of equipment. Portable equipment uses radius plates on stands placed under each front tire. It also uses stands under each rear tire to level the vehicle. Portable alignment heads are attached to the hub or wheel rim where the readings are taken.

Permanently installed equipment measures the alignment while the vehicle is on a *level* platform. The vehicle may be driven up ramps to the platform, ramps may be raised to provide a level platform, (Figure 9-4), or the vehicle may be driven onto alignment equipment at floor level. A "pit" or recessed area in the floor (Figure 9-5) provides space for the

FIGURE 9-5 Vehicle on a pit-type wheel aligner.

technician to work under and around the front of the vehicle. The alignment platform, often called the *alignment rack*, may be used with either portable alignment heads or with alignment heads that project light beams forward of the vehicle onto a screen that is marked with a scale for the reading. Still other alignment heads feed information into a computer that has a digital readout on a CRT screen. The equipment either raises the vehicle from the floor surface or provides room beneath the vehicle where the technician can make under-the-vehicle adjustments.

FIGURE 9-6 Lock pins being removed from a radius plate.

9-3 WHEEL ALIGNMENT MEASUREMENT

Vehicles with independent rear suspensions must have the rear suspension in correct alignment before the front suspension is aligned. This will ensure that the track of the vehicle is correct before the front wheels can be aligned correctly. Modern alignment equipment has alignment heads for each wheel. The alignment of both front and rear suspensions as well as track can be measured without moving the vehicle. Older and economy alignment equipment have only two alignment heads. It is necessary to back the vehicle on the rack and align the rear suspension first. Vehicle specifications are read backwards. For example, a toe-in reading would actually be toe-out on the rear suspension when the vehicle is backwards on the alignment rack. The vehicle is then turned around and the front suspension is aligned.

The alignment angles measured on rear wheels are some of the same angles measured on front wheels. They are measured in the same way. Therefore, we are going to discuss front-wheel alignment first. This will give you a chance to understand how to measure all the alignment angles. You will then know how to measure the rear suspension angles, too. Remember, when doing an alignment job, the rear suspension is aligned first.

Put the vehicle on the alignment rack with the front wheels centered on the radius plates. The vehicle must be level and as straight as possible. Remove the locking pins from the radius plates so they can rotate (Figure 9-6). Place a brake pedal depressor so it will hold the service brakes. Typical locking positions are shown in Figure 9-7. Record all alignment readings you measure before making any adjustments.

Camber. The first angle to be measured on either rear or front suspensions is camber. Camber is measured with a bubble-level gauge. The gauge compares the centerline of the wheel with true vertical. This is easily measured because the bubble-level gauge will always rest at zero if the gauge is attached to a true vertical plane. A bubble gauge is shown in Figure 9-8. Camber is always measured with the

FIGURE 9-7 Pedal and steering wheel lock positions.

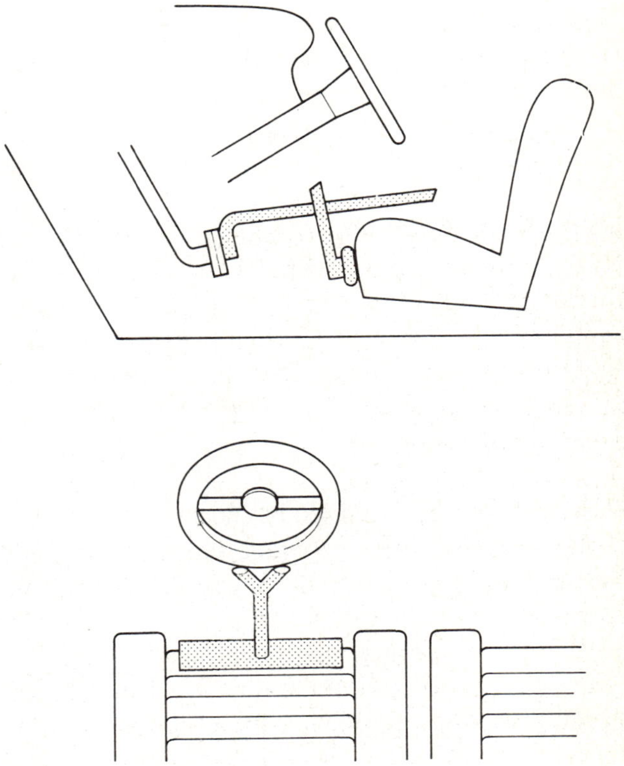

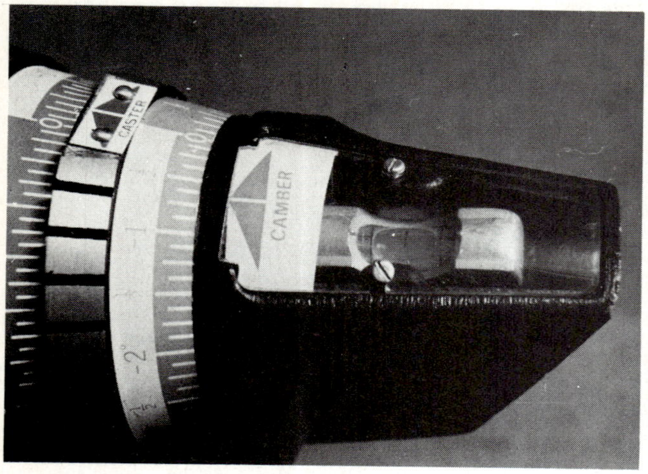

FIGURE 9-8 Bubble-level gauge in an alignment head connected to a wheel hub.

wheels straight ahead. The gauge will show the degree to which the wheel is inclined from the vertical. Camber is positive if the tire is inclined outward at the top or negative if the tire is inclined inward at the top. This is illustrated with a rod connected to the spindle of the suspension shown in Figure 9-9. When light-beam alignment is used, the angle of the light beam is compared to a graduated scale on the viewing screen to indicate the degree of camber. This is shown in Figure 9-10.

The vehicle should be jounced to settle the suspension. The camber will be shown on the readout. Depending on the type of alignment equipment, you will read camber on the scale at the center of the bubble, where a light beam crosses a scale, or on a digital readout. Record the camber readings you see. Have someone get on the bumper to compress the spring while you read the camber change. Next, have the bumper lifted to expand the spring to see the camber change. You should see very little camber change, especially when the suspension has MacPherson struts. A major change in camber is caused by faulty parts. The chamber of the rear wheels is measured the same way.

Caster. Caster is measured by the change in camber as the wheel is turned through a specified arc angle. You can see that you do not measure caster on the rear wheels. This arc may vary from 40 degrees on a typical bubble-gauge alignment head to as low as 20 degrees on a light-beam or digital alignment head. Turning the wheel through an arc produces a camber

(a)

(b)

(c)

FIGURE 9-9 Camber is shown by a rod connected to the spindle of a suspension. The rod is compared to the horizontal line on the background: (a) the rod slopes downward, indicating positive camber; (b) the rod is level, indicating zero camber; and (c) the rod slopes upward, indicating negative camber.

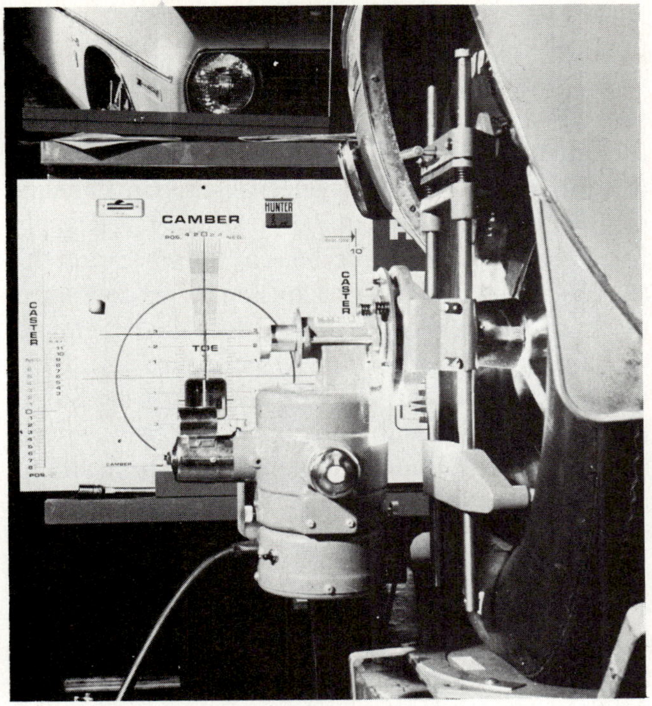

FIGURE 9-10 Wheel alignment machine using a light beam projecting on a viewing screen scale.

change and a spindle height change as a result of caster. This is illustrated in Figure 9-11. The alignment head senses the camber change produced as the wheels are turned through the specified arc. The change in camber is converted through scales to indicate caster. The procedure may vary between manufacturers, although the common method of measuring caster is through camber change. Caster is measured by turning the wheel inward, zeroing the scale, and then turning the wheel outward a specified number of degrees. The caster can then be read on the caster indicator scale.

Steering Axis Inclination. SAI is measured in a manner very similar to caster. The wheel is turned inward 20 degrees (10 degrees on some machines). Then, the caster and camber scales are referenced and locked to the spindle or wheel. The wheel is then turned outward 40 degrees (20 degrees on some models) and the SAI is read directly from the SAI indicator scale. The camber scale is not re-zeroed as was done when measuring caster. SAI produces an up-and-down rotation of the wheel spindle as illustrated in Figure 9-12. The change is sensed by the alignment head to indicate the SAI. Recall that SAI

(a)

(b)

FIGURE 9-11 The rod connected to the spindle of a front suspension shows the camber change that results from positive caster: (a) the wheel is turned inward 20 degrees and the rod slopes downward, indicating positive camber; (b) the wheel is turned outward 20 degrees and the rod slopes upward, indicating negative camber.

(a)

(b)

(c)

FIGURE 9-12 View of the left front wheel (front is to the left of this illustration). A rod is connected to the spindle and a second rod is placed so that it is parallel to the wheel. Notice that the spindle is tipped down when the wheel is turned inward (a); it is level when the wheel is pointed straight ahead (b); and it is tipped downward, but not as much, when the wheel is turned outward (c).

is the angle between the steering axis and true vertical, while camber is the angle formed between the wheel centerline and true vertical when both are viewed from the front. These angles will change as the upper or lower ball joints are moved in or out to adjust camber during a wheel alignment. The sum of the SAI and the camber angles or the angle formed between the centerlines of the wheel and the steering axis will always be constant. This combined angle, called the *included angle*, does not depend on the position of the ball joints (see Figure 8-8). During wheel alignment, the SAI reading is measured along with the camber reading for each wheel. Camber and SAI angles are added algebraically to determine the included angle. The measured angle is compared to that specified. If SAI is specified, camber is subtracted from the measured included angle to determine the vehicle SAI so this value can be compared to the alignment specifications. Because the alignment head is usually attached to the hub or wheel rather than the spindle, any rotation of the hub will alter the SAI. Having the brake pedal depressor firmly set will prevent wheel rotation. This will remove any slack that may exist in the brake system when measuring SAI as the wheel is turned 20 degrees inward from straight and then turned outward 20 degrees past center. SAI, as measured during an alignment check, is used to detect a bent spindle, steering knuckle, or strut.

Toe-Out on Turns. Toe-out on turns (turning radius) is checked with the front wheels on the radius plates. The wheels are set in the straight-ahead position with the radius plates set on zero. One wheel is turned outward 20 degrees and the turning angle is read on the radius plate degree scale under the other wheel. The turning radius on the inner wheel is normally between 21 degrees and 22.5 degrees when the outer wheel is turned 20 degrees. Toe-out on turns is the difference between the degree reading on the inner wheel radius plate and the outer wheel radius plate. Toe-out on turns is not adjustable. It is used in service to detect bent or incorrect steering arms and misaligned linkage or steering racks.

Toe. Toe is measured in a variety of ways, depending on the manufacturer of the alignment equipment. All methods eventually result in a measure corresponding to the difference in distance between the centerlines at the front and the rear of the tires at their spindle height (see Figure 8-11).

Mechanical measurement techniques include jacking each wheel and chalking a band around the tread surface of the tire. A scribe is then used to mark a fine continuous line in the chalk as the wheel is rotated, as shown in Figure 9-13. This will prevent any runout in the tire-wheel combination from affecting the toe measurement. The tires are then lowered to the radius plates and the vehicle is jounced to settle the suspension. The distance is then measured at spindle height between the scribed lines at the rear and at the front of the tires. If the distance at the front of the tires is less than the distance at the rear, the vehicle has toe-in. The difference in the measured distances is the amount of toe-in.

Another mechanical method used to measure toe is to mark a point on the inside front of each wheel rim at an equal distance above the floor at the spindle height. The distance between the two points is measured while full vehicle weight is on the tires. The tires are rotated so the mark is at the rear at the spindle height. The distance between the front and back measurements is the toe.

FIGURE 9-13 One end pointer of a toe bar aligned on a line scribed on chalk put on a tire rib.

Electronic measurement of toe includes a light beam projected across the front of the vehicle between alignment heads attached to the wheel spindles. The amount of toe is read on a scale illuminated by the light-beam line in inches or degrees.

Another method used to measure toe is to project a light beam forward from each wheel alignment head. The beam is then reflected rearward to a scale on the projector head. The total toe of the vehicle is the sum of the toe measurements read at each wheel.

Finally, toe can be read digitally in degrees from the vehicle centerline on new alignment equipment. The results are usually shown on a CRT screen. This measurement method is the best to use to measure the toe of independent rear suspensions.

Toe change can help you determine suspension problems. Measure the toe as weight is placed on the bumper in the same way you did when checking camber. Also, check the toe as the bumper is lifted. Too much change in the toe indicates a suspension or steering linkage problem.

Track. The rear wheels must be aligned with the vehicles. This is measured with strings or light beams between the alignment heads on the front and rear or a scale on the rear. Some four-wheel alignment equipment with CRT readout will show track. Follow the instructions that came with the alignment equipment you are using to check track.

9-4 FRONT-END ALIGNMENT ON VEHICLES WITH UNEQUAL LENGTH CONTROL ARM SUSPENSION

All the alignment angles are checked before any adjustments are made. Camber and caster adjustment may be made in the same or separate ways. Camber adjustment is made by shims, cams, slots, and eccentric bushings. Caster adjustment is made by shims, cams, slots, and adjustable-length control arm brake struts. Both camber and caster are affected by the position of the ball joints. Adjusting the ball joint position to make a change in either angle will affect both angles, especially on vehicles using shims, cams, or slots for adjusting both camber and caster.

The effect on camber is minimized when the caster adjustment is made by adjusting the brake

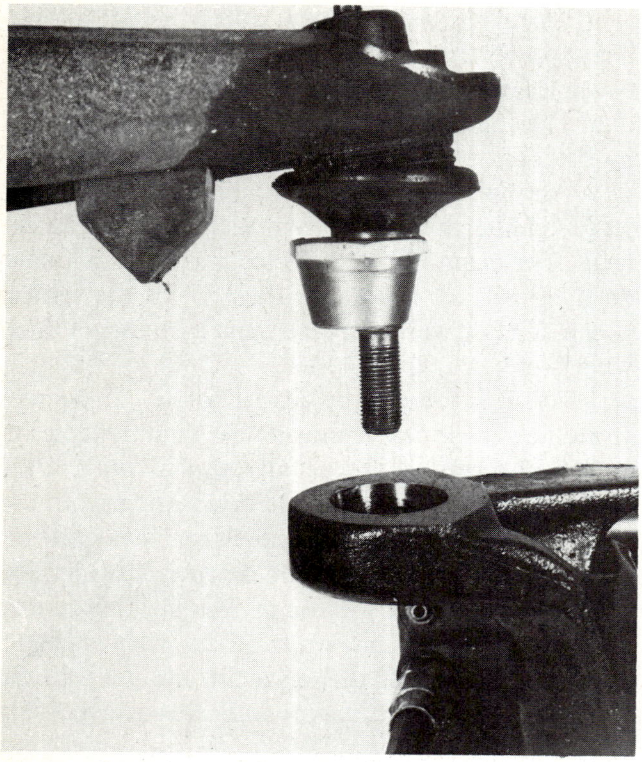

FIGURE 9-14 Eccentric ball joint bushing used for alignment.

strut. As a result, camber is adjusted first on vehicles that use strut adjustment for caster. Both camber and caster are adjusted at the same time on vehicles using shims, cams, or slotted bolt holes on the inner end of the upper control arms. Before making any change, decide which adjustment can be used to correct both camber and caster. Shims and cam adjustments may be used on the inner end of either upper or lower control arms. Slot adjustment is found only on the inner end of the upper control arm. Eccentric bushings on the upper ball joint, as shown in Figure 9-14, have been used to make the camber adjustment on some unequal length control arm suspensions and on some twin I-beam suspensions. Offset ball joints have also been used where the joint must be unbolted and rotated to effect a camber adjustment. This allows only a one-time adjustment in only one step.

Special tools are used to hold the suspension while making the caster and camber adjustment on suspensions with a slot-type adjustment. An example is shown in Figure 9-15. The special tool is hooked near the bolt to be loosened on the upper control arm cross-shaft. The control arm is loosened only on the end used to make the adjustment while supporting the cross-shaft with the special tool. It may be necessary to retighten the bolt and repeat the adjustment procedure on the other bolt to set the correct camber and caster.

A shim is added or removed after loosening the bolts that fasten the front upper control arm to the frame on shim-type adjusters. This adjustment location is shown in Figure 9-16. Before making the alignment adjustment with shims, decide where a shim must be added or removed and if the shim change will correct both camber and caster at the same time.

Cam adjusters, one on the front and one on the back at the inner end of the control arms, are used to adjust both camber and caster. A typical cam adjuster is shown in Figure 9-17. After loosening the lock, the cams are rotated to adjust camber and caster. Some vehicles have a cam adjustment on the lower control arm for camber adjustment and an adjustable strut bar for caster adjustment. On these vehicles, camber is adjusted before caster.

FIGURE 9-15 One type of tool used to control the suspension when the vehicle has slot adjust.

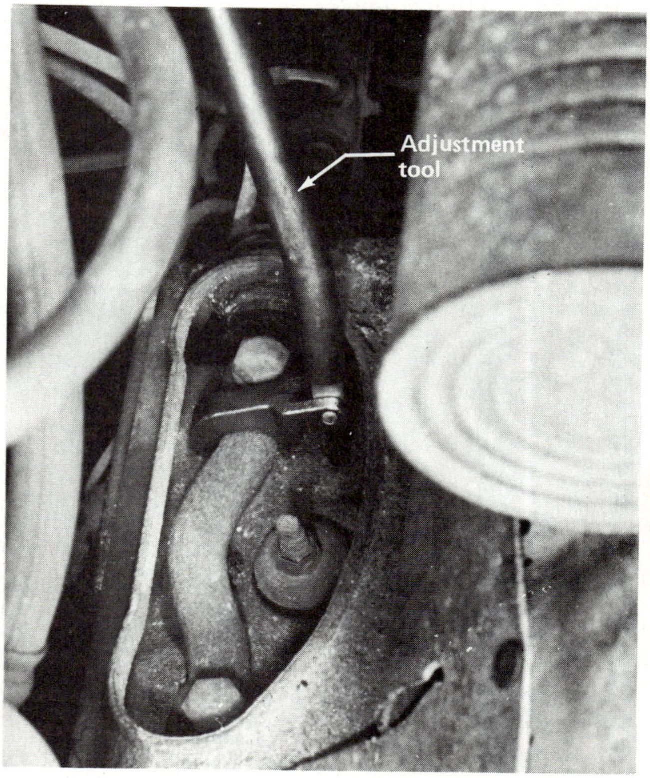

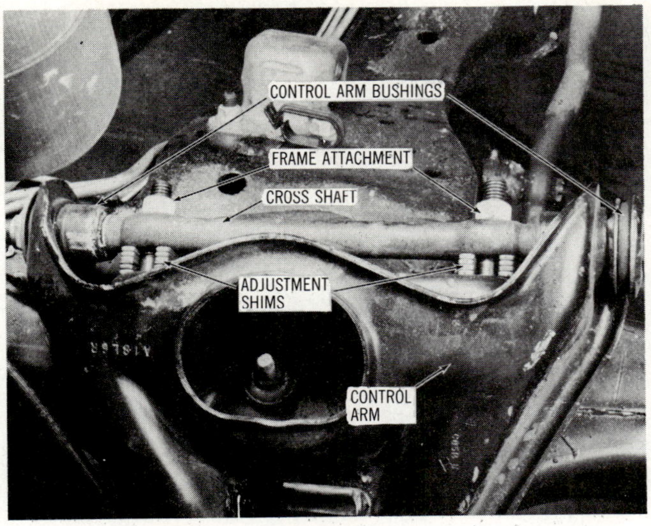

FIGURE 9-16 Adjustment shims for wheel alignment.

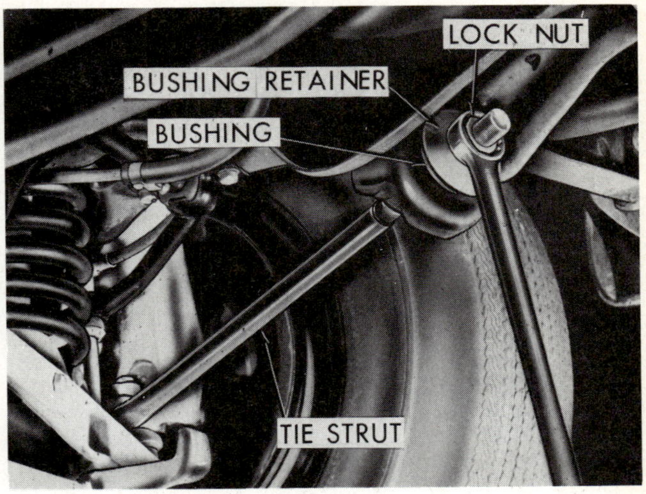

FIGURE 9-18 Strut adjustment to set caster. (Courtesy of Cadillac Motor Car Division, General Motors Corporation)

FIGURE 9-17 Adjusting cam for wheel alignment.

When the strut bar is used to change caster, the bar length is changed by increasing or decreasing the distance from the fixed point on the frame to the outer end of the control arm. The brake strut is adjustable only when there is a separate means used to adjusι ːamber and caster (Figure 9-18). Struts may be either in front of or to the rear of the control arm. Before making an adjustment, you should decide whether the strut needs to be lengthened or shortened to correct caster.

9-5 FRONT-END ALIGNMENT ON VEHICLES WITH MACPHERSON STRUT-TYPE SUSPENSION

Vehicles equipped with MacPherson strut or modified strut suspensions may or may not have adjustable camber and caster. All have an adjustment for toe. Most domestic vehicles with MacPherson suspensions have ways to adjust all three angles. Many import vehicles designed with MacPherson strut suspensions do not have adjustments for camber or caster.

Camber is adjusted on some struts by turning cams at the point where the strut connects to the steering knuckle (Figure 9-19). Others are adjusted by moving the upper strut mounting plate (Figure 9-20). Camber is also adjusted on some MacPherson struts by moving cams on the inner end of the lower control arm.

Caster is adjusted by either the brake strut or by moving the mounting plate on the upper end of the strut. The brake strut adjustment corrects caster the same way it does on unequal-length control arm suspensions.

If either camber or caster need adjustment, it is most important to check the complete suspension. Camber and caster rarely need adjustment on vehicles with MacPherson struts. If they do, worn bushings or bent or worn parts are usually caused

FIGURE 9-19 Adjusting cam on a MacPherson strut.

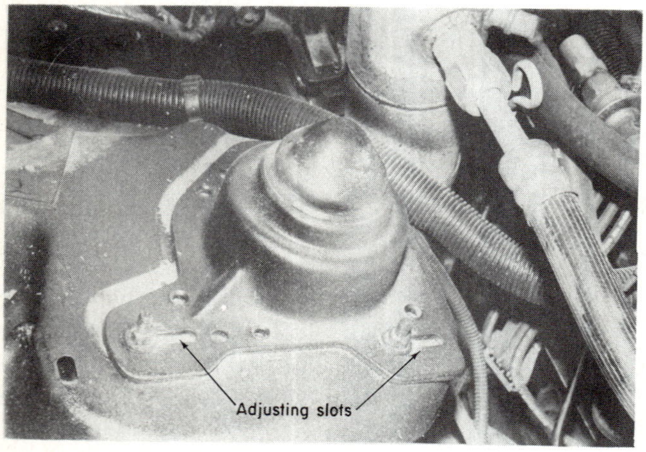

FIGURE 9-20 Adjusting plate at the top of a MacPherson strut.

by misalignment. Always make a careful inspection before aligning caster or camber on vehicles with MacPherson struts.

9-6 FRONT-END ALIGNMENT WITH TWIN I-BEAM FRONT AXLES

Light- and medium-duty trucks and vans manufactured by Ford Motor Company have had twin I-beam suspensions for many years. This suspension combines some of the ruggedness of the I-beam suspension with some of the ride qualities of an unequal-length control arm suspension.

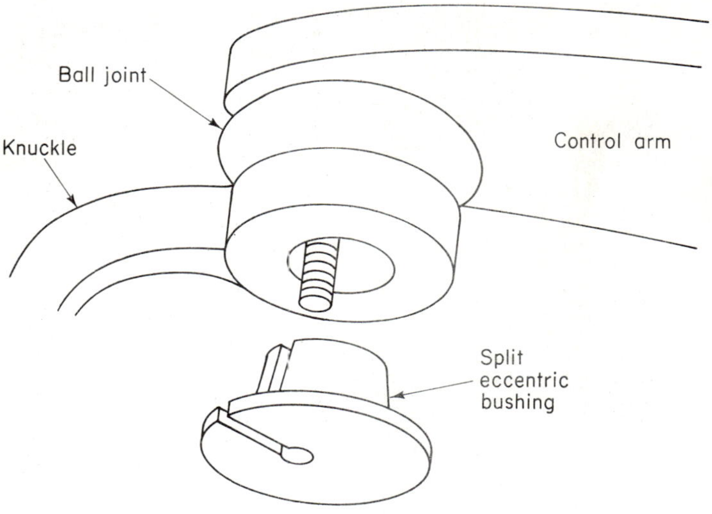

FIGURE 9-21 Twin I-beam adjustment with an offset eccentric bushing type sleeve cam adapter.

Early designs of this front suspension (see Figure 1-12) had forged steel axles. The wheel pivots were kingpins that allowed the wheels to be steered. Camber and caster adjustment required bending or twisting the axle to make the desired setting. Bending required special heavy-duty equipment. Toe was adjustable with an adjustment sleeve on the left tie rod only.

In the early 1980s, stamped steel axles (Figure 9-21) were introduced on many of these same light trucks. This design used ball joints for wheel pivots. A sleeve is used between the spindle assembly and ball joint. This sleeve, with an offset hole for the ball joint, is called a *camber adapter*. It can be repositioned or replaced with a sleeve with a different amount offset to adjust camber. This allows precise adjustment without using special bending equipment.

9-7 TOE ADJUSTMENT

Toe adjustment is made by changing the effective distance between the inner and outer tie rod ball pivots. On most domestic automobiles, the adjustment involves loosening two clamp bolts on an adjuster sleeve and rotating the adjuster sleeve, as shown in Figure 9-22, to obtain the correct toe setting. Vehicles with rack and pinion steering require a locknut to be loosened on the steering linkage next

FIGURE 9-22 Toe is set by rotating an adjusting sleeve on the tie rod.

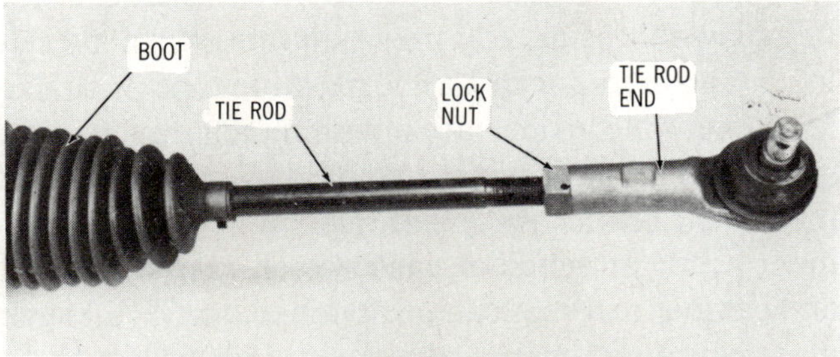

FIGURE 9-23 Lock nut is loosened to adjust the tie rod end of a rack and pinion steering system.

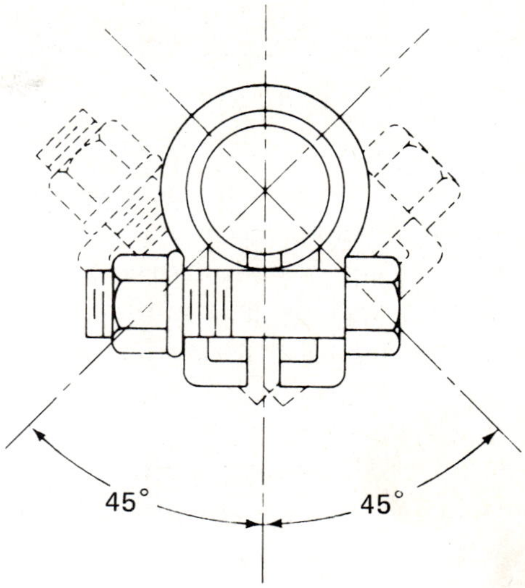

FIGURE 9-24 Typical range of the tie rod sleeve clamp positions.

to the tie rod end (Figure 9-23). The tie rod is then rotated to adjust the toe.

Adjustment of one tie rod will change the toe setting; it will also change the straight-ahead center position of the steering wheel. Rotating both tie rod sleeves an equal amount in the same direction will change the toe setting and will not change the steering wheel position. Adjustment of both tie rods, in equal amounts but in the opposite directions, will produce no toe change but will rotate the steering wheel to a new position.

It is important that the steering wheel and the steering gear be in the center position when the wheels are aimed straight ahead. The steering gear is designed to have the least amount of free play when it is centered. Steering linkages are designed to be symmetrical from side to side. If the overall length of the tie rods differs on each side of the vehicle, as would be possible with an uncentered steering gear, erratic vehicle handling will result. Proper tie rod adjustment gives the driver more positive, predictable steering control.

Steering wheel position may be checked visually or it may be measured to assure proper positioning. Using either method, the front wheels are first set in the straight-ahead position, then the steering wheel position is checked. If the wheel is not centered and the toe adjustment is incorrect, you may be able to correct both malfunctions by adjusting only one of the tie rods.

Toe is the last adjustment made during wheel alignment. The tie rod sleeves and clamps must be properly positioned to provide operating clearance and to prevent any binding in the tie rod pivots during movement of the vehicle body and steering linkage. The tie rod adjuster sleeve clamp bolts are then tightened. The proper position is shown in Figure 9-24.

Remember, in an actual alignment, the rear suspension alignment would have been checked before the front suspension. Before moving the vehicle from the alignment area, the vehicle is rechecked to ensure that all suspension and steering fasteners have been properly tightened and that the cotter pins are in place. You should test drive the vehicle to make sure that there is no directional drift, vibrations, or other alignment and handling conditions the operator might associate with alignment or suspension problems.

9-8 REAR-WHEEL ALIGNMENT MEASUREMENT

Modern automobiles with independent rear suspensions have adjustments for camber and toe. These automobiles should have four-wheel alignment to provide good handling and minimum tire wear. Vehicles using dependent rear-wheel-drive systems need to have the rear-wheel alignment checked whenever the vehicle has tire wear or a handling problem. Wear of suspension bushings and linkages can cause the dependent suspension to misalign.

Independent Suspension. Vehicles equipped with an independent rear suspension have some way to make toe and camber adjustments. On this type of suspension, toe can affect vehicle tracking when one wheel is adjusted to point inward and the other wheel is adjusted to point outward. This will cause the vehicle rear end to steer in one direction or the other, as shown in Figure 8-16. Periodic adjustment of the rear suspension angles may be required as a result of sagging springs and wear of the independent suspension parts.

Independent rear suspensions use either shim- or cam-type camber adjustments. Toe of the rear wheels may be adjusted with shims or with an adjustable tie rod (Figure 9-25). Shims are used to change the angle of the spindle on the suspension when the spindle is bolted to the suspension. The brake backing plate is loosened and the shim is placed between the spindle flange and axle flange. Do not place the shim between the backing plate and spindle. This will cause pulsing brakes. The spindle will have to be removed and cleaned if there is rust on the flanges. A shim between the top of the flanges will increase positive camber. A shim between the rear of the flanges will increase toe-in. The alignment will have to be checked after the suspension has been reassembled. Race cars often have adjustable-length control arms on the rear suspension. The adjustments are used to set rear toe, and camber. An applicable shop manual should be used to determine the adjustment procedure for the vehicle being serviced.

FIGURE 9-25 Tie rod used to adjust rear-wheel toe.

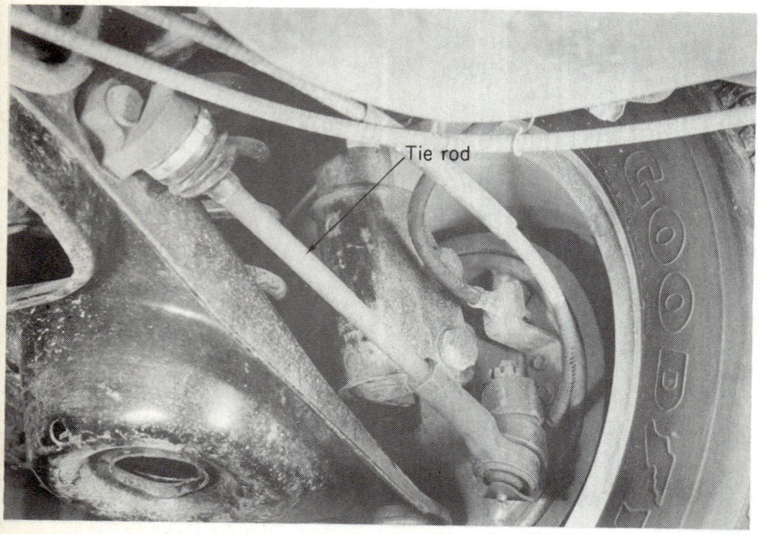

Rear suspension angles, as well as track, can be measured using alignment equipment or by using a tape measure and some masking tape or chalk. Most new electronic alignment machines are designed to check the alignment of all four wheels with the vehicle on the alignment rack. With wheel alignment heads attached at each wheel, rear-wheel toe and camber readings can be obtained quickly. Alignment equipment with only two alignment heads requires backing the vehicle onto the alignment rack and measuring the angles as described in Section 9-3. When this method is used, it is important to realize that toe-in is measured as toe-out on the alignment machine. This is caused by the reversed position of the vehicle on the alignment rack. When alignment measurements are being made, it is important that the rear tires are on the radius plates to permit free movement of the tires.

Camber readings are measured in degrees on the alignment machine; however, many manufacturers specify rear axle camber readings in inches. In this case, the camber specifications will have to be converted to degrees. This can be done by using a trigonometric value called an *arc-sine* ($\sin^{-1}$). The fractional specification is divided by the distance from the floor to the center of the wheel. The arcsine of this number is the angle in degrees equivalent to the specified distance. This is shown in Figure 9-26.

Dependent Suspension. A vehicle with a dependent rear suspension is supported on the frame so the suspension is extended when checking toe and

Standard conversion of camber specifications

Inches		Degrees
$\frac{1''}{16}$	=	$\frac{1^\circ}{4}$
$\frac{1''}{8}$	=	$\frac{1^\circ}{2}$
$\frac{3''}{16}$	=	$\frac{3^\circ}{4}$
$\frac{1''}{4}$	=	1°
$\frac{5''}{16}$	=	$1\frac{1}{4}^\circ$
$\frac{3''}{8}$	=	$1\frac{1}{2}^\circ$

Example: inches to degrees

Tire: 28" diameter = 14" radius

Specification: $\frac{1''}{8}$ positive

$$\text{Camber (degrees)} = \sin^{-1} \frac{\text{Specification (inches)}}{\text{Tire radius}}$$

$$= \sin^{-1} \frac{0.125 \text{ in.}}{14 \text{ in.}}$$

$$= \frac{1^\circ}{2}$$

Example: degrees to inches

Tire: 28" diameter = 14" radius

Specification: $\frac{1^\circ}{2}$ positive

$$\text{Camber (inches)} = \sin \frac{\text{Specification (degrees)}}{\text{Tire radius}}$$

$$= \sin \left(\frac{1^\circ}{2}\right) \times 14 \text{ in.}$$

$$= \frac{1}{8} \text{ in.}$$

FIGURE 9-26 Converting an angle to a distance and a distance to an angle.

camber. Short pieces of tape are placed on each rear wheel on one tread bar near the center of the tread. The wheels are rotated so the tape is positioned at the extreme front of the tire. The distance between the pieces of tape on the two tires is measured and the reading recorded. The wheels are then rotated until the two pieces of tape are at the extreme rear. The distance between the pieces of tape is again measured and the readings recorded. Finally, the wheels are rotated so the tape is at the bottom and a measurement is taken and recorded.

Toe is the difference between the front of the tire readings (FTR) and the rear of the tire readings (RTR). The suspension has toe-in when the reading at the front of the tires is less than at the rear. If the front of the tire reading is greater, the axle has toe-out. Thus,

$$\text{RTR} - \text{FTR} = \text{Toe}$$

For determining camber, the front of the tire reading and the rear of the tire reading are averaged. The bottom of the tire reading (BTR) is subtracted from the average. Thus,

$$\frac{\text{RTR} + \text{FTR}}{2} - \text{BTR} = \text{Camber in inches}$$

If the bottom of the tire measurement is larger than the average, the vehicle has negative camber. If the measurement is less than the average, the camber is positive.

The axle housing will have to be replaced when toe and camber readings are out of specification. Toe measurements that are out of specification result from a bent axle housing. The housing can be bent by a sharp rearward blow to the tire and wheel as a result of a pothole in the road surface or from an accident. Negative camber on both rear wheels is usually caused by severe overloads that slightly bend the axle housing. In either case, straightening a axle housing is not recommended. A bent housing must be replaced. Heavy-duty and off-the-road vehicles designed for rough use may be equipped with struts connecting the left and right sides beneath the differential carrier to add strength. Rear-wheel camber and toe can be normal and the track can be incorrect if the suspension bushings have deteriorated.

Alignment in vehicles with dependent and independent suspensions can be checked in another mechanical way. The vehicle must be raised and supported at the frame to mark the tire with chalk. This is done by rotating the tire and marking a straight chalkmark around each tire on one tread rib. After the chalkmark is made, it is scribed with a sharp tool as the wheel is rotated. This produces a fine, smooth sharp line in the chalk. Measurements can be taken from this line without rotating the tires. Toe and camber on independent suspension vehicles can be measured easily from these lines. The vehicle must have its full weight on the radius plates of the alignment machine when measurements are taken. After measuring, the same calculations need to be done to determine toe and camber as described above. Toe and camber are automatically measured and converted on four-wheel alignment machines.

Track Alignment. Track measurement is easily checked with electronic alignment equipment that uses light beams. Measurement is started on each

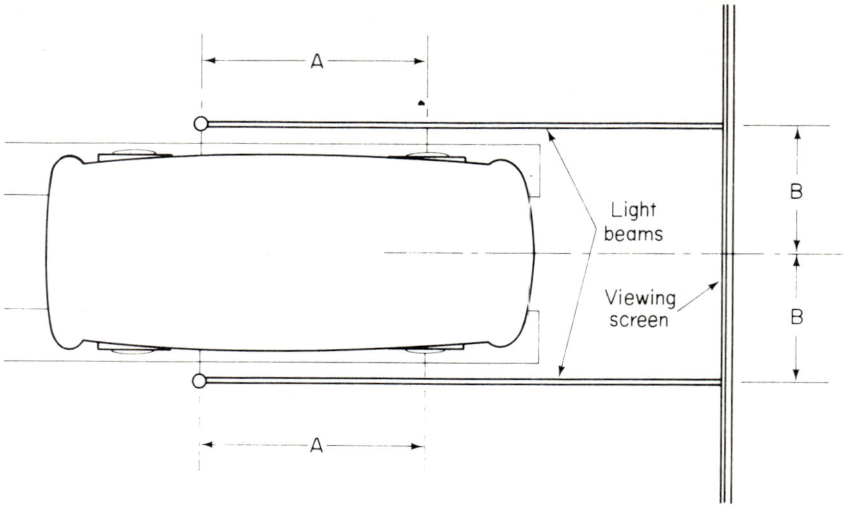

FIGURE 9-27 Checking vehicle track. The vehicle track is correct if the distances marked A are equal and the distances marked B are equal.

side of the vehicle at a point an equal distance back from the front wheels. The light projects forward from these points on each side of the vehicle. The vehicle will track properly if the measurements shown in Figure 9-27 are equal.

Other rough methods can be used to find out if the vehicle is tracking properly. One of these is to sight forward across the sidewall of the tire so the front and back of one rear tire sidewall appear as a single plane while an assistant holds a yardstick in a horizontal position, touching the frame at a specific distance back from the front wheels. The assistant moves a marker across the yardstick until the plane of the tire and the marker line up. The position of the marker is noted and the procedure is repeated on the opposite side of the vehicle using the corresponding frame point. The measurements should be identical when measured from identical points on each side of the frame. If the measurements differ, track is incorrect. Toe must be set before adjusting independent rear suspension to set track. Another rough means of checking track uses a long, straight board or similar object placed against the side of the tire. The distance is measured to a frame or body point at an equal distance forward on each side of the vehicle. Again, the measured readings are compared. The best way to measure track is to use a four-wheel alignment machine. In all cases, track is correct when the thrust line falls on the vehicle centerline.

Rear tires have the least wear when the rear wheels have a little positive camber and there is slight toe-out on rear-drive cars. Power driving the wheels produces toe-in. The normal tendency of the vehicle weight is to reduce camber. Working together, these conditions make the running toe and camber near zero. During braking, toe-out is produced and this tends to make the vehicle less stable. To compensate for this, some vehicles specify toe-in for the static setting. Slight toe-in is normally set on independent rear suspension with front-wheel drive. The vehicle pulling the rear wheels tends to make them move backward slightly so they run with zero toe.

If all rear-wheel angles are within specifications, rear tire wear should be normal. It is relatively uncommon to find excessive tire wear produced by incorrect rear-wheel angles on vehicles with dependent rear suspensions.

9-9 PROBLEM DIAGNOSIS

After alignment, it is possible that the operator may notice problems that appear to be caused by improper wheel alignment. It is possible that alignment settings were made incorrectly and are causing the problem. If so, these should be corrected.

Directional Drift. When a vehicle has directional drift, the camber and caster should be rechecked. If camber and caster are within specifications, the tires and wheels should be re-inspected to make sure they match.

Tire wear caused by incorrect camber will have unequal tread wear across the tread width. After camber adjustment, this unequal wear will move the effective tread centerline away from the center of the tire and this will change the effective scrub radius. With an unequal scrub radius on each tire, the wheel will pull the vehicle in one direction while operating on a level road surface. This pull will be exaggerated during braking.

Drift from wheel pull can be caused by problems within a belted tire. If a belt shifted during tire construction, the stiffness across the tire tread width is not symmetric. This becomes critical with radial tires. The unsymmetric loads in the tire footprint will create the same effect as incorrect camber settings. If this is suspected, it can be checked by replacing the tire you suspect is faulty with the spare tire and road testing the vehicle. It can also be checked by changing the tires, left to right, on the front of the vehicle and road testing. If the pull changes to the opposite direction, replace the faulty tire.

Directional drift can result from a problem within the steering gear or steering linkage. It can also be caused by an upper spring support bearing on MacPherson strut vehicles. If the bearing binds, the spring will wind up slightly, causing steering pull. A preloaded rubber-bonded socket (RBS) tie rod end, introduced by Ford on the 1983 Ranger pickup, or a malfunctioning power steering gear will act to steer the vehicle to one side. Be sure you tighten rubber bonded joints while they are in the neutral position.

Wheel Pull While Braking. Pull during braking may be the fault of either the front suspension or vehicle brakes. The problem must first be isolated. The brakes can be inspected after all visual checks of the tires, wheels, and suspension have been made. Pull during braking can be caused by a different scrub radius on each side. The scrub radius will not be the same when different wheels are used on each side of the vehicle. Caster can cause wheel pull during braking, even when the caster is within specifications. This can happen when the caster on one wheel is positive and the other is negative. To minimize this problem, manufacturers recommend a maximum cross-caster specification, usually 1/2 degree. While setting alignment, it is always a good policy to avoid going across zero on caster (+ 1/4 degree on one wheel and − 1/4 degree on the other wheel, for example) even though all adjustments are within specifications.

Wander. *Wander* is the tendency of a vehicle to be directionally unstable. The driver is not sure which way it will drift. The first thing to check when a vehicle wanders is the tire pressure. If this does not correct the wander, recheck the wheel alignment angles. Pay particular attention to toe and caster settings. The vehicle can wander when it has excessive caster and when the wheels toe-out while operating under dynamic loads. Center-link nonparallelism may also cause wander, especially when the vehicle is driven over an uneven road surface. The linkage is not parallel when the heights of the inner tie rod sockets are unequal. This problem will exist if the steering gear, idler arm, or rack and pinion steering gear housing are out of position. If the distance between the linkage pivot point differs from side to side by more than 1/2 inch (13 mm), an adjustment is required. Unequal-length tie rods can cause problems similar to nonparallel steering linkage. This will cause a different toe change, from wheel to wheel, while making a turn. These toe changes will produce steering pull. The steering wheel must be centered or wander will occur due to looseness in the off-center section of the steering gear. This is especially noticeable when the rack and pinion steering gear housing is not mounted securely.

Front-Wheel Shimmy. Front-wheel shimmy is often blamed on front-wheel misalignment. Common factors contributing to shimmy include a faulty tire, loose steering linkage, incorrectly adjusted wheel bearings, weak shock absorbers, and, possibly, wheel alignment. Most shimmy problems are related to tires and wheels that are either out of balance or out of round. Shock absorbers, if excessively worn, may not control wheel motions and this can allow wheel tramp which will shake the front of the vehicle. Excessive caster under certain conditions will allow the wheels to shimmy as it applies a strong inward turning force. If caster is not excessive, it should be ruled out as a possible cause of shimmy.

STUDY QUESTIONS

Section 9-1

1. What loads must be in the vehicle before it is aligned?
2. List the things to be checked before aligning wheels.
3. Why is wheel bearing adjustment critical to wheel alignment?

Section 9-2

4. What methods are commonly used to measure camber?
5. What is used to measure turning angle?
6. Why is the vehicle above the work area while checking alignment?

Section 9-3

7. Why should track be corrected before the front wheels are aligned?
8. What is the first wheel alignment angle to be measured?
9. How can you tell if the wheel has positive camber?
10. How is caster measured?
11. Why is caster only measured on the front wheels?
12. What is the included angle?
13. Define SAI.
14. Why is a brake depressor used when measuring SAI?
15. Why is it necessary to check SAI?
16. Toe-out on turns is not adjustable. Why is it necessary to measure it while checking wheel alignment?

Section 9-4

17. What alignment angles can often be corrected by making one adjustment?
18. What type of adjustment requires the use of special tools?
19. What should be done before making an adjustment with shims?
20. What alignment angle is adjusted with a strut bar?

Section 9-5

21. Where can alignment adjustments be made on MacPherson struts?
22. If caster or camber need to be adjusted on a vehicle with MacPherson struts, what should you also check?

Section 9-6

23. What two methods are used to align twin I-beam suspensions?

Section 9-7

24. What must be done to center the steering wheel?
25. What is the last adjustment made when aligning wheels?
26. After alignment is completed, what should be done before moving the vehicle?

Section 9-8

27. What results from incorrect rear-wheel toe?

28. What results from incorrect rear-wheel camber?

29. Why is it necessary to convert from inches to degrees when checking rear-wheel alignment?

30. What parts must align when the vehicle has normal track?

31. Why isn't the rear suspension set with zero camber and zero toe?

Section 9-9

32. List three causes of directional drift.

33. If the brakes are normal, what can cause wheel pull while braking?

34. What faults are common to wander and shimmy?

10

Front Suspension Service

The primary purpose of the steering and suspension system is to provide maximum vehicle control under all operating conditions. Suspension system parts wear gradually and wear will cause a slow change in the vehicle ride quality, handling, and steering response. Because this change is slow, the operator of the vehicle will usually become accustomed to the change and accept it as normal.

As an automotive service technician, you must be able to recognize a vehicle's abnormal handling characteristics. The general area of the suspension that is causing the operator's complaint can usually be identified during a road test. You should be able to experience the complaint described. Often, the problem will be reflected as excessive or abnormal tire wear. You must be able to relate tire wear to suspension and steering angles or to faulty parts that cause this type of tire wear.

Suspension parts require service or replacement when their wear makes the vehicle either unsafe or difficult to handle. Rough road conditions, neglected maintenance, high mileage, and age cause parts to wear. These things will cause the springs to sag, control arm bushings to deteriorate, and ball joints to loosen. Wear of any of these parts will cause changes in the wheel alignment.

A thorough inspection done in an orderly sequence, so you do not to overlook any part, is important to a complete and satisfactory suspension repair. The vehicle can be operated safely when careful inspection and correct repair procedures are properly completed.

10-1 INSPECTION OF THE SUSPENSION

A visual inspection to identify worn parts should always be done before doing any alignment. Raise the vehicle by placing it on a lift or support it, both front and rear, with safety stands. This will allow you to examine the suspension and steering system parts. It is important to check the condition of the rear suspension as well as the front suspension because the condition of the rear suspension also affects vehicle handling. Details of rear suspension service are covered in Chapter 11.

Inspection Procedure. During the visual inspection, check for broken, bent, or worn components. You can usually see faulty suspension parts from beneath the vehicle. Look for broken springs, worn bushings, broken or leaking shock absorbers, stabilizer bar link or bushing failure, and bent control arms or struts. The steering system inspection should include a check for loss of fluid from the power steering gear and hoses. It also includes checking for Pitman arm damage or wear, idler arm wear, and tie rod or linkage damage or wear. With rack and pinion steering, the rack mountings should be carefully inspected and the rack moved through its travel to make sure it is not bent. You can locate any excess play as you watch the linkage while shaking the wheels vigorously from side to side. Excess play may also show up as you grasp each part and attempt to move it in both vertical and horizontal directions. Tie rod ends and other steering linkage pivot points should have zero clearance with no play. There should be no wheel bearing looseness while shaking the wheel.

A thorough inspection of the tires and wheels is part of the visual inspection while the vehicle is raised from the floor. Check the tire and wheel assembly for tire tread runout. While making these inspections, listen for any abnormal noises coming from the bearings.

The vehicle is lowered to the floor surface for further checks. These checks include the front suspension upper control arms and bushing, ball joints, vehicle ride height, and any suspension noises or binding, both as the vehicle is moved up and down and as the steering wheel is turned. You can find an excessively worn or internally damaged shock absorber while the vehicle is cycled through jounce and rebound. A good shock absorber should stop the up-and-down motion of the vehicle body in approximately two cycles. It should be noted that this is not a positive test of the shock absorber condition. Any questionable shock absorber should be removed from one end or completely removed and rapidly extended through its full travel by hand to determine its effectiveness.

The condition of the upper control arm shaft bushing can be seen through the engine compartment after raising the hood (see Figure 2-13). On MacPherson strut suspensions, the upper spring seats should be inspected for free rotation as the wheels are steered. The front and rear ride height of the vehicle can now be measured. Typical measurement points (Figure 10-1) may be referenced between the inner and outer control arm pivots and the floor for the front suspension, and between the axle housing and frame for the rear suspension (Figure 10-2). The ride height of other vehicles is measured from specific body points to the ground (see Figure 9-2). Check an applicable shop manual for the specific measurement points and dimensions for the vehicle being measured. At the correct ride height, the lower control arm is usually slightly higher on the inside end than on the outside end. The ride-height measurement, when compared with specifications, will allow you to determine if the springs need to be replaced. Small differences in ride height are expected over a period of time. Good judgment should be used to determine whether spring replacement is necessary. Some older rear drive Chrysler vehicles use adjustable torsion bars. With these vehicles, torsion bars are adjusted to set the front ride height before alignment begins.

FIGURE 10-1 Typical ride-height measuring points on a front suspension.

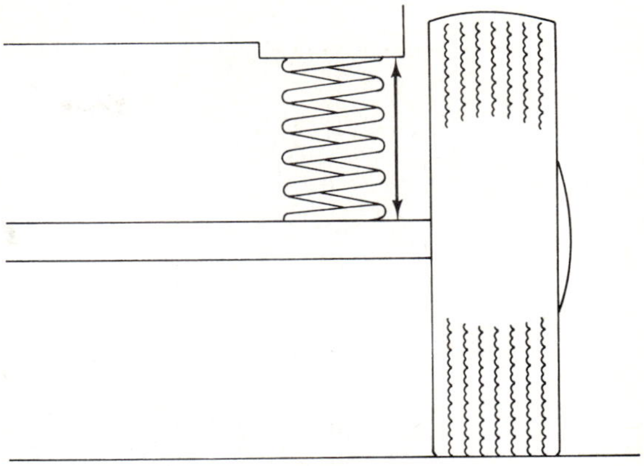

FIGURE 10-2 Typical ride-height measuring points on a dependent rear suspension.

Ball Joint Inspection. The condition of the ball joints must be checked during the inspection of the front suspension. Each ball joint is inspected for lateral (sideways) and axial (up-and-down) movement. The vehicle must be raised, one side at a time, to relieve all vehicle weight and spring force from the load-carrying ball joint. You can determine where to jack by noting whether the upper or lower ball joint is the vehicle-loaded joint. The *loaded ball joint* is the joint on the control arm to which the spring is attached. On MacPherson strut systems, the spring force acts directly through the lower part of the strut and spindle. The ball joint is a preloaded ball joint; it should have no free play.

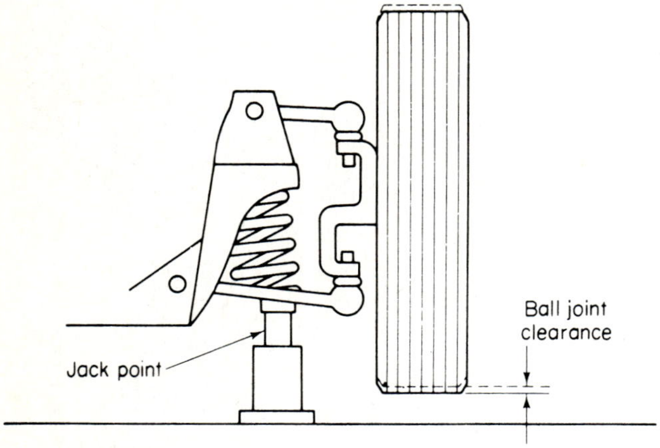

FIGURE 10-3 Jacking point for checking the clearance of a lower loaded ball joint.

If the lower ball joint is the loaded ball joint, the jack should be placed under the lower control arm as shown in Figure 10-3. When the upper ball joint is the loaded ball joint, a spacer block is placed between the upper arm and frame; then the jack is placed under the vehicle frame. This can be seen in Figure 10-4. This block will hold the force of the spring and will keep the upper control arm in its normal position to ensure a true ball joint clearance measurement.

With a MacPherson-strut-type front suspension, the jack is placed under the frame (Figure 10-5). This allows the strut to fully extend to take the load off the joint. Full extension of the MacPherson strut holds the force of the spring. This takes all of the load off the ball joint. On the modified strut system (Figure 10-6), the spring acts on the lower control arm and, therefore, this system requires the jack to be placed under the lower control arm. If you have any doubt about which method of inspection to use, consult the applicable shop manual. With the vehicle raised, a dial indicator or ball joint wear indicator

FIGURE 10-4 Jacking point for checking the clearance of an upper loaded ball joint.

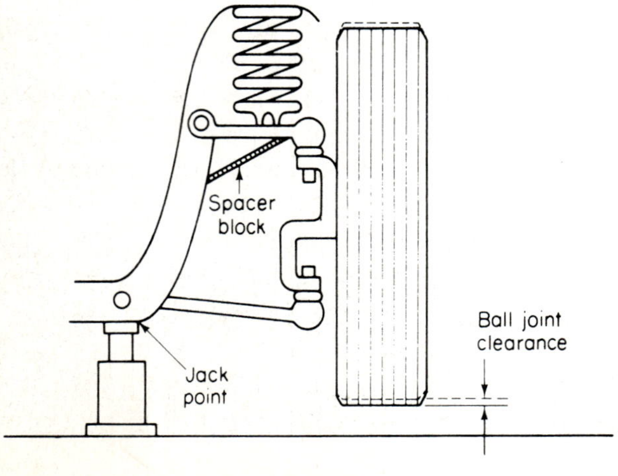

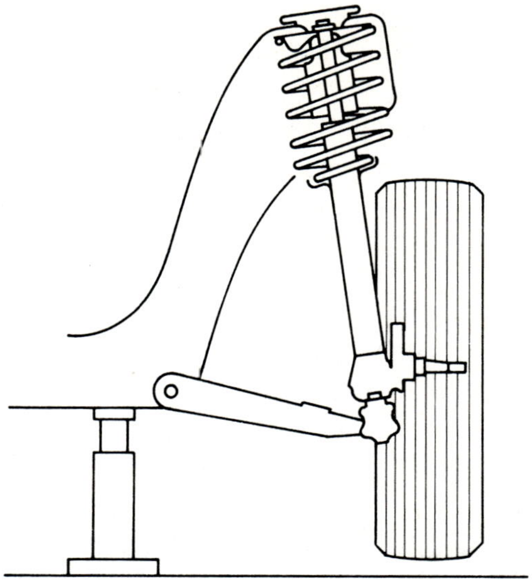

FIGURE 10-5 Jacking point for checking the clearance of a lower ball joint used with a MacPherson strut.

is used to determine if the ball joints are satisfactory for continued service.

A ball joint wear indicator makes use of a dial indicator. The dial indicator is fastened to the control arm with the indicator anvil positioned against the steering knuckle, as shown in Figure 10-7, to measure any movement between the two parts. A bar is placed beneath the tire of the suspension being checked. The bar is lifted and moved sideways

FIGURE 10-6 Jacking point for checking the clearance of a lower ball joint used with a modified MacPherson strut.

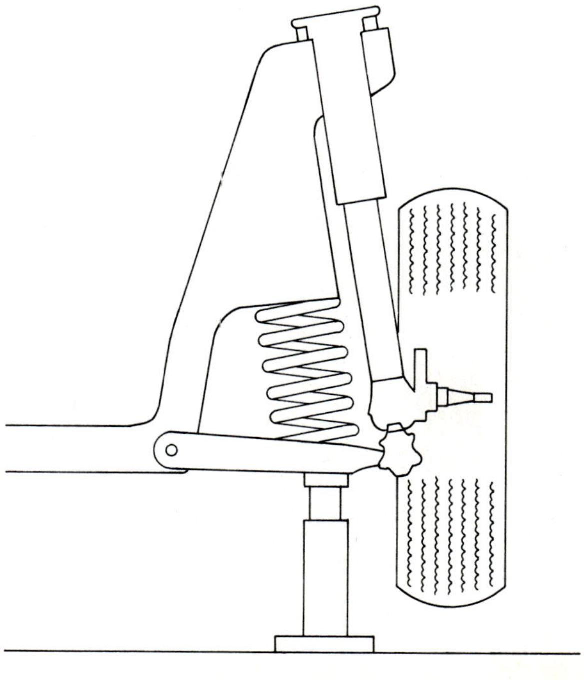

as readings are taken to check maximum lateral and axial movement of the joint. The friction-loaded or preloaded ball joint should have zero lash while the vehicle-loaded ball joint should have a specified clearance. You will need to check the specification chart to determine the normal ball joint clearance. A ball joint with clearance greater than that specified should be replaced so you can properly set alignment angles and minimize the chance of premature failure.

One manufacturer's procedure for checking the friction-loaded ball joint includes separating the joint from the steering knuckle. The ball joint is then rotated in its socket with a torque wrench and the amount of torque noted. If the torque falls below a minimum specified value, the ball joint should be replaced. If a faulty ball joint is kept in service, it could fail like the ball joints pictured in Figure 10-8.

Many manufacturers use a ball joint wear indicator incorporated into the ball joint. This provides a quick and simple way to inspect the joint while the vehicle is raised. Special jacking or measurement are not required. A visual inspection of the protrusion of the grease-fitting nipple boss will indicate excessive wear. The nipple recedes into the ball joint housing as the joint wears. The ball joint should be replaced when the nipple boss is recessed below the lower portion of the housing. Figure 10-9 shows an external view of the protruding boss. A section view of the ball joint was illustrated in Figure 6-37(a).

Control arm bushings are inspected for looseness while the vehicle is jacked for the ball joint inspection. The load-carrying control arm will have to be supported to relieve all loads so you can check the bushings for excessive wear and deterioration.

FIGURE 10-7 Indicator set up to measure ball joint wear.

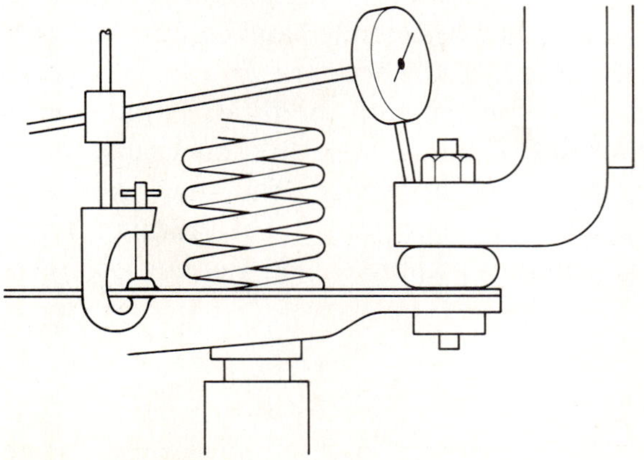

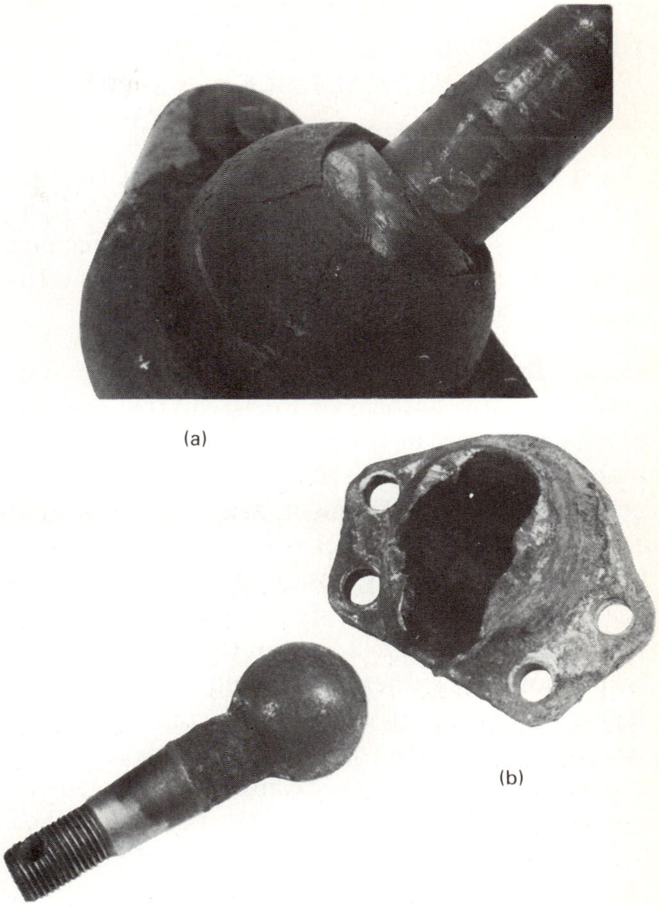

(a)

(b)

FIGURE 10-8 Failed ball joints: (a) cracked and (b) separated.

FIGURE 10-9 Grease fitting boss indicating the amount of ball joint wear.

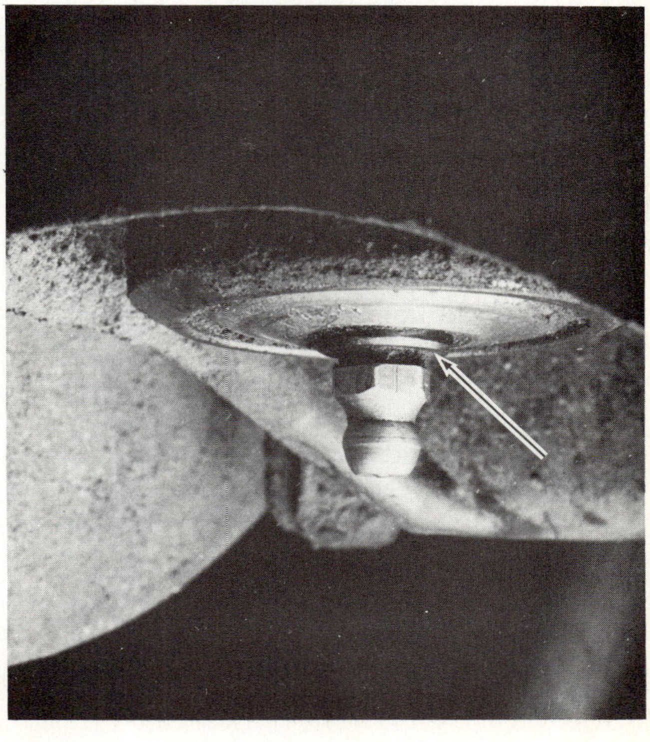

10-2 UNEQUAL-LENGTH CONTROL ARM SUSPENSION SERVICE

When inspection of the suspension and steering system indicates that a repair is required, an orderly procedure should be followed. A systematic approach to repair will save you, the service technician, time and the vehicle owner money.

Shock Absorbers. The shock absorbers are removed before any front suspension work requiring removal of the coil springs or control arms can be done. This is necessary because the shock absorber is connected between the vehicle frame or body and one of the control arms. Front shock absorbers on unequal-length control arm suspensions are usually located inside the coil spring.

Shock absorbers are one of the more common suspension replacement items and they are often replaced even though they are still working properly. Two of the more common, although often unnecessary, reasons for shock absorber replacement are slight fluid seepage around the ram and seal or noise produced during jounce and rebound. Slight fluid seepage is a normal condition and necessary in shock absorber operation to lubricate the seal and shaft. Noise may be produced by aeration of the fluid within the shock absorber. This air can be purged from the fluid. When replacement is required, shock absorbers should always be replaced in pairs to provide uniform suspension control.

Shock absorbers are removed in much the same way for all current-model passenger vehicles. One exception is the MacPherson strut suspension that combines the strut, shock absorber, and spring into one unit. MacPherson strut suspension service is covered in Section 10-3.

Servicing Conventional Shock Absorbers. The front shock absorber is connected between one of the control arms and the frame or body. You can reach the top shock mount from within the engine compartment. You can loosen the lower shock absorber mount on the suspension arm from under the vehicle. Full shock absorber extension limits suspension rebound on some vehicles. The suspension on these vehicles is jacked to support the spring force before the shock absorber is disconnected.

The upper stud end shock absorber mount is the most difficult to disconnect. The spike has small flats to hold with a wrench or a special tool as the nut is loosened with a second wrench. The nuts usually rust to the stud in the "rust belt" area of the United States. This makes removing the nut very difficult. There is not enough surface on the stud flats to hold much torque when trying to loosen the nut. Sometimes, penetrating oil will loosen the rust. The rusted nut can be split with a "nut cracker" to remove it from the stud. Both of these procedures will save the shock absorber, but they take time. Technicians seldom remove a shock absorber just to check it. It is removed because it is faulty. Often, the technician will slip a piece of close-fitting pipe over the stud above the nut. The stud is bent back and forth until it breaks off, then the lower end of the shock absorber is unbolted and the shock absorber is removed.

The vehicle is raised to a good working height and supported with safety stands or a hoist so you can work on to the suspension parts. If suspension fasteners are corroded, spray penetrating oil on the threads of all fasteners requiring removal. Due to the potential danger of releasing a coil spring from its seat and harming you or a fellow employee, always use safe procedures. For example, place a jack under the lower control arm to limit the suspension travel while the shock absorber is being removed. (*Note:* The coil spring is highly preloaded. Be sure to follow all precautions to prevent its uncontrolled release.) The fasteners holding the upper and lower shock end mountings are disconnected and the shock absorber is removed from the vehicle as shown in Figure 10-10. On some vehicles, it is necessary to remove a plate or bracket along with the shock absorber as it is removed from the vehicle. The plate is then removed from the shock absorber.

If the shock absorber has been removed for a bench test, it should be carefully mounted in a vise in the same direction it was mounted on the vehicle. It can then be extended and compressed and its action noted. If the action is erratic, it is possible that air has mixed with the fluid. Purging will get the air out of the fluid so the shock absorber can be placed back in service. The shock absorber is purged by fully extending it, then turning it upside down and fully compressing it. Purging is illustrated

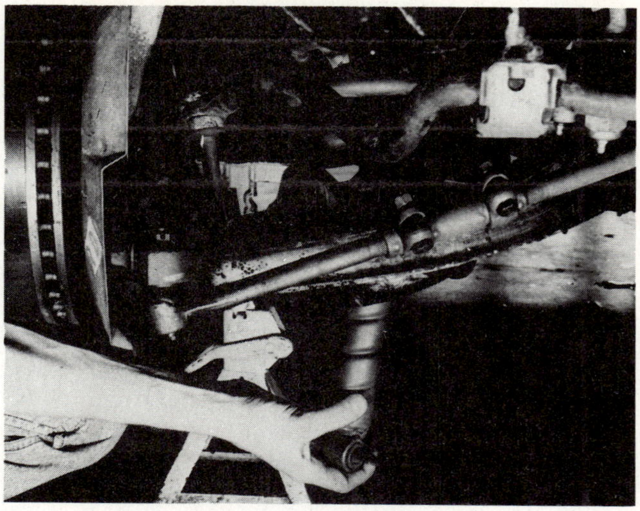

FIGURE 10-10 Front shock absorber being removed through the lower control arm.

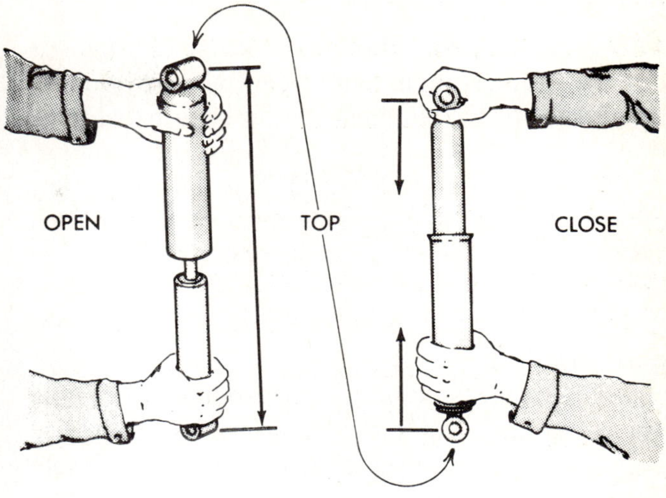

FIGURE 10-11 Purging air from a shock absorber. (Courtesy of Chrysler Corporation)

in Figure 10-11. This cycle is repeated at least three times. Replace the shock absorber in its normal position in the vise and extend and compress it as fast as possible. If the operation of the shock absorber is still erratic, the shock absorber should be replaced. Purging and bushing replacement are the only service repairs made on conventional shock absorbers. The shock absorbers must be replaced to correct any other type of abnormal operation.

Shock absorbers may have a spike end, on which one bushing is used on each side of the mounting hole. Shock absorbers may fit over a stud with a bushing fitted on each side of the shock eye. Sometimes, a single rubber bushing is molded in the eye. It may have a flanged end with a rubber mount made as a part of the shock absorber. These mounting styles, shown in Figure 10-12, are made to keep the road vibration and noise from being transmitted into the vehicle frame. For this reason, it is important that the shock absorber be mounted with insulators in such a way that there is no metal-to-metal contact between the shock absorber and its mount.

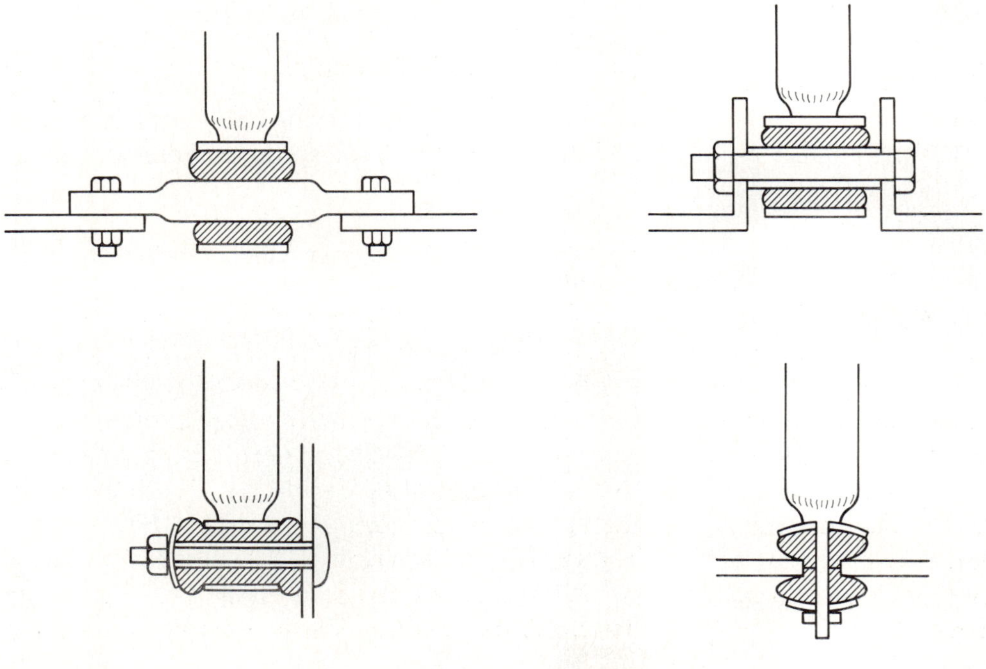

FIGURE 10-12 Typical shock absorber mountings.

Ball Joint Removal. Replacement is necessary whenever the ball joint has excessive wear or damage. Wheel alignment should always be done after replacing the ball joints because the actual position of the joints determines the alignment angles.

Ball joint removal begins by jacking the front of the vehicle and supporting the vehicle frame on safety stands. The stands should be positioned behind the front wheels and suspension members so you have full access to the suspension parts that require removal.

One wheel stud and a corresponding lug hole are marked as a guide for replacement and the tire and wheel assembly is removed. The disc brake caliper and rotor normally require removal to allow access to the bolts holding the ball joint to the steering knuckle. The removal procedures discussed in Section 3-4 should be followed.

Separating the Ball Joint from the Steering Knuckle. Bolts or rivets hold the lower ball joint to the steering knuckle. You can reach them after the brakes are removed. The steering arm is part of the lower ball joint on some vehicles. On these vehicles, the tie rod end must be separated from the steering arm. To do this, first remove the cotter pin and loosen the castellated nut. The tie rod end is then separated from its socket using a tie rod spreader (Figure 10-13) or by hitting the tie rod socket of the steering arm with a hammer (see Figure 12-2). Once the tie rod stud has been separated from its socket, the castellated nut is removed and the tie rod is lowered.

FIGURE 10-13 Removing a tie rod end with a tie rod spreader.

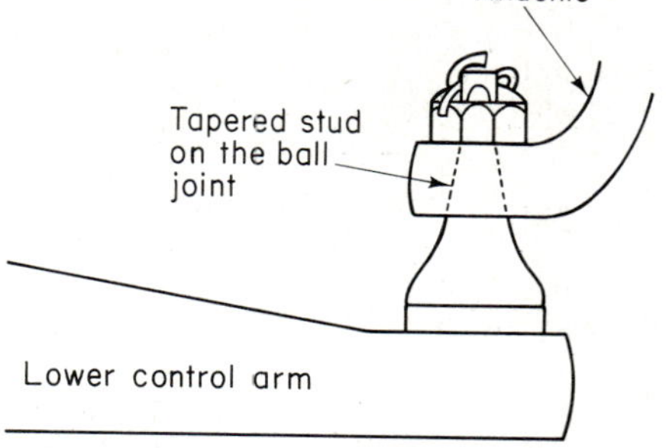

FIGURE 10-14 Tapered ball joint stud fits into a tapered hole in the steering knuckle.

The nut that holds the ball joint stud to the steering knuckle is loosened. If both upper and lower ball joints are being removed, both nuts may be loosened at this time. *Caution:* Do not remove the nuts completely. If the nuts were removed completely, the ball joint and steering knuckle might suddenly separate. The sudden release of the spring force and the control arm is likely to cause injury. The ball joint fastens to the control arm. The ball joint stud has a taper that fits into the steering knuckle, as shown in Figure 10-14. You saw this stud in Figure 6-37. As an alternative design, the ball joint is made part of the steering arm and the ball joint stud fits into a bracket in the lower control arm (Figure 10-15). Also see Figure 12-9(c).

To separate the ball joint from the steering knuckle, the suspension load must be off the ball joint. Use a jack to support the spring force as you did to check the load-carrying ball joint (see Figure 10-3 through 10-6). A coil spring compressor can be

FIGURE 10-15 Ball joint is part of the steering arm.

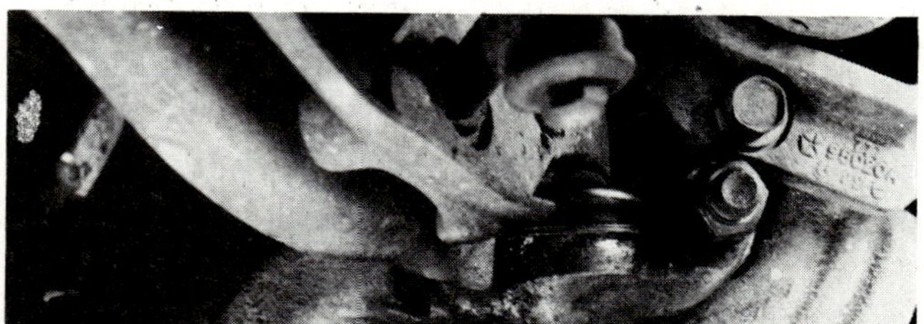

FIGURE 10-16 Spring compressing tool in place to hold the spring as the ball joint is separated.

used to hold the spring force. The extent of the repairs determines whether to use the spring compressor or the jack. If control arms are to be removed, it is easier to use the spring compressor and remove the spring. Figure 10-16 shows one type of spring compressor. If repairs are limited to ball joint replacement, the jack can be used. It should be positioned toward the outside of the vehicle under the lower control arm, as shown in Figure 10-17. The jack is raised until the rebound snubber clears the frame. This releases all loads from the ball joints. Remember, the vehicle must be supported on safety stands.

Various methods are used to separate the ball joint stud from its socket. A ball joint tool that applies a force between the upper and lower ball joint studs may be used to separate them. This is pictured in Figure 10-18. If the ball joint is not to be used again, a ball joint spreader, often called a *pickle fork* by mechanics, may be used. It is shown in Figure 10-19. The spreader will usually damage the seal and, for this reason, this tool is not recommended if the ball joint is to be reused. Another method, one that must be used with care, is to tap the ball joint stud seat with a hammer while backing it with another hammer, as illustrated in Figure 10-20. With the ball joint loosened, the tapered socket and the spring forces work together to cause the ball joint stud to separate from the knuckle. Never use heat to free ball joints; it will change the heat treating of the metal to weaken the parts.

FIGURE 10-17 Jack position to lower the suspension as the ball joint is separated. The vehicle frame is supported on safety stands.

FIGURE 10-18 Tool used to push the ball joints out of the knuckle. (Courtesy of Chrysler Corporation)

FIGURE 10-19 Separating a ball joint from the knuckle with a ball joint spreader.

The knuckle must be supported to prevent damage to the flexible brake line, steering linkage, or brake assembly. Some vehicles may require the separation of the brake hose from the brake line to allow the suspension to move far enough to provide clearance to work.

With the steering knuckle assembly out of the way, the ball joints may be removed. Ball joint removal is most often done by removing rivets that attach the joint to the control arm. A few lower ball joints are bolted to the steering knuckle and the ball stud fits into a socket in the lower control arm. Other ball joints are a part of, riveted to, or press fit into the control arm. The control arm and ball joint of the nonseparable type are replaced as a unit. Check the applicable shop manual if you have any question about the type you are servicing.

FIGURE 10-20 Loosening a ball joint by hitting the knuckle with a hammer while it is backed up with a second hammer. This distorts the hole shape enough to release the ball joint stud.

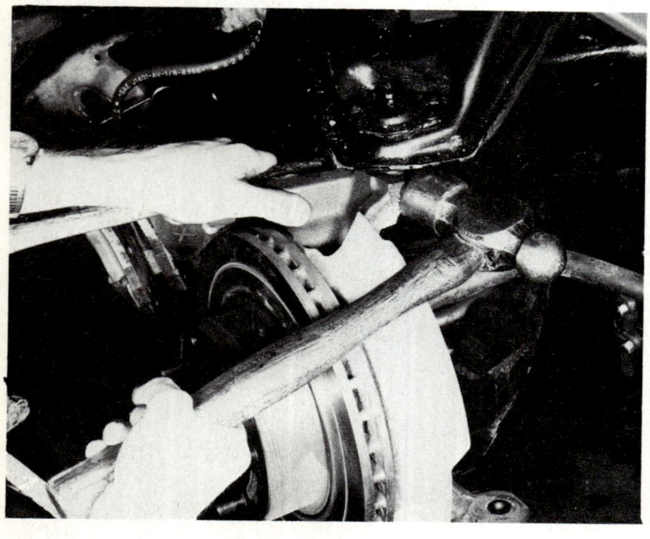

When the riveted type is used, the rivet head is chiseled off or the rivet is drilled out. Care must be taken when drilling or chiseling to avoid damage to the control arm. Replacement ball joints use bolts to hold the joint to the control arm. The bolts provide a solid and secure mounting for the service replacement joint.

When reassembling the suspension, care should be taken to prevent damage to the boot seals on the new ball joints. They are needed to retain grease. After assembly, suspension bolts and nuts are installed. They are torqued to specifications with a torque wrench after weight is on the suspension. The correct torque is important to properly seat the tapered stud used on most ball joints. All cotter pins that have been removed must be replaced with new ones.

Control Arms and Bushing. Control arms and A-frames are removed in much the same way as ball joints. The disassembly procedure for ball joint removal is followed, up to the point of removing the ball joint from the arm. The ball joint does not have to be removed from the control arm when servicing bushings.

Upper Control Arm Removal. If the unloaded upper control arm is the only suspension member to be removed from the vehicle, the spring-loaded lower control arm can be supported on a jack. It is not necessary to remove the coil spring or the torsion bar to service the unloaded arm. The required disassembly would look like that shown in Figure 10-21. Conversely, on vehicles using a spring force on the upper control arm, the coil spring does not have to be removed if the lower control arm is the only suspension member that requires service.

Spring Removal on Long-Short Suspensions. The coil spring or torsion bar must be removed to service the spring-loaded lower control arm. First, the shock absorber is removed and the stabilizer bar is disconnected from the control arm. If the arm is to be removed, the coil spring is compressed using a

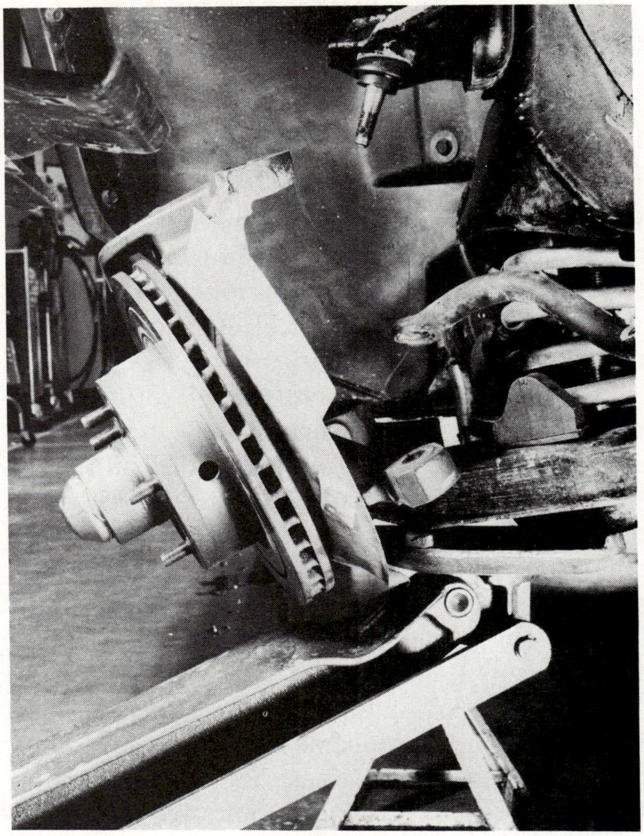

FIGURE 10-21 Knuckle removed from the ball joints with the spring and control arms still attached to the frame.

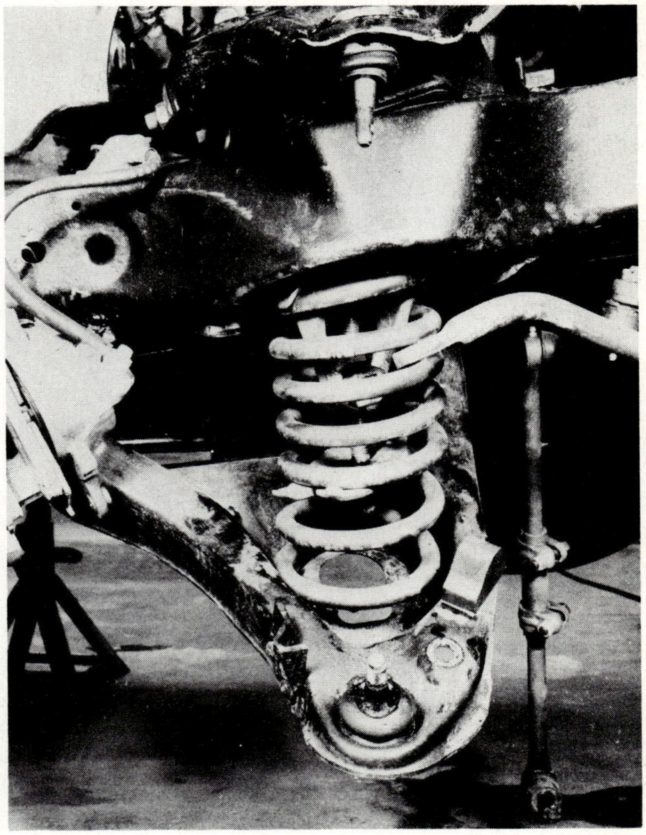

FIGURE 10-22 Spring held with a spring compressor so the lower control arm can swing clear.

spring compressor. The spring compressor fits up through the coil spring and attaches to the highest and lowest accessible coils of the spring. The center bolt of the compressor is tightened. This will shorten the effective distance between the attaching points to compress the spring. Figure 10-22 shows a spring being held with a compressor ready to be removed.

Vehicles using spring force on the upper control arm must have the spring removed before the spring-loaded upper control arm is removed. The spring is compressed and removed in the same way using the spring compressing tool.

Control Arm Removal. The lower control arm is disassembled by first removing the brake strut. It may be attached to the frame with a pivot held by two or more bolts. If the bolts attaching either upper or lower control arms to the frame are equipped with eccentric adjustments for camber, the adjuster position should be marked to speed alignment after reassembly. If shims are used, they are taped together as a shim pack and marked to prevent loss or mixing with shims from the opposite side. Marking a slot adjustment will also aid realignment during assembly.

Control arms are connected to a cross-shaft inserted through the bushings. The shaft is bolted to the frame (see Figure 9-16). When all of the load is off the control arm, the bolts holding the shaft to the frame are removed to free the control arm and shaft assembly.

Bushing Replacement. Once the arms and shaft assembly are removed, the bushings in the control arms are easily replaced. Worn bushings are pressed from the control arm (Figure 10-23). The complete bushing is removed. Care should be taken to prevent damage to the control arm during removal. Control arms using a cross-shaft are serviced with either new bushings or an entire bushing and cross-shaft assembly.

It is important that the bushing be neutralized

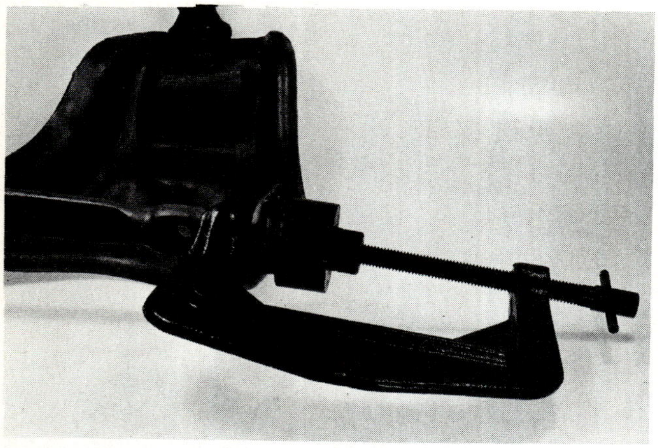

FIGURE 10-23 Clamp-type tool used with adapters to remove and install control arm bushings.

during assembly. The purpose of neutralizing the bushing is to allow the bushing rubber to be relaxed when the vehicle is standing with normal weight on the suspension. It will twist within its natural limits during operation. If the bushing is installed and tightened with no reference to the control arm position, the bushing will have to be twisted to connect the control arm to the ball joint. This is called *preloading a bushing*. In operation, the bushing will exceed its elastic limit and rupture. Once this occurs, the bushing is destroyed. To prevent shearing the bushings, the control arms are placed on the vehicle in the center of their normal travel. The end bolts that press on the inner sleeves are then tightened. For bench assembly, service manuals will give reference angles that must be used to set the neutral position for optimum service life.

Some suspension systems use asymmetric bushings. This allows a different amount of compliance in each direction. The asymmetric bushing must be installed in a specific direction.

Assembly. The upper and lower control arms are properly positioned on the vehicle. The bolts and any shims that were present on disassembly are reinstalled. The control arms are placed at their normal ride-height position and then the bolts are tightened. Eccentric cams or bolts in slots are positioned as they were before disassembly.

On suspension systems having the coil spring above the upper control arm, the steering knuckle is refastened to both the upper and lower ball joints. The brake strut is then fastened to the lower control arm. The spring is inserted in place and the spring compressor removed. *Note:* The coil end fits into a specific recessed area in the spring seat. A hole in the seat is often used to correctly locate the end of a coil spring.

Vehicles that have the coil spring seat in the lower control arm are assembled in a slightly different way. After the control arms are mounted on the vehicle, the coil spring is positioned in the spring tower. The lower control arm is raised into position with a jack. The steering knuckle is connected to the upper and lower ball joints, making sure the nuts are finger-tight. Make certain the coil spring is properly located in its seat. Again, a hole in the spring seat will help you locate the end of the coil spring. The jack and spring compressor can now be removed. When a spring compressor is not used, the lower control arm is carefully jacked into position as the spring is correctly seated. The steering knuckle can then be connected to the upper and lower ball joints with the nuts finger-tight.

Vehicles using torsion bar suspension are assembled in much the same way. After connecting the ball joints to the steering knuckle, the torsion bars are installed, with the left and right bars placed on the correct side.

The brake struts, when used, are connected to the lower control arms. Jack the suspension so the weight of the vehicle is on the suspension. With the suspension in its normal position, all bolts and nuts used for assembly are correctly torqued. New cotter pins are installed where required. If the tie rod end was disconnected during disassembly, it is reconnected. Make sure the steering wheel is straight ahead, then tighten the nut to the correct torque and install the cotter pin. Shock absorbers and stabilizer bars are installed and their fasteners are secured. *Note:* Care must be used to align the stabilizer and shock absorber bushings in their holes to prevent metal-to-metal contact.

If front wheel bearings require repacking with grease, follow the procedure in Section 3-4. The brake drum or rotor is then installed and the bearings are properly adjusted. With the spindle nut saftied, the dust cover is replaced. The brake caliper is installed on the rotor, then the assembly bolts are put in place and torqued to specifications. The tire and wheel assemblies are installed and the wheel nuts are properly torqued. The suspension pivots are lubricated as required, the safety stands removed, and the vehicle is lowered to the floor to complete the assembly procedure. Finish torquing any bolts or nuts that must be torqued after full vehicle weight is on the suspension.

Wheel alignment should follow assembly to complete the repair. This is most important due to the close relationship of the suspension members to wheel alignment angles.

10-3 MACPHERSON STRUT SUSPENSION SERVICE

MacPherson strut suspension differs from the unequal-length control arm system in a number of ways. You will recall from studying Section 6-5 that it has one pivoting joint on the lower control arm and one sliding linkage (the strut itself). The strut is a rugged shock absorber. The suspension spring is on the strut. The lower spring seat is a part of the strut housing. The upper spring seat is on a bearing plate attached to the strut ram. The strut and spring assembly is removed from the vehicle (Figure 10-24). The spring can then be removed from the strut. There is only one ball joint at the bottom of the strut on each side of the vehicle. The strut upper pivot bearing allows the spring to rotate with the turning strut. The following discussion describes procedures unique to MacPherson strut suspension service.

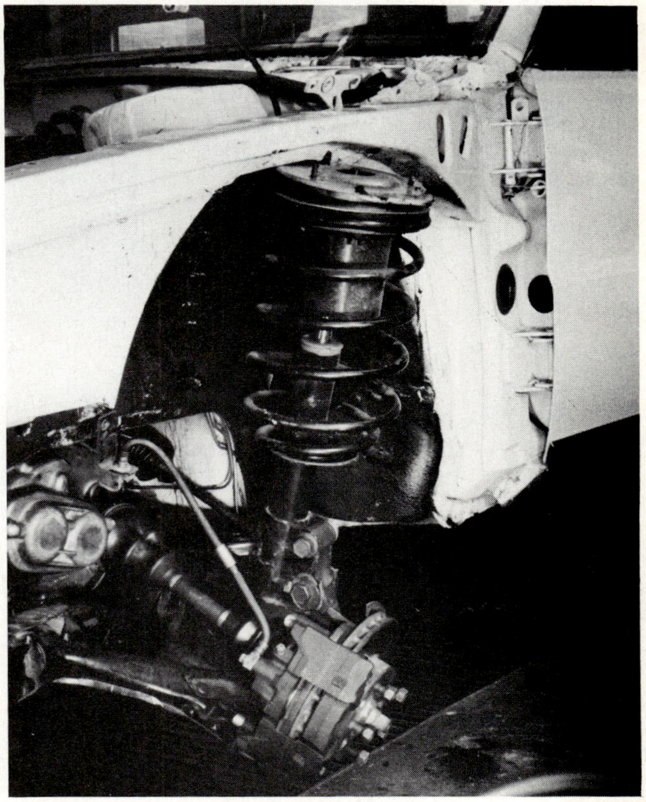

FIGURE 10-24 The MacPherson strut and spring is removed as an assembly.

Servicing MacPherson Strut Shock Absorbers. The replacement of the MacPherson strut shock absorber differs from replacement of conventional shock absorbers. The bottom of the strut assembly connects to the steering knuckle or lower control arm. The upper end connects to the fender housing. The strut can be removed when the vehicle is raised and supported so that the front suspension is fully extended (see Figure 10-5).

A modified MacPherson strut has a load-supporting coil spring between the lower control arm and the frame (see Figure 10-6). Removing this type of strut is done with the vehicle raised and jacked under the lower control arm to hold the spring under compression.

The lower attaching bolts may have a camber adjustment in either type of strut. Mark the adjusting cam position (Figure 10-25) before disassembly to aid in alignment after reassembly. The lower clamp bolts or ball joint can then be freed from the knuckle or lower control arm.

On the upper end of the strut, remove only the attaching plate bolts that connect the plate to the upper fender housing (Figure 10-26). Do not remove the upper shock absorber rod nut. If the rod nut is removed, the coil spring will be released suddenly from the shock absorber. This could cause personal injury and damage the vehicle. MOOG makes tools that can be used to remove the spring and replace the strut cartridges with only the top of the strut removed (see Figure 10-24).

The spring can be easily removed from the strut after the strut has been removed from the vehicle. The coil spring is compressed, as shown in Figure 10-27, using a spring compressor to release the spring load from the strut. Remove the rod nut, upper bearing plate, spring seat, and dust boot; then remove

FIGURE 10-25 Camber adjustment marked before removing the strut.

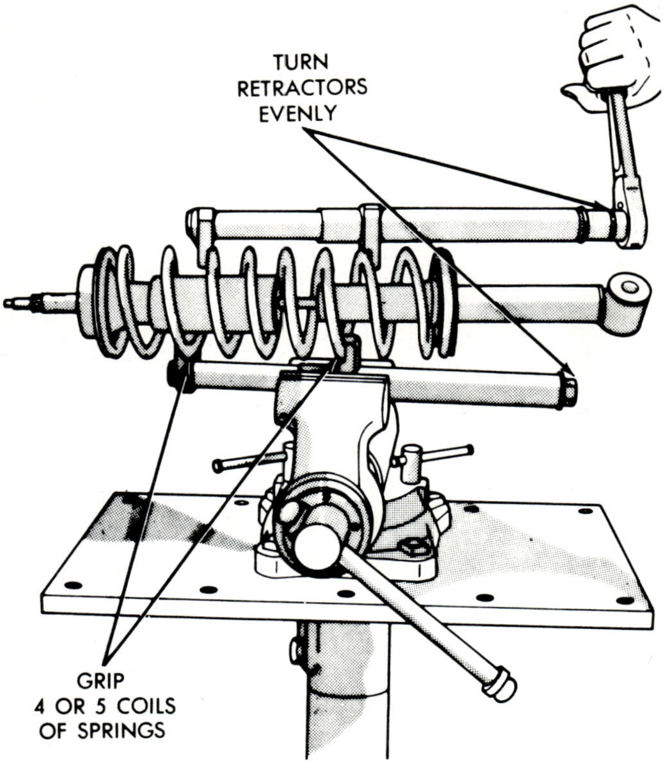

FIGURE 10-27 Coil spring retractor (compressor) used to compress the spring on a MacPherson strut. (Courtesy of Chrysler Corporation)

the spring (Figure 10-28). Keep it compressed for reassembly. After the spring is removed, the shock absorber part of the strut assembly may either be repaired, rebuilt, or replaced, depending on the problem and the strut design.

Repairing and rebuilding a MacPherson strut requires that the strut be completely disassembled.

FIGURE 10-26 Nuts to be removed to disconnect the upper end of a MacPherson strut from the shock tower.

Strut repair kits are available for some types of struts. These kits include new seals, cups, valving, piston rod, and the required fluid. The strut replacement parts are installed following the instructions that apply to that strut. These instructions are packaged in the repair kit. Both parts and the labor required to recondition the strut are charged to the customer. This has led some manufacturers to market replacement cartridges for the strut shock absorber assembly. The strut is disassembled and the cartridge assembly is inserted in the strut in place of the many separate shock absorber parts removed from the strut. The cartridge assembly costs somewhat more than the repair kit, but the labor cost is less. Some of the companies that market replacement shock absorbers produce only complete replacement shock strut assemblies. New replacement shock struts are more costly than the cartridge repair, but the labor cost is the lowest of the three reconditioning methods described.

If the shock strut is to be repaired, the piston rod nut is removed with an adjustable spanner wrench or, on some types, a large pipe wrench. In

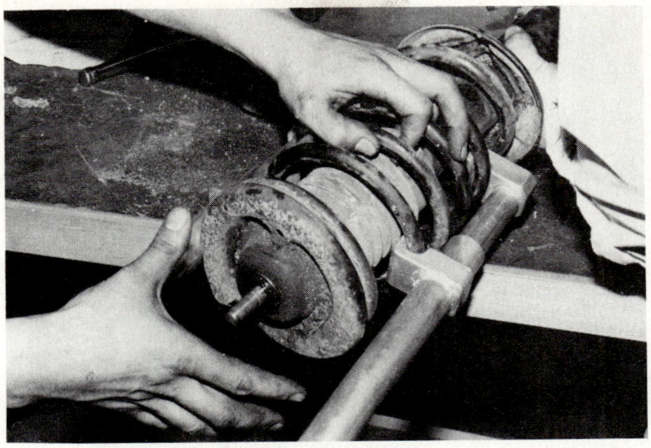

FIGURE 10-28 Removing the upper bearing plate, upper spring seat, coil spring, and dust boot.

practice, it is often removed with an air chisel (Figure 10-29). Turn the strut upside down to remove the fluid and all parts. Follow the instructions packed in the repair kit if the shock absorber is to be rebuilt. Discard all parts if a replacement cartridge is to be used. Slide the new shock absorber cartridge into the strut. Fill the space around the cartridge with shock absorber oil to help transfer heat from the cartridge to the strut body where it can be cooled by air. Fasten it in place with a new piston rod nut (Figure 10-30).

To assemble the shock strut, pull the rod out of the shock absorber as far as possible. This will make it easier to assemble the rest of the strut parts. Make sure the flats on the strut rod align with the upper spring seat. Reassemble the strut by installing the parts in the reverse order of disassembly. Take care to ensure that the spring is properly seated

FIGURE 10-29 Removing the piston rod nut with an air chisel. The nut will be replaced with a new nut.

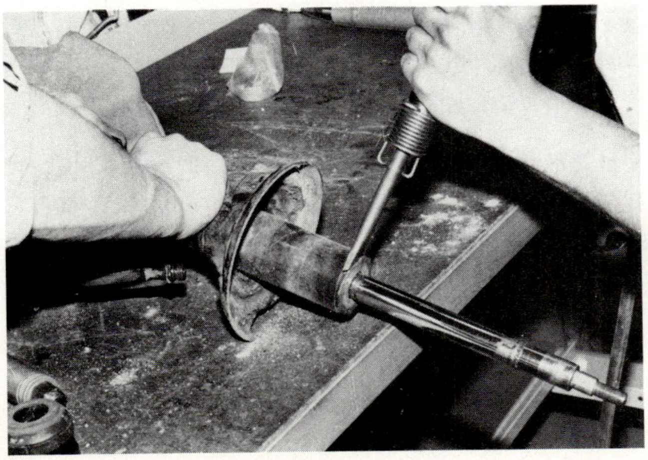

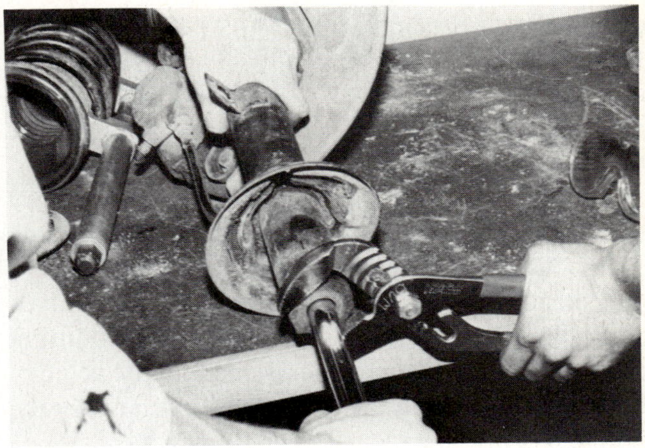

FIGURE 10-30 The new cartridge is fastened in place with a new nut.

as the spring tension is gradually released. The assembled MacPherson strut is ready to be installed on the vehicle. Align any marks you made before removing the strut. Install the strut, reversing the removal procedure. The wheels should be aligned to ensure proper installation and trouble-free operation.

Ball Joint Replacement. Ball joints are relatively easy to remove from a MacPherson strut suspension. First, the vehicle is raised and supported from the frame to allow the suspension to extend to its full length. The wheels are removed as previously discussed in Section 10-2. Brake components are not normally removed to remove the ball joint on a MacPherson strut.

The ball joint stud connection is removed from the knuckle assembly. On this type of suspension, many manufacturers are using a bolt that locks and clamps the stud in position (Figure 10-31). This eliminates the need for a tapered stud as is commonly used in other suspensions. The ball joint is separated by removing the lock bolt and sliding the stud from its mounting hole.

The ball joint mounting to the control arm is similar on all suspensions. It may be riveted, bolted, pressed, or not serviceable as a separate unit but replaced as an assembly with the lower control arm. If it is serviceable, the procedures in Section 10-2 should be followed when replacing the ball joint.

Lower Control Arm Replacement. The lower control arm on MacPherson strut suspensions is removed by following the procedure for separating the ball joint from the knuckle assembly. The lower control arm can be an A-frame shape (Figure 10-32) or of the design using a strut to constrain the fore and aft motions (Figure 10-33).

Support the vehicle frame on a hoist or safety stands. If the lower control arm is of the A-frame design, the front stabilizer bar is disconnected from the control arm. The lower control arm frame pivot points are then disconnected from their mounting points. The arm can now be lifted from the vehicle. On some vehicles, the lower control arm has a brake

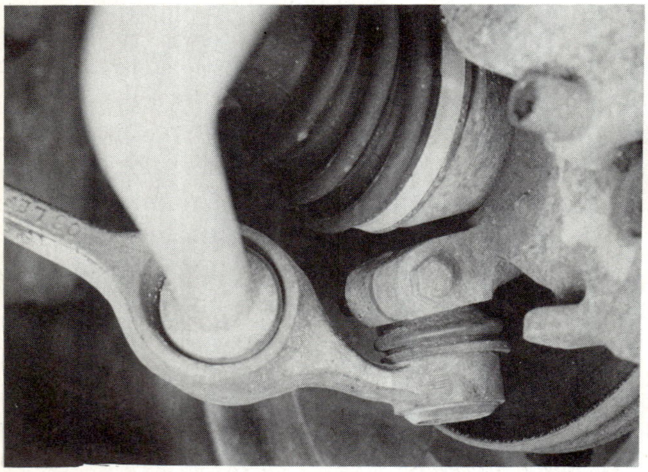

FIGURE 10-31 Bolt clamps the knuckle around the ball joint stud.

FIGURE 10-32 A-frame-type lower control arm used with a MacPherson strut.

FIGURE 10-33 Single-arm lower control arm is kept from moving forward and rearward by the stabilizer bar acting as a brake strut.

strut or uses the stabilizer bar that also functions as a brake strut. The stabilizer bar and lower control arm pivot points are disconnected and the arm is removed.

Control arm bushings are removed as discussed in Section 10-2. The remaining suspension parts are serviced as they would be on an unequal-length suspension system.

10-4 COMMON ELEMENTS FOR ALL TYPES OF SUSPENSION

Service of the stabilizer bar and brake struts are common to all types of suspensions. Control arms are similar; however, as previously discussed, separation of the ball joint may differ. In this section, the suspension-related items that are common to independent suspension systems will be discussed.

Stabilizer Bar Link and Bushing. The most common repairs on the stabilizer bar are link and insulator replacement. The outer link and insulators are used on many vehicles to connect the stabilizer bar to the control arm. They will fail in time when the vehicle is frequently driven over rough road surfaces.

FIGURE 10-34 Removing the stabilizer bar link.

The link insulators will wear and age (see Figure 6-32) and the link itself may break and require service. Occasionally, the stabilizer bar will break or bend in service due to a vehicle accident and so it will have to be replaced.

The outer link is replaced as an assembly including the insulators, retainers, through bolt, and sleeve. To replace the outer link, first remove the through bolt; then remove the sleeve and the four large insulators (Figure 10-34). The link assemblies are normally replaced in pairs so unequal sleeve lengths will not preload the bar.

Stabilizer Bar Replacement. To replace the stabilizer bar, remove the outer connections at the control arm. The connection may either be a through bolt or a clamp fastened on the strut bar near the lower control arm. The stabilizer-bar-to-frame mountings are then disconnected. These mountings may be a bracket that is bolted to the frame holding a bushing around the bar. Other mounts have a link with insulators that holds the bar to the frame. In either case, the frame mounting is disconnected and the bar removed. It is necessary on some vehicles to remove one front tire and wheel assembly to provide clearance for the stabilizer bar to be removed from the vehicle.

Stabilizer Bar Frame Bushings. Stabilizer bar frame bushings can be replaced without removing the complete bar. The frame mounts are disconnected and the brackets removed. The bushings are then removed from the bar. The new bushings can then be installed over the stabilizer bar and the brackets positioned and bolted into place.

Reassembly. The first step in reassembly is to install the bushings on the stabilizer bar. The bar is fitted beneath the vehicle and the mounting brackets are correctly positioned. The assembly bolts should be finger-tight. The outer links are then fastened and their assembly bolts are torqued to specifications. The tire and wheel assembly is installed and properly torqued. The vehicle is lowered so the weight is supported by the wheels. Finally, the frame bracket mounting bolts should be torqued to specifications. This prevents preloading the bushings in an unnatural position. Preloading would shorten the bushing service life. Wheel alignment is not affected by repair of the stabilizer bar.

Strut Bar, Strut Bushing, and Radius Arm. Strut bar, strut bushing, and radius arm replacement require removing the bar or arm from the vehicle. Bushing replacement is the most common strut repair, although a vehicle collision may cause a bent bar or arm, requiring its replacement. These suspension parts are highly stressed in use; therefore, the strut should never be straightened.

The strut bar or radius arm is disconnected from the vehicle frame by removing a nut and washer from the end mounting, as shown in Figure 10-35. On most vehicles, the bolts that fasten the strut to the lower control arm are removed to disconnect the

FIGURE 10-35 Typical location of strut bar suspension bushings.

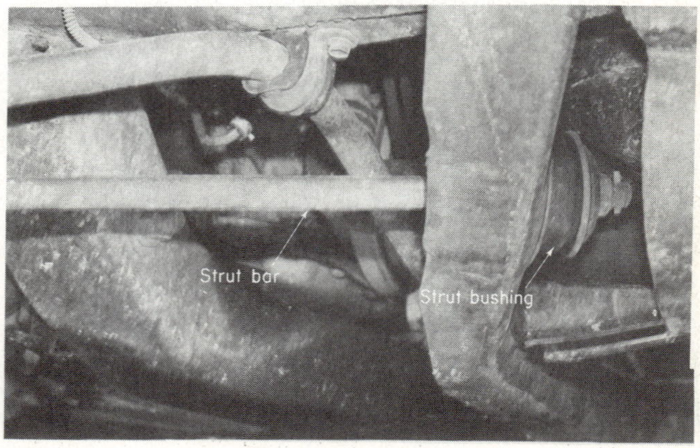

strut. On some vehicles, the lower control arm and strut are removed as an assembly in order to have the clearance necessary to remove the frame end of the bar from the vehicle. The bar is separated from the control arm for service. A radius arm on an I-beam suspension is removed by disconnecting the frame end of the bar and removing the bolt that fastens the radius arm to the axle. The arm can then be lifted from the vehicle.

Bushing designs vary according to manufacturer, although replacement of all bushings is quite similar. Strut bushings are designed to prevent metal-to-metal contact between the bar and frame. The bushing has an internal sleeve to provide the correct amount of compression when installed. A typical worn suspension bushing is pictured in Figure 10-36. The bushing is positioned either on the bar or into the support before the bar is put in place. The fasteners are installed and torqued to specifications. All required cotter pins are installed. After assembly, the vehicle suspension must be aligned to correct any change in camber or caster setting that occurred during reassembly.

Ride-Height Adjustment. The position of the control arm in relation to the vehicle frame has a bearing on front suspension angles. Because of this, the ride height of the vehicle must be measured; it may require adjustment before the wheels are aligned.

Before measuring the ride height, remove any excess cargo from the vehicle. Tire pressures and tire sizes are inspected and changed if incorrect. The fuel tank must be filled for an accurate ride-height measurement. The ride height is measured, following manufacturer's recommendations, and compared to specifications.

FIGURE 10-36 Worn suspension arm bushing.

On vehicles using coil spring front suspension, shims can be placed beneath the coil spring seat to obtain a maximum of one-half-inch (13 mm) change in the ride height. If the suspension height is more than one-half inch out of specification, both coil springs on that end of the vehicle will have to be replaced.

Torsion bars with height adjusters are reset for correct ride height. To adjust height, the vehicle is positioned on a level floor surface. Each adjusting bolt is turned alternately until the specified measurement height is achieved.

Checking Air Springs. The linkages or suspensions with air springs are serviced like suspensions with coil springs. The air spring control in the trunk is turned off or the battery ground cable is disconnected; then the vehicle is supported on a hoist or safety stands. The air must be exhausted so you can remove the air spring bag. The faulty suspension parts can then be removed for repair. The air suspension system can be checked to see that it is working correctly.

The air suspension system will normally re-adjust for 1 hour after the ignition is turned off. The system has a 45-second delay after the load has changed before changing the suspension height. In this way, the suspension does not adjust for momentary jounce and rebound. The suspension system will not adjust while the door is open. This keeps the door from jamming on some object as the passengers get in or out. The system does not adjust when the brake is applied. This keeps the suspension from adjusting for nose-down transfer of weight while braking.

The air spring system should adjust immediately after the doors are closed and the ignition is turned on to correct for the load, even if the brake is on. Up corrections have priority over down corrections. Rear corrections are large so they have priority over front corrections. There is no leveling correction when electricity does not get to the microprocessor, if a leak makes the system try to correct in over 3 minutes, or any time the sensor is out of range. Normal venting is limited to 3 seconds in the front and 10 seconds in the rear.

You can make sure the system is operating correctly by using the following procedure. Open the driver's window for access to the ignition key. With no weight in the vehicle, turn the ignition on. One flash of the failure warning lamp indicates normal circuits. A problem is indicated by any other flashing or lack of a flash. For further checking, connect a charger to the battery. Set the charging rate low to supply the electricity needed by the air pump. Turn the ignition on for 5 minutes then turn it off. Ground the diagnostic wire pigtail that hangs next to the air suspension switch on the left pad in the trunk; then turn the ignition on again while watching the failure warning lamp as you perform the following sequence.

Start the test by opening and closing the driver's door. This first raises, then lowers, the suspension to the trim height. This should be accompanied by a glowing failure warning lamp. There is a problem if the lamp flashes. If this test is normal, open and close the driver's door again to make the next test in the sequence. Open and close the door twice to bypass the faulty test and to continue the test sequence. The test sequence checks the rear air springs, each front air spring, compressor, and brake switch. The electrical circuits can be checked with a test light using a #194 bulb only. Do not use any other type of test light bulb.

If the system accidently looses air pressure, it can be pumped up using the following procedure. Take the weight off the suspension, if possible, by jacking the vehicle and supporting it on safety stands. Connect a charger set on low to the battery. Turn the trunk switch on and open the driver's door. Turn the ignition switch on for 5 seconds, then turn it off. Ground the diagnostic pigtail next to the trunk switch. Hold the brake pedal down and turn on the ignition, but do not start the engine. Close and open the driver's door. After 6 seconds, the rear air springs will fill. Again, open and close the driver's door. After 6 seconds, the front air springs will fill. After the suspension is refilled, check the system for air leaks with a brush and soapy water.

STUDY QUESTIONS

Introduction

1. What causes the front wheel alignment to change?
2. What is necessary before you can do a complete and satisfactory suspension repair?

Section 10-1

3. What abnormal conditions do you look for on a visual inspection of the suspension?
4. What abnormal conditions do you listen for on an inspection of the suspension?
5. How can you check the control of a shock absorber?
6. What measurements do you make on an inspection of the suspension?

Section 10-2

7. What part is removed first when disassembling unequal-length control arm suspensions?
8. Why are shock absorbers replaced in pairs?
9. What safety procedures should you use when removing shock absorbers?
10. How is a shock absorber checked after it has been removed from the vehicle?
11. What service repairs are done on shock absorbers?

12. What precautions should be taken when installing shock absorbers?
13. When should ball joints be replaced?
14. What should be done after the ball joints have been replaced?
15. Why are the ball joint stud nuts loosened only?
16. How is the suspension jacked to remove the spring force from the vehicle-loaded ball joint?
17. How much is the spring compressed before the ball joints can be removed?
18. When should you not use a ball joint spreader?
19. Why should heat never be used on front suspensions?
20. Some ball joints are riveted to the control arm. How is the replacement ball joint held to this arm?
21. What is different about servicing the upper and lower control arm bushings?
22. How are bushings held in the control arm?
23. What precautions should be taken before torquing the suspension pivots?
24. Why are the wheels aligned after servicing the front suspension?

Section 10-3

25. Describe the procedure used to remove a MacPherson strut from a specific vehicle. Name the vehicle.
26. Describe the procedure used to remove the spring from the strut in question 25.
27. What three methods are used to correct faulty MacPherson struts?
28. How is a cartridge installed in a strut?

Section 10-4

29. How does stabilizer bar bushing replacement affect wheel alignment?
30. When are the brackets tightened around new bushings on a stabilizer bar?
31. What is the most common strut repair?
32. How much of a bend can be straightened on a strut bar?
33. What holds the strut bar to the lower control arm?
34. What should be done after the strut bar bushings are replaced?
35. How is ride height adjusted?
36. What switches keep the air suspension from leveling?
37. What can be done if all air leaks from the air suspension?
38. How can you find leaks in the air suspension system?

11

Rear Suspension Service

The rear suspension plays an important role in vehicle handling. Vehicle ride depends as much on the rear suspension as it does on the front suspension. Most of the useful load the vehicle carries is supported by the rear springs. The amount of load being carried will change the ride height. Ride-height change at the rear will affect the front-end alignment angles as discussed in Chapter 8. Rear suspension wear can allow the wheel positions to change relative to the body and/or frame. If the position of the wheels changes, this can cause a steering tendency resulting in wheel pull, tire wear, or an uncomfortable ride. The movement is used in four-wheel steering. Vibrations from drive line misalignment can also result from a misaligned dependent rear-drive suspension. Because differences do exist between independent and dependent rear suspensions, these repairs are discussed separately in Sections 11-2 and 11-3.

11-1 REAR SUSPENSION INSPECTION

The rear suspension should be carefully inspected before doing any repair. A thorough inspection is performed with the vehicle raised and supported with safety stands placed under the frame structure. This positioning allows the suspension to fully extend, removing the static weight from the suspension and making it easier to find worn parts. Be sure to turn off the air spring suspension control before lifting a vehicle with air springs. Rear suspension bushing wear is often difficult to see. With the rear suspension hanging free, a small pry bar can be used to lightly pry on the parts to determine if the bushings are worn.

As in any suspension-related inspection, the tires should be carefully examined for abnormal wear or signs of impending failure. Slowly spin the tires while on the vehicle as you watch for bent wheels or out-of-round tires. These problems lead primarily to vibration complaints. On rear-drive vehicles, the universal joints should be visually inspected and checked for looseness. Looseness could be the source of a suspension complaint.

Rear shock absorbers or strut assemblies should be visually inspected for signs of leakage or binding. With conventional shock absorbers, the mounting should be inspected for security and the condition of the bushings. Worn or loose bushings usually cause noise. Noises can also occur in shock absorbers, both conventional and strut types, due to seal drag, aeration, and other problems as discussed in Section 10-1.

11-2 REAR-WHEEL-DRIVE DEPENDENT SUSPENSION SERVICE

Dependent rear suspension systems are used on many domestic passenger cars and most trucks (see Figure 1-13). This type of suspension system uses a Hotchkiss drive differential. It uses an open drive line and two rear springs. This type and similar-type suspensions are the focus of this section.

Shock Absorbers. Rear shock absorbers are usually replaced while the rear axle is positioned at or near its normal ride height. When the vehicle is raised and supported so the suspension holds the vehicle weight, you will still have enough room to remove the shock absorbers. If the vehicle uses gas-charged shock absorbers, the pressurized gas will tend to force the shock absorber to extend. It may be easier and safer to allow the suspension to lower until the gas-charged shock absorbers are fully extended. This can be done by supporting the frame with safety stands and lowering the axle with a stable jack to a point where the shock absorbers are extended fully.

Removing Rear Shock Absorbers. Fasteners for the upper end of the shock absorber mounting may be accessible from beneath the vehicle or through access holes placed either in the trunk or behind the rear seat back. The fastener for the lower end of the shock absorber mounting is accessible from under the vehicle. Typical shock absorber mounting ends are shown in Figure 10-12. The lower end of the rear shock absorber is disconnected first, so the shock absorber can be cycled when there is a question about its effectiveness. This procedure is the same as testing and purging front shocks as described in Section 10-2.

A "nut cracker" becomes an effective tool for removing stubborn shock absorber nuts. It can be used when the shock absorber is to be replaced. With this tool, as pictured in Figure 11-1, you can remove the nut from the end of the shock absorber without breaking the mounting stud or spike. If the shock absorber is to be replaced, regardless of the condition, the stud can be broken off for rapid shock absorber removal. The upper end can then be disconnected and the shock absorber removed from the vehicle.

FIGURE 11-1 Upper shock absorber mount nut being split with a "nut cracker."

Installing Rear Shock Absorbers. To prepare shock absorbers for installation, place the required retainers and insulator bushing over the stud or spike end of the shock, or into the eyes of the shock absorber, depending on design (see Figure 10-12). The upper end is mounted and the insulators, retainers, bolts, and nuts are installed finger-tight. The lower end is then connected using the proper retainers and insulators. The upper and lower connections are tightened to complete the job.

Stabilizer Bar. Rear stabilizer bar service includes the replacement of links, bushings, and the entire stabilizer bar. Do not attempt to straighten a damaged stabilizer bar.

Rear stabilizer bars, similar to front stabilizer bars, may be bolted to the trailing arms (see Figure 6-30), mounted to the frame with links connected to the suspension, or mounted to the suspension with links connected to the frame or underbody (Figure 11-2).

Rear Stabilizer Bar Service. For service, the vehicle is raised and supported on a hoist or safety stands so you have room to work on the stabilizer bar. The position of the suspension is not critical while servicing the stabilizer bar.

Stabilizer bars bolted to the trailing arms do not have bushings or linkages. Removing fasteners connecting the bar will free the bar from the suspension.

Stabilizer bars mounted to the body or axle

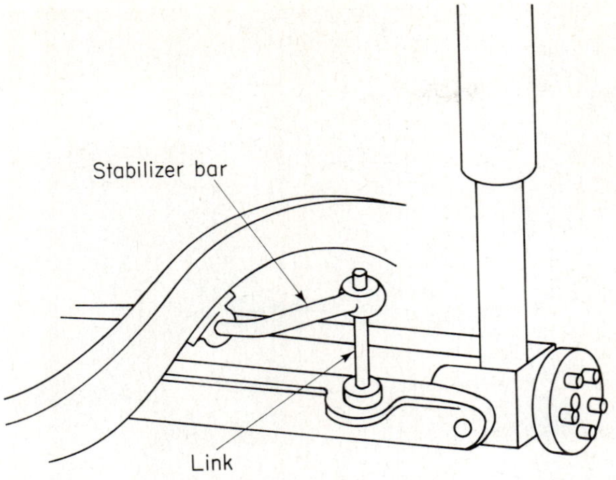

FIGURE 11-2 Stabilizer bar bolted to rear control arms and fastened to the body with a link bolt going through a bracket attached to the underbody.

housing use links with insulators and bushings, sometimes called *grommets*. Broken linkages, damaged bushings, and worn insulators cause poor vehicle handling. These faulty parts will cause rattles or noise as the vehicle maneuvers. Connecting linkage insulators and retainers can be replaced while the stabilizer bar remains mounted on the vehicle. A through bolt holds the assembly of insulators, retainer, and spacer (see Figure 11-2). If the pivot bushings are to be replaced, the bar is disconnected by removing the bolts holding the bushing bracket. It may be necessary to slide the bushings in from the end of the bar.

The bushings are placed on the bar before the stabilizer bar is installed. The bar is placed in position and secured loosely until the weight of the vehicle is supported by the springs. The fasteners are torqued to specifications with the suspension at its normal ride height. Jouncing the vehicle will ensure that there are no binding or noise problems in the installation.

Track Bar Service. A track bar, sometimes called a *Panhard rod*, is used with some coil spring rear suspensions to prevent excess lateral body movement during turns (Figure 11-3; also, see Figure 6-18). Dangerous handling, uncomfortable ride, and noise will result from the sideways body movement if the bar breaks or the bushings wear. In extreme cases, this movement will allow the tires to rub the fender wells.

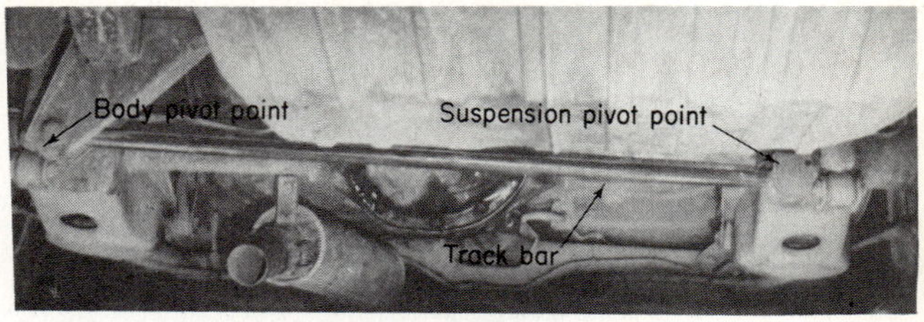

FIGURE 11-3 Typical rear track bar.

The track bar can be serviced by removing the fasteners from both ends of the bar with the vehicle raised and supported on the rear axle.

The bushings in the bar ends are replaceable. The old bushings are pressed out and the new bushings pressed in. If the bar is bent or broken, it is replaced with a new one with new bushings already in place.

The bar is placed over the frame lug first; then, by pushing the body slightly, the lug on the rear axle is aligned. The retainer washers and nuts are installed. The vehicle is then lowered to the floor and the nuts are torqued to specifications.

Recall that the shock absorbers, stabilizer bars, and track bar have no effect on ride height. Springs establish the ride height. If the rear ride height is out of specification, all parts of the suspension should be inspected even though you may see damage to only one of them.

Spring Replacement. Spring service is required whenever the ride height is not within specification, or when you can see damage to the spring. It is sometimes necessary to remove a coil spring to correct damage or wear to the trailing arms.

Coil Spring Service. To remove the coil spring from a dependent rear suspension, the vehicle must be raised on a frame contact hoist or have safety stands placed under the frame forward of the axle assembly. A floor jack is positioned beneath the center of the rear axle or axle housing. You may find it easier to disassemble and reassemble one side at a time. If you do, the jack is positioned on the side being serviced. This procedure will eliminate any chance of placing the parts on the wrong side of the vehicle and will maintain major component alignment.

The axle is raised to support the rear axle weight, but not enough to lift the vehicle from the safety stands. The lower end of the shock absorbers

FIGURE 11-4 Coil spring being removed from a rear suspension.

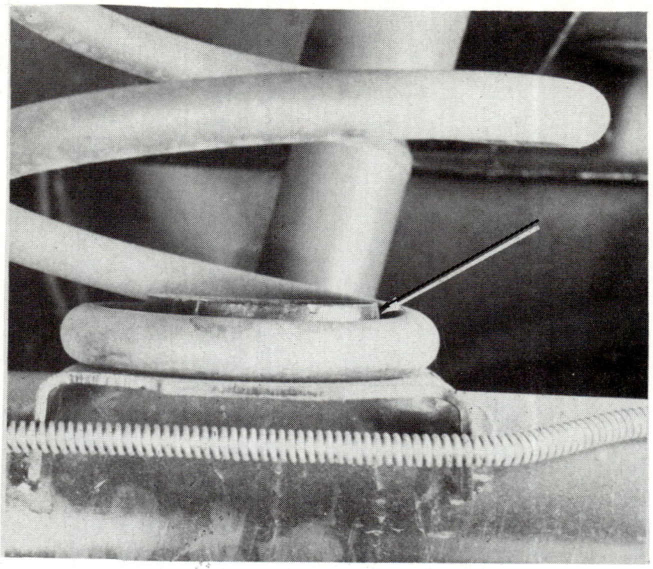

FIGURE 11-5 Typical coil spring locator on a rear axle housing.

are disconnected. The jack is then lowered, allowing the rear axle to drop below its normal travel limit to relieve all of the coil spring force as illustrated in Figure 11-4. The coil spring positioner, if used, is unbolted from the center of the coil spring. On many vehicles, the coil spring may then be removed from the vehicle by lifting it from its spring seat. If the coil spring remains under load after the rear axle assembly is lowered, a spring compressor will have to be used (see Figure 10-16). Compress the coils so the spring can be removed. If the springs are to be reused, mark each spring so it can be returned to its original location.

Coil springs may be shimmed to adjust the side-to-side height when there is less than one-half-inch (13 mm) difference. Coil springs that vary more than this will have to be replaced. When replacement is required, coil springs are always replaced in pairs to ensure a uniform, level ride height. The insulators used on the springs are replaced if they show any sign of damage.

When necessary, the coil spring is compressed for installation. The insulator is then placed on the top of the coil spring and the spring is positioned on the lower spring seat locator (Figure 11-5). The location of the end of the top coil is critical. It must be positioned to line up with the recess in the spring seat. Some spring seats have a drilled hole where the spring should end. Look in the hole to make sure the spring is seated correctly. The spring compressor can then be removed. The opposite-side coil spring is installed the same way. The rear axle housing is raised into position and the shock absorbers are reconnected.

Leaf Spring Service. The vehicle is raised and supported on the frame to remove a leaf spring. Safety stands are placed on the frame ahead of the springs. A floor jack is used to lift the rear axle housing close to the spring being removed. It is recommended, when working with leaf springs, that one side be serviced at a time. The assembled side will give support to the suspension and aid in locating the axle housing on the spring being serviced.

The axle housing is raised slightly and the lower end of the shock absorber is disconnected. The jack is then lowered so the leaf spring does not place any force on the rear axle housing. The U-bolts or bolts holding the isoclamps together are removed as shown

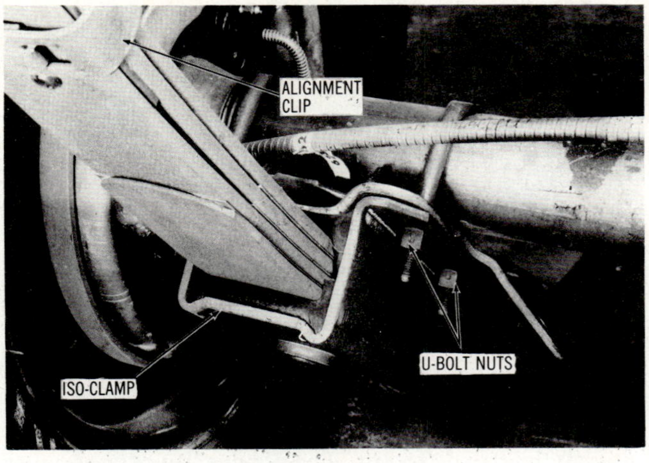

FIGURE 11-6 U-bolts are loosened to free a leaf spring.

in Figure 11-6. The lower spring clamp plate is removed and the axle housing is jacked up from the leaf spring. The rear spring shackle nuts are then removed. The spring must be supported to keep it from dropping before removing the rear shackle (Figure 11-7) or the front spring eye bolt or bracket. The spring front attachment may have a mounting bracket bolted to the underbody. It is removed from the frame to free the front of the spring. An alternative design is to have the spring front eye bolt pass through the vehicle frame. The eye bolt is removed to free the front of the spring. When both ends of the spring are free, the spring can be lowered to the floor.

FIGURE 11-7 Removing the spring shackle while the spring is supported with a jack.

Leaf spring service includes replacing the spring eye bushings, the interleaf separators, and the center bolt. With the spring on the workbench, eye bushings are readily pressed out of the main leaf and new bushings are pressed in. In most cases, the complete spring assembly is replaced if the spring requires more than new spring eye bushings.

Rear Leaf Spring Installation. The repaired or replacement leaf- type spring is positioned under the vehicle and the front of the spring is connected. On some vehicle spring designs, a spring spreader is required to aid in connecting the rear spring shackle to the spring. The spreader, as shown in Figure 11-8, increases the span between the spring eyes to allow the spring shackle to be connected. In some installations, the front bracket is connected and the rear shackle bracket is pulled rearward to align with the frame mounting bolt holes as shown in Figure 11-9. If a spreader is not available, the front end of the spring is connected. Then the rear axle housing brackets are positioned so the housing can be fastened to the spring. Be sure to position the head of the center bolt in the hole in the axle housing. The center of the spring is jacked upwards and the rear end of the spring is pulled down to decrease the spring arch. Continue this until the shackle bolt will pass through the spring eye and frame hole.

Check again to make sure the axle housing is properly centered over the spring center bolt. The U-bolts used for clamping the spring to the axle

FIGURE 11-8 Spring spreader used to align the leaf spring eye with the spring shackle.

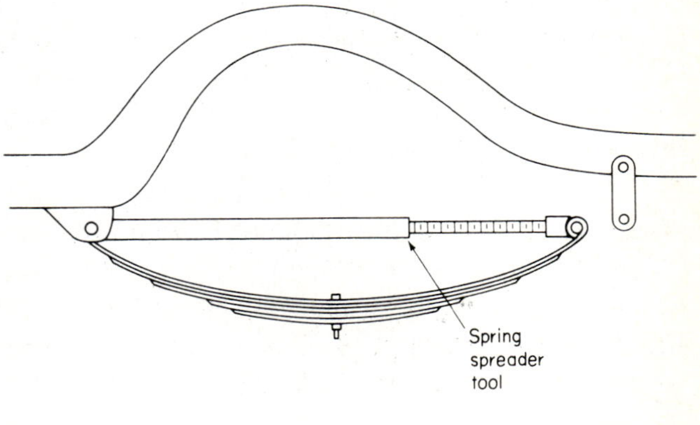

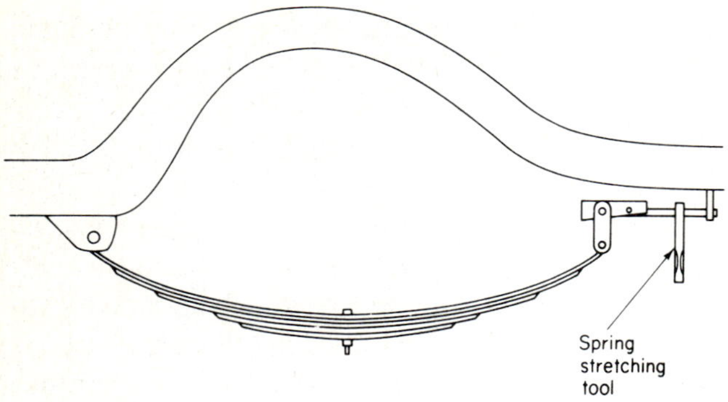

FIGURE 11-9 Spring stretching tool used to align a leaf spring shackle bracket with bolt holes in the frame.

housing are installed with the isoclamps and clamp plate in place. The nuts are installed and torqued to specifications. Overtorquing the nuts on the U-bolts may result in excessive noise transfer or damage to the axle housing or spring. The shock absorbers are reconnected. The wheels and tires, if removed, are now replaced and the wheel covers installed to complete the repair.

Drive Line Working Angle. Many rear-wheel-drive vehicles have a way to adjust the drive line working angle. This adjustment establishes the angle through which the universal joints must operate. The working angle is important because too large a working angle will result in vibration and rumble on acceleration. If the working angle is zero, the U-joints will rapidly wear or brinell. This makes grooves across the bearing race as the needle rollers keep operating in the same place.

The working angle of the universal joints on the drive shaft is always checked after the rear suspension has been serviced to minimize the possibility of drive line vibration and noise. The *rear working angle* of the drive line is the angle between a centerline drawn through the drive shaft and a centerline drawn through the rear axle drive pinion (Figure 11-10). The *front working angle* is the drive shaft centerline compared to the angle of the transmission main shaft centerline.

On a rear-wheel drive vehicle, the speed of the drive shaft will increase and decrease twice each revolution when the universal joint is working at an angle. The greater the working angle, the greater this speed change will be. The speed or velocity change results in a vibration when the working angle is too great or if the front and rear universal joint working angles are different. The service life of the universal joint is greater when it has a small working angle of 1 to 5 degrees to move the bearings in the universal joints.

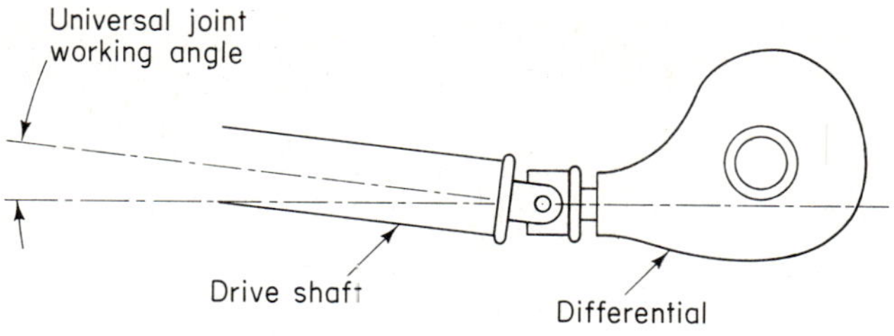

FIGURE 11-10 Rear universal joint working angle.

A constant velocity joint, as shown in Figure 11-11 and used on some rear-drive vehicles, does not produce a speed change. The working angle is not as critical on the rear constant velocity joint. As a result, there may be no way to make a working angle adjustment when these joints are used.

After rear suspension work is completed, the nose of the differential may have a different angle than it had before disassembly. This angle must be corrected so that it is within limits to ensure minimum drive line vibration. To check the rear universal joint working angle, the vehicle is raised and supported by the rear axle housing and the front suspension so the entire vehicle is level. This may be done

FIGURE 11-11 Typical constant velocity rear universal joint.

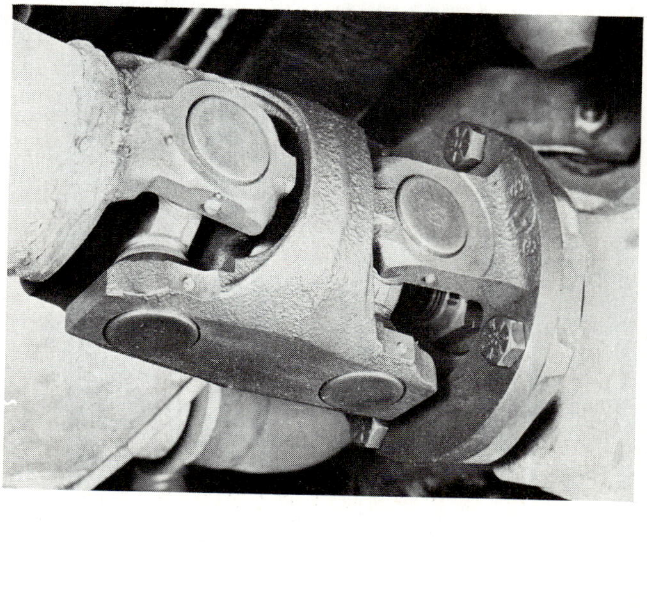

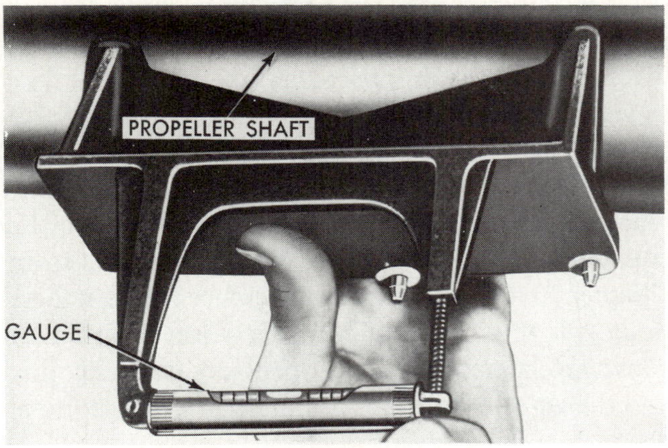

FIGURE 11-12 Inclinometer designed to measure the working angle of the universal joints. (Courtesy of Chrysler Corporation)

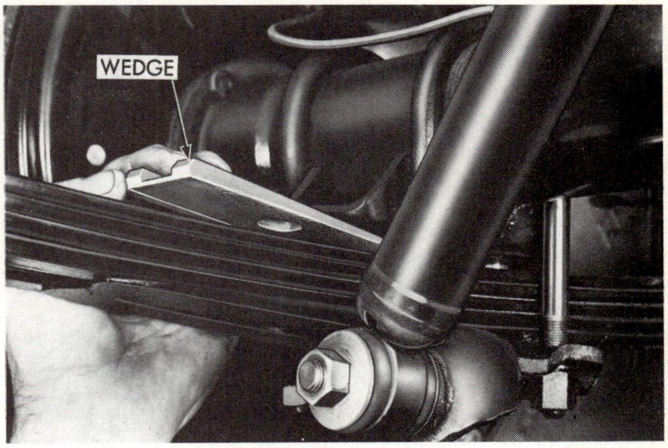

FIGURE 11-14 Location of a taper shim on a leaf spring. (Courtesy of Chrysler Corporation)

easily on an alignment rack or on a twin-post hoist. The rear axle housing must be at its normal ride height in relation to the vehicle frame. An inclinometer that measures the angular degree from level is held against the drive shaft to measure the drive shaft angle. One type of inclinometer is shown in Figure 11-12. The inclinometer is then held against the rear universal joint cup with one bearing cup positioned at the bottom to measure differential angle. The working angle is the difference between the drive shaft angle and the differential angle. If the working angle is out of specification, it will have to be adjusted so that it falls within the specifications.

Adjustment of the Drive Line Working Angle. Adjustment of the working angle, when leaf springs are used, is done using tapered shims placed between the spring and the axle housing pad. Shims are readily available for one, two, and four degrees. A typical taper shim is shown in Figure 11-13. The amount of angle change required is found by comparing the measured angles to the specifications. To install

FIGURE 11-13 Typical taper-shaped leaf spring shim used to adjust the working angle of the rear universal joint.

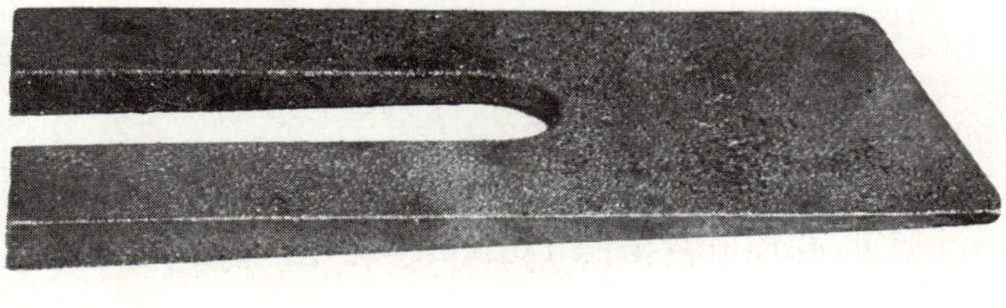

shims, the vehicle must be supported on the frame. The rear axle is then jacked just enough to hold the axle so the U-bolts can be loosened. The taper shims or wedges are positioned between the axle housing and spring to give the desired angle change, as pictured in Figure 11-14. The U-bolts are retorqued and the vehicle is lowered, allowing the axle to again support the vehicle weight in a level position. The body of the vehicle is jounced to ensure that the suspension is settled. The measurements of the working angle are checked again.

The universal joint working angle on the coil spring rear suspension is adjusted by lengthening or shortening the effective length of the upper control arms. Adjustment can be made with slots, shims, or eccentric bolts. The shims space the upper control arm bracket from the frame to change the working angle. The slots or eccentric adjustment provide a change in distance from the axle housing to the frame along the upper control arm to change the differential angle. After the required adjustments are made, all bolts on the control arms are retorqued to specifications. The vehicle is test driven to ensure vibration-free operation.

Rear Wheel Tracking. Rear suspension alignment problems can be divided into two types: problems that result in excessive tire wear and problems that cause a vehicle to track improperly. Track was discussed at the end of Section 8-1. Suspension alignment problem detection, analysis, and correction depend on whether the rear suspension is dependent or independent.

You will remember from Section 9-8 that alignment checks on rear-wheel dependent suspension vehicles include toe, camber, and track. Tire tread scuffing and rapid wear result from toe problems. The rear tires may show uneven wear across the surface of the tread. This usually identifies a camber problem. Tracking problems occur when the axle centerline is not perpendicular to the centerline of the vehicle as shown in Figure 8-15.

On dependent suspension vehicles that use coil springs, tracking problems are most often the result of loose or damaged control arms or failed bushings. The control arms hold the axle position on these suspensions.

Broken rear leaf springs will cause tracking problems on dependent rear suspensions using leaf springs. Track on a leaf spring rear suspension is maintained by the leaf springs themselves. The axle is held in position by the center bolt fitting into a hole in the spring pad on the axle housing. Failure of either the spring or center bolt will allow the axle to move on the spring. This will cause the track to change.

11-3 REAR-WHEEL-DRIVE INDEPENDENT SUSPENSION SERVICE

Front-engine, mid-engine, and rear-engine positions can be used with rear-wheel-drive and four-wheel-drive independent suspensions (Figure 11-15). Rear-wheel-drive vehicles with independent suspension are gaining in popularity. Service procedures for independent rear suspensions are similar to those followed for front-wheel-drive suspensions. In some mid-engine vehicles, the rear drive system is the front drive system moved to the rear with the tie rods fixed at the inner end (see Figure 9-25). This section discusses the repair procedures common to these types of independent rear suspension systems.

FIGURE 11-15 Independent-strut-type rear suspension used with a front engine.

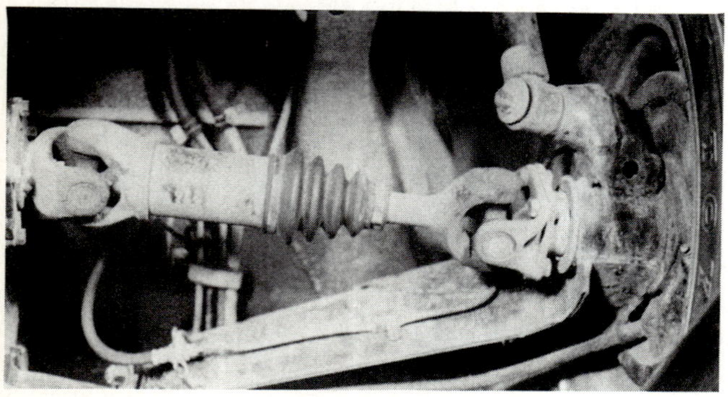

Shock Absorbers. Rear shock absorbers for rear-wheel-drive independent suspension systems may be conventional or strut type. You will recall from Chapter 10 that the primary difference between these designs is that the strut controls wheel position. A conventional shock absorber provides damping only and the wheel position is maintained by the control arms.

Conventional shock absorbers on independent rear-drive suspensions are serviced and replaced in the same way as those on dependent rear-drive suspensions. The vehicle is raised and the suspension is supported at or near its ride position. In many suspensions, the shock absorber limits rebound. Supporting these suspensions holds the load so the shock absorbers are free from any load for easy removal. Care must be used when removing gas-charged shock absorbers because they extend to their full travel after they are removed.

The strut-type suspension on rear-wheel-drive vehicles is nearly identical to the strut on the front-wheel-drive system. Differences may include the lower control arm arrangement and the lower pivot, which may not be a ball joint. Since the rear strut does not turn, as is required of the front strut, lower pivots and upper spring pads can be simplified. For rear strut service, follow the guidelines given in Section 10-3.

Tracking Adjustment. Out-of-track can result in poor handling, reduced fuel economy, and possible tire wear. The vehicle tracks correctly when both rear toe and symmetry of toe are proper. Rear suspension toe can be adjusted with shims, tie rod struts, or slots.

11-4 REAR SUSPENSION SERVICE WITH FRONT-WHEEL DRIVE

Front-wheel drive is the most popular drive arrangement on new automobiles. The rear suspension systems used with front-wheel-drive vehicles are quite varied in design (see Figures 1-15 through 1-17).

Designs include the trailing arm beam suspension, trailing arm strut suspension with integral sway control, trailing arm modified strut suspension with torsion bars, a beam axle with leaf springs, and a control arm tie rod suspension in a modified strut arrangement. This section will discuss repair procedures for the nondriving rear suspension systems.

Conventional Shock Absorbers. On most vehicles, conventional rear shock absorbers can be removed with the suspension at ride height. This may be necessary because most rear shock absorbers provide the extension limit for the suspension. If the suspension is allowed to fully extend, the suspension may drop when the shock absorbers are disconnected.

Strut Systems with Remote Springs. Many rear strut systems have suspension springs in a location other than on the strut (Figure 11-16). To remove the strut, the vehicle support must hold the suspension so the spring holds the weight of the vehicle. This will relieve all loads from the strut assembly.

Strut Systems with Integrated Springs. Some independent rear suspension systems use a spring on the struts (MacPherson strut) to carry the vehicle weight. The rear strut assembly is simplified, compared to the front strut, in that it does not use an upper support plate or have a bearing to allow for wheel steering. Rear strut service is performed the same way it is done on the front strut assemblies.

FIGURE 11-16 Rear strut suspension with the remote spring between the lower control arm and frame.

11-5 COMMON ELEMENTS IN REAR SUSPENSIONS

Many rear suspension systems use control arms to control the axle or the individual wheel positions (Figure 11-17). The control arms may require bushing replacement within the lifetime of the vehicle. Because their replacement is similar on all arms, the repair methods will be covered in general terms. If the bushings are worn, the usual cause for complaint is noise during braking, a tracking problem, or, on independent suspensions, tire wear.

Control arms on dependent rear suspensions are either upper and/or lower control arms. If the control arm has the spring load placed on it, the spring must first be removed. The spring is removed when the suspension is fully extended. With the suspension extended, the control arm can be easily removed and serviced one side at a time. Servicing one side of the vehicle at a time is safer as this prevents axle rotation caused by the spring loads and will make it easier for you to realign the parts when you reassemble them. Axle rotation could result in damage to the vehicle and injury to you. Control arm bushings are replaced on serviceable control arms by removing the defective bushing with a suitable puller or press. The new bushing is properly positioned and pressed into place. Some service bushings and bolts have an eccentric adjustment so you can adjust the drive line angle and track after service, as shown in Figure 11-18. The repaired con-

FIGURE 11-17 Typical rear control arm on a dependent suspension with a rear coil spring.

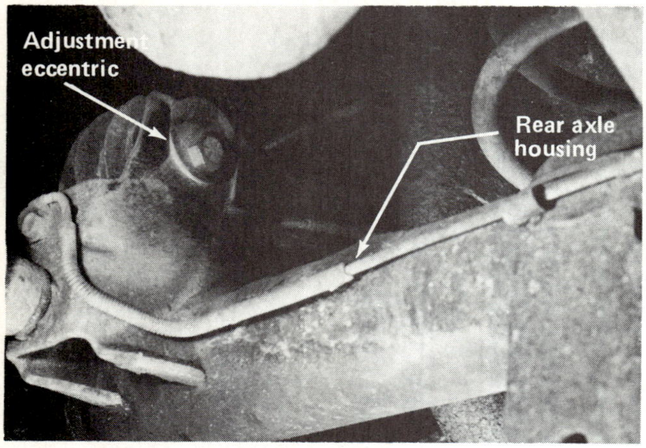

FIGURE 11-18 Typical rear control arm eccentric adjustment.

trol arm is put in place in the vehicle and the fasteners are loosely installed. The service is repeated on the control arm located on the other side of the vehicle, if required. The bolts and nuts are properly torqued after the entire vehicle weight is on the springs.

Control arms on MacPherson-strut-type independent rear suspensions (see Figure 1-16) are serviced with the weight fully removed from the wheels. The vehicle is supported from the frame with the suspension fully extended. Vehicles with the springs on the lower control arms (see Figures 1-17 and 11-16) will require the use of a spring compressor to compress the spring. An alternative method is to use a floor jack under the outboard end of the control arm. The jack will support the vehicle enough to allow disconnecting the attachment point at the upright. Lowering the jack will allow the control arm to move away, releasing the weight on the spring. The control arm attachment points at the frame can then be disconnected and the arm removed. The arm and bushings can be serviced and reassembled.

11-6 SUMMARY

The rear suspension is a very important, often overlooked, part of vehicle handling. The many different rear suspension systems being manufactured make it difficult to detail each one separately in one general textbook. This chapter introduced basic rear suspension designs and basic repair procedures. It should not be considered a replacement for but a supplement to the vehicle service manual.

STUDY QUESTIONS

Introduction

1. Why is the chassis lifted so the suspension extends when inspecting the rear suspension?
2. What faulty suspension parts are likely to cause a noise?

Section 11-2

3. Why is the lower rear shock absorber mount disconnected before the upper?
4. When would a "nut cracker" be used to remove a mount nut rather than breaking the spike?
5. What is done to prepare the shock absorber for installation?
6. When would a rear stabilizer bar be replaced?
7. Why are the stabilizer mounts tightened after the vehicle weight is supported by the springs?
8. What driving conditions indicate failed track bar bushings?
9. When should the rear springs be replaced?

10. Why should springs be marked when they are removed?
11. Why is one rear spring serviced at a time?
12. What is included in leaf spring service?
13. What problems might be encountered when installing a leaf spring?
14. Why is the universal joint working angle important when working on rear suspensions?
15. How is the rear universal joint working angle adjusted when the vehicle has leaf springs?

Section 11-3

16. How is an independent rear suspension adjusted?
17. What service procedures are followed when working on independent rear suspensions?

Section 11-4

18. What precautions should be taken to remove a strut from an independent rear suspension with a remote spring?
19. What service procedures should be followed when servicing rear struts?

Section 11-5

20. What is the usual complaint that results from worn control arm bushings?
21. What is the difference in the way control arms are serviced between those with MacPherson struts and those with the spring on the control arm?

12

Steering Linkage Service

The steering linkage plays an extremely important role in vehicle handling. Steering linkage problems no longer are just problems with the tie rod ends. Steering linkage misalignment, improper rack centering, and nonparallel rack or center link are common faults causing handling problems. This chapter discusses these and other problems to provide you with a greater background for diagnosing problems on the modern lightweight, responsive automobiles.

The vehicle should be inspected to verify the cause of the steering linkage problem. The steering linkage is repaired with the vehicle raised. After any work is done on the steering linkage, the toe should be adjusted and, for most satisfactory operation, the vehicle should be given a complete wheel alignment.

12-1 TIE ROD SERVICE

Tie rod ends and tie rod assemblies are often replaced as a result of wear or bending. For replacement on parallelogram linkage, the tie rod adjusting clamp is loosened (see Figure 9-22). The rack and pinion linkage uses a jam nut against the outer tie rod end to lock it to the tie rod (see Figure 9-23). The inner tie rod repair on a rack and pinion linkage will usually require rack removal unless you have some of the special tools designed especially for this work.

Spray penetrating oil on all fasteners you plan to remove. This will make it easier to disassemble the parts. Loosen the tie rod adjusting sleeve clamp. Remove the cotter pin and loosen the castellated nut at the tie rod ball stud. Use a tie rod spreader (see Figure 10-13), puller (Figure 12-1), or the two-hammer method (Figure 12-2) to free the tie rod stud from its tapered socket. With the two-hammer method, use a steady force to help separate the stud from the steering arm as the stud socket is hit.

When the entire tie rod assembly is to be replaced, both ends of the tie rod are removed. When only one end is being replaced, the tie rod end stud is removed from the steering arm or linkage. To aid reassembly, measure the length of the tie rod from pivot to pivot or count the number of turns required to remove the tie rod end. The tie rod end is unthreaded either right or left handed from the adjuster sleeve.

The replacement tie rod end is threaded into the sleeve with approximately the same number of threads on both inner and outer tie rod ends (Figure 12-3). Adjust the distance between pivots. The tie rod ball stud is placed into its tapered socket and the nut is installed. The steering wheel is positioned so that the wheels point straight ahead. The tie rod stud nut is torqued to the correct specifications and a new cotter pin is installed.

The tie rod ends must be positioned with each ball stud in the center of its free movement to prevent binding. This is especially important with tie

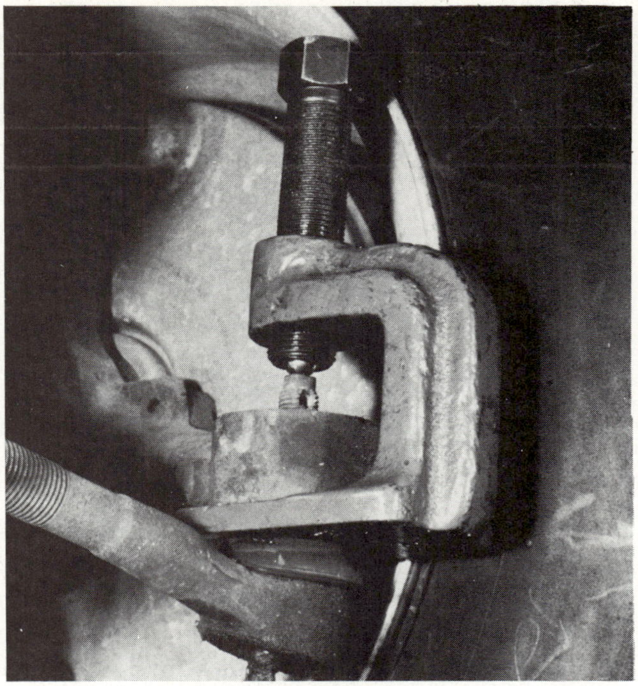

FIGURE 12-1 Puller used to remove a tie rod end.

FIGURE 12-2 Loosening a tie rod end from the steering arm with two hammers, one supports the arm and the other is used to hit the arm.

rod ends that have rubber bonded studs (RBS) held to the tie rod end with molded rubber. Toe should be checked before locking the adjusters. The adjuster sleeve clamp, if used, is positioned on the adjuster sleeve so that it does not touch anything as it moves and it is in the best position to hold the sleeve. This was shown in Figure 9-24. The clamp bolts are then tightened. The joint is lubricated, if required, and

FIGURE 12-3 Tie rod adjuster is equally spaced between the tie rod and tie rod end.

checked to make sure that there is no interference with the steering linkage as it moves through its full travel.

12-2 PARALLELOGRAM LINKAGES

The center link, idler arm, and Pitman arm are used only on parallelogram linkages along with a recirculating ball steering gear (Figure 12-4). Although most current vehicles use rack and pinion steering gears, a number of vehicles are still in service and some are being manufactured using recirculating ball steering gears.

FIGURE 12-4 Typical parallelogram linkage used with a recirculating ball steering gear.

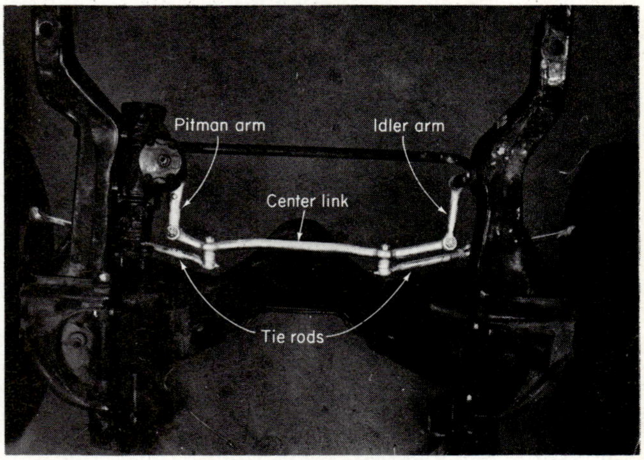

FIGURE 12-5 An idler arm with a rubber bushing pivot.

Center Link Replacement. To replace the center link, disconnect both inner tie rod ends from the Pitman arm and the idler arm. Most often this is done by taking out a cotter pin and loosening the castellated nut from each end and then separating the joint using a spreader (see Figure 10-13). If any of the tie rod ends or other connections are to be reused, a puller similar to the one shown in Figure 12-1 is used to prevent damage to the seal at the joint. A tie rod spreader will damage the tie rod end seal so it should not be used if the joint is to be placed back in service. An alternative design may be found on the idler arm end, which uses a rubber bushing, as shown in Figure 12-5, eliminating the need for using the spreader to separate the connection. All loosened nuts and the center link can be removed from the vehicle.

Install the replacement link by making all the connections, centering the steering wheel, and torquing the nuts to specifications. The cotter pins are replaced and all grease fittings are lubricated. The linkage is inspected for interference anywhere through full travel.

Idler Arm Service. Idler arm service may consist of only bushing replacement or it may require replacing the complete assembly, depending on the idler arm design. The serviceable idler arm has a replaceable press-fit bushing. The nonserviceable idler arm is usually of the ball-and-socket design, which incorporates bearings. Both types require complete assembly replacement. Some idler arms may be serviceable on one end but may require replacement if wear occurs on the other end. The idler arm is removed by disconnecting the center link stud. The bolts are removed to disconnect the arm from the frame (see Figure 12-5).

The bushings on serviceable arms are removed using a suitable puller or by carefully driving the bushing out with a suitable driver. The new bushing is pressed in, using properly fitting adapters that push on the outer bushing sleeve.

The new or repaired idler arm is now installed. The wheels must be in the straight-ahead position while the nuts are being tightened to the correct torque to prevent preloading the new bushing. New cotter pins are installed in all studs equipped for their use. Any lubrication fittings are greased to ensure proper service life.

Pitman Arm Service. The Pitman arm does not normally require replacement. Causes for replacement would be accidental damage or, if equipped with a bearing stud joint, wear of the joint. The Pitman arm is removed by first disconnecting it from the steering linkage. To do this, take out the cotter pin, remove the castellated nut, and then use a suitable puller to separate the joint stud from its socket. The arm is disconnected from the Pitman shaft by first removing the nut and lock washer. The Pitman arm is then pulled from the shaft with a puller similar to the one shown in Figure 12-6. Do not hammer

FIGURE 12-6 Pulling the Pitman arm from the Pitman shaft.

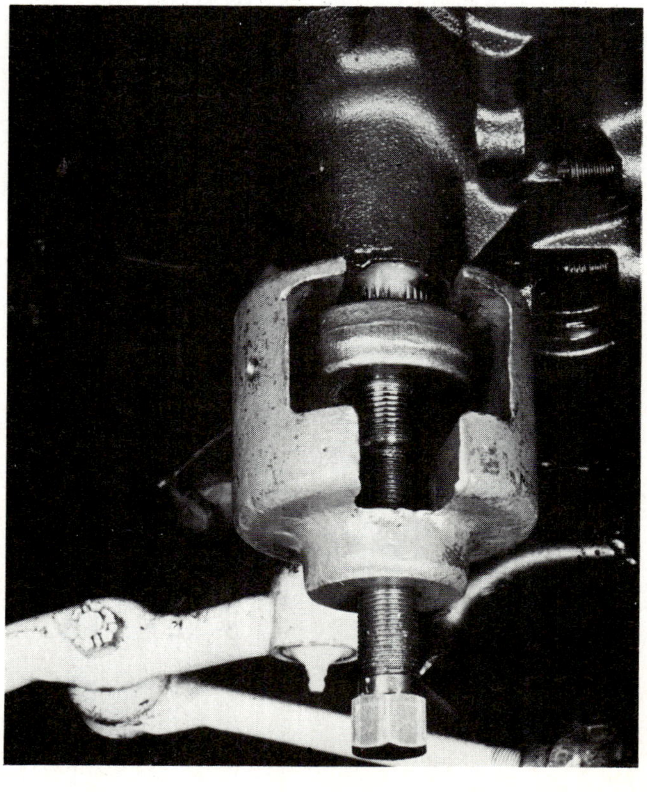

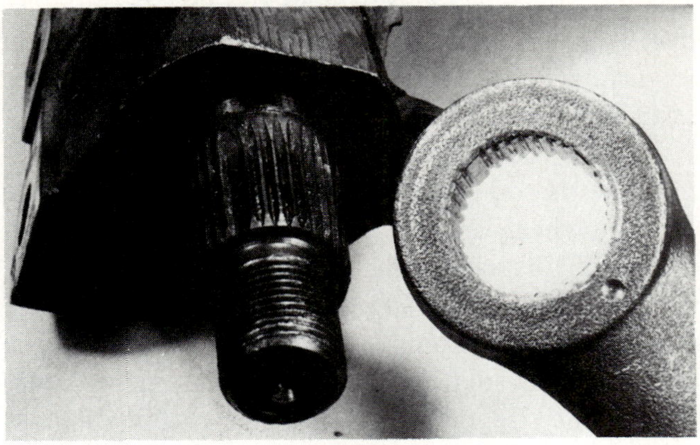

FIGURE 12-7 Pitman arm and Pitman shaft indexing splines.

on the arm or shaft; hammering is likely to damage the steering gear.

When replacing the Pitman arm, install it on the Pitman shaft with the indexing splines correctly aligned. You can see indexing splines in Figure 12-7. The lock washer and nut are installed and torqued to specifications. With the wheels in the straight-ahead position, connect the steering linkage and then install and torque the castellated nut. The cotter pin is replaced and the linkage is checked for binding.

12-3 STEERING ARM DIAGNOSIS AND REPLACEMENT

The steering arm is the lever at the knuckle or strut (Figure 12-8) that turns the wheel as the steering linkage moves. The steering arm may be made of forged or cast steel or, on many struts, stamped steel.

FIGURE 12-8 Steering arm on the steering knuckle.

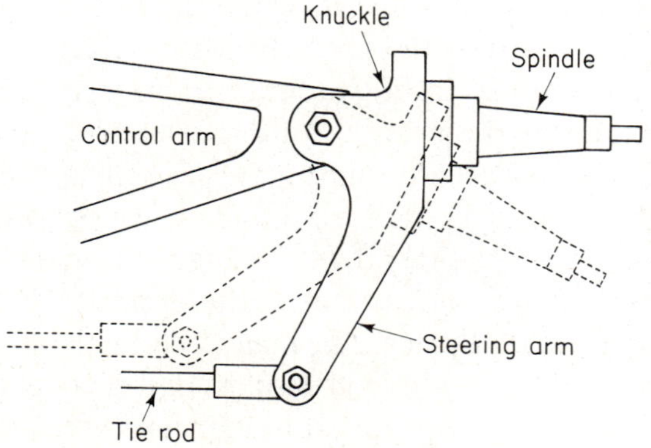

The arm could bend as a result of high loads placed on the steering arms when the wheel is forced to turn with the vehicle stationary, striking potholes, or a collision.

Steering Arm Diagnosis. Steering arm problems may first shown up as excessive toe and tire wear. However, if the vehicle has already been realigned, it is more difficult to determine if the arm has been bent. A bent steering arm may also show up as a customer complaint of excessive tire squeal on turns or as drifting while the vehicle is driven over rolling, hilly roads.

To detect a bent steering arm, the vehicle should be positioned with the front wheels on alignment radius plates. A visual inspection is performed with the rack or linkage centered. Next, check to make sure the wheels point straight ahead. In extreme cases, you can see a bent arm. If everything appears to be normal, toe-out on turns is checked. The inside wheel is turned to a specified angle and the outside wheel turning angle is read. The procedure is repeated, turning the wheels in the opposite direction. If the toe-out on turns reading is not the same in each direction, one steering arm may be bent.

Steering Arm Replacement. The steering arm should be replaced if, during the inspection, the arm shows visible damage. An out-of-specification reading for toe-out on turns may indicate a bent steering arm.

Depending on the manufacturer, steering arms may be bolted to the steering knuckle, may be a part of the steering knuckle, or may be a part of the lower ball joint. These are shown in Figure 12-9. Due to the high forces placed on steering arms in service, heat must not be used to straighten them. Heat will alter their heat-treated strength. The vehicle is raised and the wheels and brakes are removed to remove the steering arms. The tie rod is then removed.

Steering arms that are not an integral part of the knuckle or the ball joint are now readily replaced. After disconnecting two bolts and nuts passing through the steering arm and knuckle, the arm is removed. The brake may have to be disassembled to reach the bolt heads. The new steering arm is positioned and the bolts are inserted. The nuts are installed and torqued to specifications.

If the steering arm is an integral part of the knuckle or ball joint, the procedure for ball joint

(a)

(b)

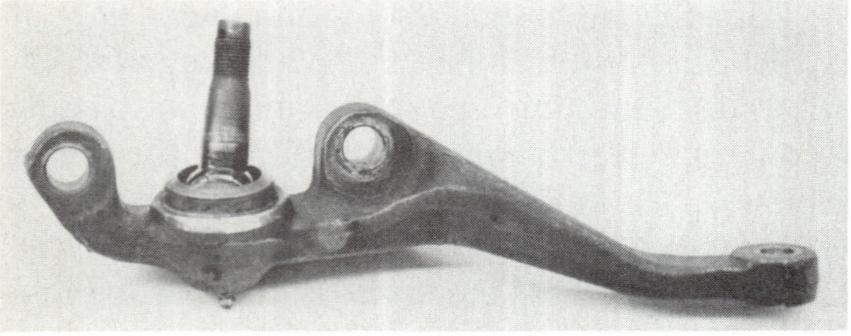

(c)

FIGURE 12-9 Typical steering arm: (a) bolt on, (b) integral, and (c) bolt on with a ball joint.

and steering knuckle replacement described in Section 10-2 should be followed. When the bent steering arm is part of the MacPherson strut, the whole strut must be replaced.

After the steering arm has been installed, reconnect the tie rod end to the steering arm, center the steering wheel, and torque the nut. The wheel bearings are inspected for wear and repacked if necessary. The hub and rotor, or hub and drum, are installed on the spindle and the wheel bearing is properly adjusted. Install the cotter pin and dust cover. The tire and wheel assembly is installed using the lug and wheel alignment marks to locate them in the same position they had before disassembly. The wheel nuts are torqued to specifications and the wheel cover is installed. The vehicle is lowered to the floor. A wheel alignment must follow the repair of any steering linkage part to ensure correct alignment angles, steering, and vehicle handling.

12-4 LINKAGE ALIGNMENT

Today's vehicles have more responsive steering than many of their early counterparts. Good response has been accomplished, in part, by steering geometries, steering ratios, and the use of radial ply tires. This steering responsiveness has also made these same vehicles more sensitive to steering geometry problems. This section discusses basic linkage geometry as well as problems that lead to poor handling.

Steering Linkage Characteristics. The position, symmetry, and angularity of the steering linkage can cause the wheels to steer in or out as the suspension is moved through jounce and rebound. These characteristics were described in Section 7-2. Notice that, in Figure 12-10, the angle of the tie rod slopes down toward the outside. This will cause the steering arm to steer outward (with front steering linkage) as the suspension moves through jounce. In Figure 12-11, the tie rod slopes upward toward the outside. As the suspension moves into jounce, the steering arm will steer inward during jounce (with front steering linkage). Also notice that, if the linkage is in back of the wheel spindles (rear steer), the opposite steering effect will take place. The manufacturer carefully designs the steering linkage lengths and pivot locations to provide the proper roll steer characteristics.

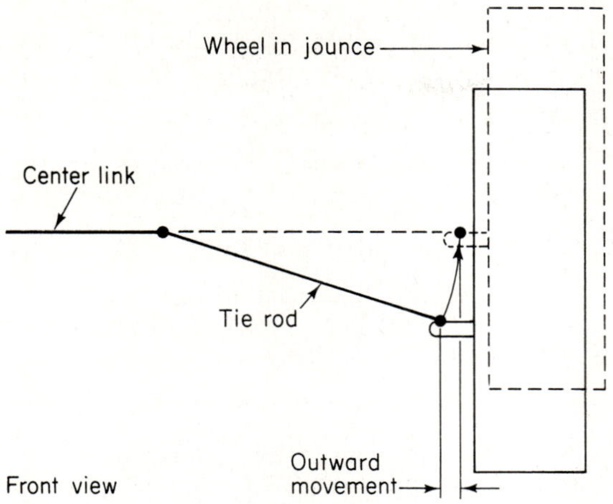

FIGURE 12-10 The downward slope of the tie rod will cause the wheel to turn outward during jounce.

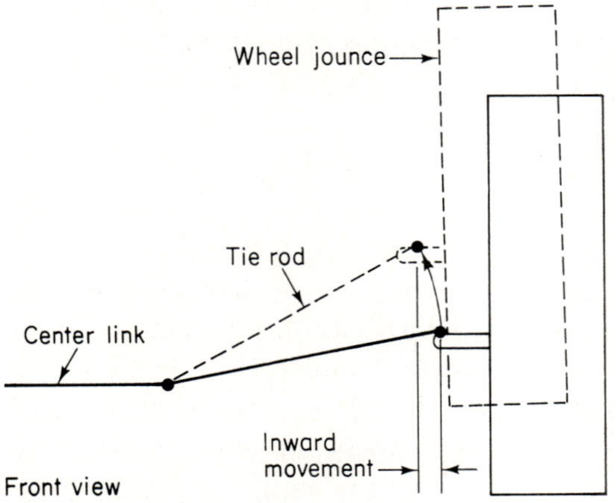

FIGURE 12-11 The upward slope of the tie rod will cause the wheel to turn inward during jounce.

Most are designed to steer the outside wheel outward while making a turn (in jounce) and the inside wheel inward (in rebound), producing a slight roll understeer condition. Excessive roll steer leads to wheel fight as described in Section 7-2.

Steering Rack or Center Link Centering. The position of the steering rack, with the wheels pointed straight ahead (Figure 12-12), will affect the steer tendency of the vehicle as it encounters bumps. *Note:* For this discussion, the steering rack of rack and pinion and the center link of parallelogram linkage serve the same purpose, that of supporting the inner tie rod end. As a result, similar problems and solutions result from both types of steering linkage.

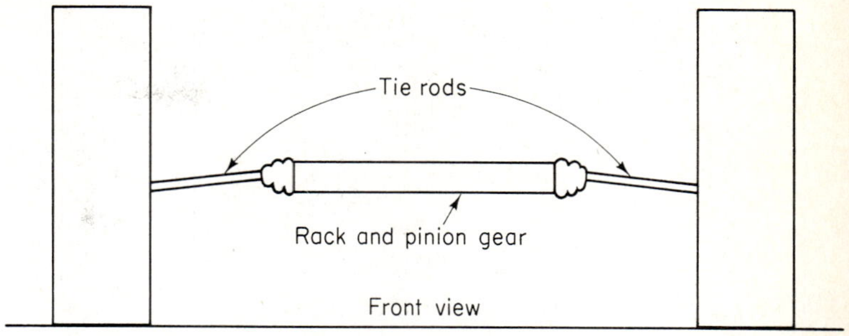

FIGURE 12-12 Typical mounting of a rack and pinion steering gear.

Remember that the lengths of the tie rods determine the swinging arc as the wheels move up and down. The arcs predict the roll steer. The position of the arms is critical. If the rack is not centered left to right, the tie rod lengths would be adjusted to set the correct toe. A vehicle with unequal tie rod lengths will have a tendency to roll steer while being driven on a rolling, hilly road. With equal jounce travel on each wheel, one wheel will have a greater steer tendency and will result in a slight directional pointing of the wheels.

If a steer tendency complaint is received, make sure the tie rod lengths are within 1/4 inch of each other. With the lengths equal and the steering wheel centered, the rack (or center link) should be centered when the wheels are straight ahead. If it is not centered, the rack mounting points may be out of position or loose, or the frame may be bent. With parallelogram linkage, the Pitman and idler arms should point straight forward or rearward when the wheels are straight ahead when the tie rods are equal in length. Correcting these problems will usually correct bump steer complaints. See Section 12-6 for a discussion of bump steer.

Steering Rack or Center Link Horizontal Positioning. The steering rack housing or center link may also be out of position relative to the lower control arm pivot height (Figure 12-13). This causes steer characteristics similar to those of improper centering. First, measure ride height to make sure it is correct, side-to-side, for the vehicle. The tires must have the correct air pressure. Then, take measurements from the floor to the lower control arm inner pivot and from the floor to the inner tie rod position. The differences between these two measurements for each side of the vehicle are recorded. The two recorded

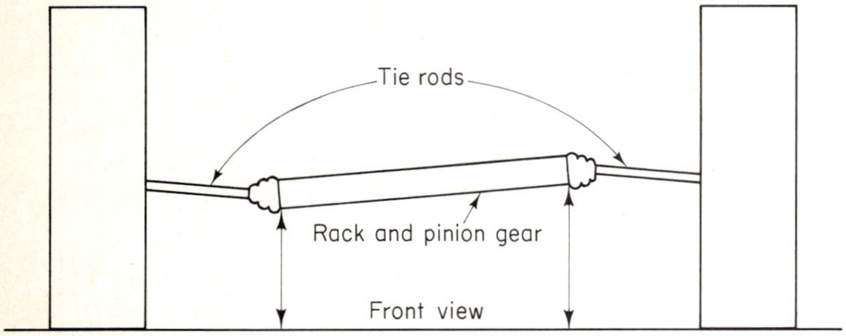

FIGURE 12-13 Misaligned rack and pinion steering gear will affect toe changes during jounce and rebound.

measurements should differ by less than 1/2 inch. Best ride and handling will result when the side-to-side readings differ by less than 1/4 inch. If the readings are out of specification, a check of rack and frame mounting position may be in order. On parallelogram linkage, the causes of ride and handling problems can be an incorrect idler or Pitman arm, a loose gear, worn out idler arm bearing, or other related problems.

Steering Rack or Center Link Angularity. Although this occurs relatively infrequently, the rack or center link may be out of position when viewed from directly above (Figure 12-14). Misalignment usually results from an accident. This problem can also cause a self-steering tendency over rolling bumps and incorrect toe on turns. The angular position difference can result in both a difference in the tie rod lengths and the angular relationship between the steering arm and the tie rods. The different angular relationship is the cause of incorrect toe-out on turns. The out-of-position rack or center link can be detected by using a carpenter's plumb bob to locate the position of the rack. Measure back to the control arm pivot on each side. Inspect the frame mounting points and frame position to check for anything out of specification. On parallelogram linkage, an incorrect-length idler arm or Pitman arm will cause the center link to be out of position when viewed from above.

FIGURE 12-14 Misaligned rack and pinion steering gear as viewed from above.

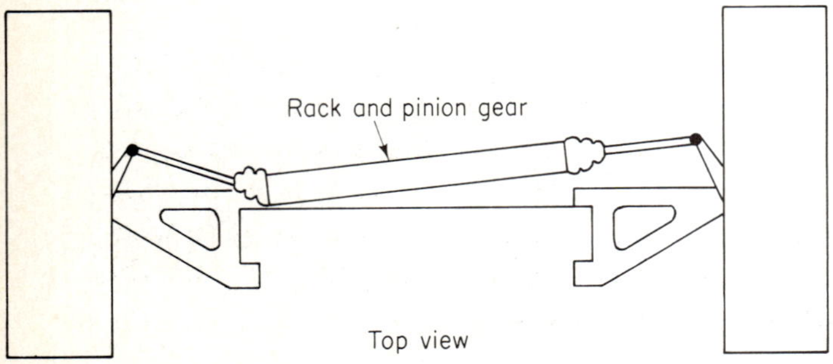

12-5 WHEEL SETBACK

Wheel setback is a condition where the wheelbase is different on each side of a vehicle. This creates an angle between the thrust line and the vehicle centerline. Wheel setback can lead to nonresponsive steering as the vehicle enters a turn. It may also lead to a wheel pull that is difficult to diagnose, incorrect toe-out on turns, or a bump steering tendency. Wheel setback may be caused by frame damage, control arm or bushing problems, or out-of-specification cross-caster.

Measuring Wheel Setback. Some electronic four-wheel aligners measure wheel setback automatically. However, in most cases, it must be measured manually. Wheel setback must be correct if the vehicle is to be aligned correctly.

Wheel setback can be measured with a tape measure from the front side of the rear wheel rim to the front side of the front wheel rim (Figure 12-15). Each side of the vehicle is measured and the measurements are compared. When using this technique, it is extremely important that the wheels are precisely straight ahead or you will get an incorrect reading.

An alternative and more precise method of measuring setback is to use a plumb bob to project

FIGURE 12-15 One way to measure setback.

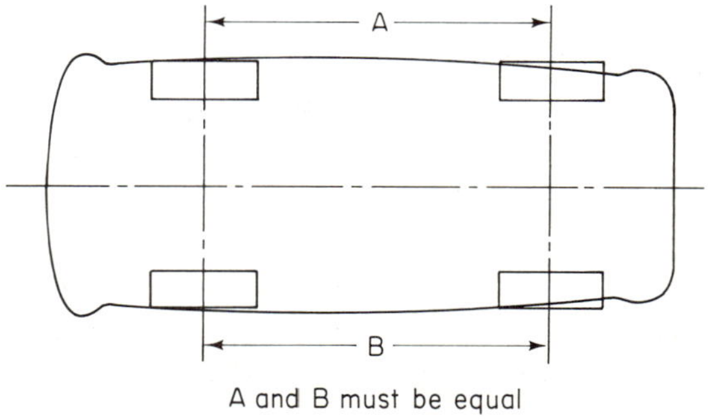

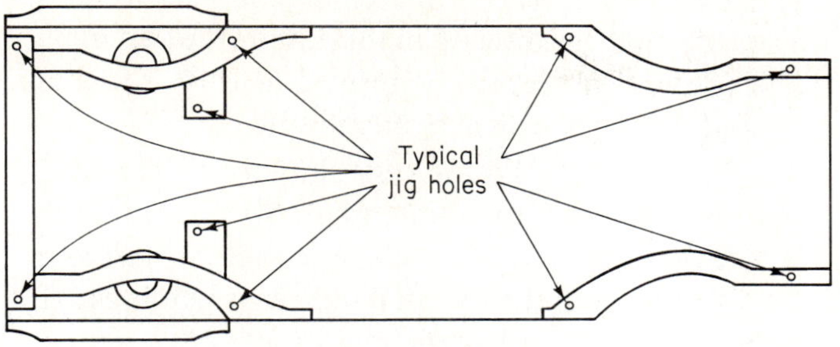

FIGURE 12-16 Jig hole used to locate critical parts and frame realignment.

the exact ball joint position onto the floor and mark the floor at the plumb bob point. Then, use the plumb bob to project critical frame jig holes to the floor. All framed, as well as unibody, vehicles have jig holes (Figure 12-16). Jig holes are used during assembly to locate critical parts. They are also used by frame-straightening specialists for frame realignment. Measure between the ball joint and jig hole points you marked on the floor on each side of the vehicle. The measurements are compared to determine setback. Combine the two methods of measurement to determine whether the rear-wheel or front-wheel position is incorrect.

Wheel setback should be less than 1/2 inch. Excess caster difference from side to side, or cross-caster, will position the wheel spindles to cause wheel setback. Bent frame mounting points, frames, or control arms, as well as bushing problems can lead to wheel setback. Careful visual inspection of the suspension and suspension-mounting points will usually lead to identification of the cause of the problem. Once found, the problem can be corrected.

12-6 ROLL STEER

Roll steer, or *bump steer*, is the steer tendency caused by each front wheel as the suspension moves through jounce and rebound. Although this does not normally cause a problem on production vehicles, it must be considered by those who are involved in performance vehicles, road racing, and oval track racing. They need to know how bump steer is both measured and changed. In this section, we discuss topics related to bump steering a vehicle. For those interested in studying handling, this will provide information on predicting how a vehicle will behave when it enters a turn. This can be done as a laboratory project as well.

Characteristics of Roll Steer. Most production vehicles are designed to produce a certain amount of toe-out during jounce and toe-in during rebound. This toe change produces a roll understeer condition as the vehicle enters a turn. Roll understeer is very predictable so it is used on average passenger vehicles with average drivers. The same type of vehicle may be used for racing performance road courses, solo events, and even drag racing. It can be modified to provide more precise handling and/or reduced wheel fight tendency by changing roll steer.

Roll steer is measured without the dynamic suspension loads that occur during cornering. As a result, it is important to understand how other factors influence roll steer when the suspension is dynamically loaded. One of these factors is *compliance steer* or the tendency of the wheel to steer in or out due to compliance of the suspension parts. Remember, *compliance* is the rearward movement of the suspension as the tire encounters road irregularities. Because the control arm bushings are quite compliant during cornering on current automobiles, the ball joint and spindle on the outside of a turn tend to move in toward the center of the vehicle. If the ball joint end of the steering arm is held in one position by the steering linkage, compliance will affect the steer angle of the wheel. The direction it is steered depends on whether the steering linkage is ahead of or behind the spindle. With the linkage ahead (front steer), compliance will result in a tendency toward roll understeer for the vehicle. With the linkage behind (rear steer), the vehicle will tend toward roll oversteer as a result of compliance.

It is important to understand the effects of steer caused by suspension compliance. When a vehicle has zero static roll steer with the steering assembly ahead of the axle, the compliance tends to produce roll understeer, which is safe. However, if the steering linkage is behind the axle with zero static roll steer, front suspension compliance would cause the vehicle to roll oversteer, which is unsafe. Caution should be used in either case when changing what the manufacturer has carefully designed into the suspension and steering.

Roll Steer Measurement. Roll steer can be measured in a number of ways, although the basic principle is the same: detecting the amount the wheel steers as compliance lets the suspension move. This measurement is complicated because wheel camber changes as the wheel moves through jounce and rebound. This camber change must be measured at both the front and rear of the tire from some fixed reference point.

To measure roll steer, the suspension must be free to move into both jounce and rebound. The springs must be removed to do this. The suspension is reassembled without the springs. The vehicle must be supported from the frame and be level. In addition, the steering wheel must be locked in a straight-ahead position and the wheels must have the proper toe setting. A method of easily raising and lowering the suspension from its normal ride-height position should be available. A small hydraulic jack or scizzors jack often can be used to move the suspension.

One method is to make and bolt a large smooth rectangular plate onto the wheel hub. The plate should be approximately the same width as the tire diameter. A stand with a dial indicator on one end, pointer on the opposite end, and a weight to force the pointer to maintain contact with the plate is used. The weight holds the pointer against the plate while the dial indicator measures any change in distance from the pointer to the indicator (Figure 12-17). This device then cancels out any change in camber and reads toe changes directly in thousandths of an inch. The suspension should be moved at least 3 inches up and 2 inches down from ride height.

A second method is very similar to the above method; however, two dial indicators are mounted to a fixed arm. The suspension is then moved and the differences in the two dial indicator readings are recorded for the roll steer measurement.

The third method is easy to do. Clamp a straight bar with a length equal to the tire diameter to the wheel hub. An extended tape measure is fastened so it can pivot at a distance 6 to 10 feet from the wheel hub (Figure 12-18). The measurements are taken at each end of the attached bar. The suspension is moved to a new height and the measurements are taken again. The first measurements for each end of the bar are subtracted from the second. The difference in the subtracted figures is the toe change. If the front measurement changes less than the rear measurement when the suspension moves, the suspension is producing a toe-out change. If this toe-out occurs with the suspension moving up (jounce), the vehicle would tend toward roll understeer, ignoring compliance. Many racing vehicles try to use suspension pivots with minimal compliance, front steer, and zero roll steer for optimal vehicle handling.

The roll steer of a chassis can be altered by slightly changing the steering linkage pivot locations. On race cars, spherical ends are used for the tie rod

FIGURE 12-17 Special purpose indicator for measuring roll steer.

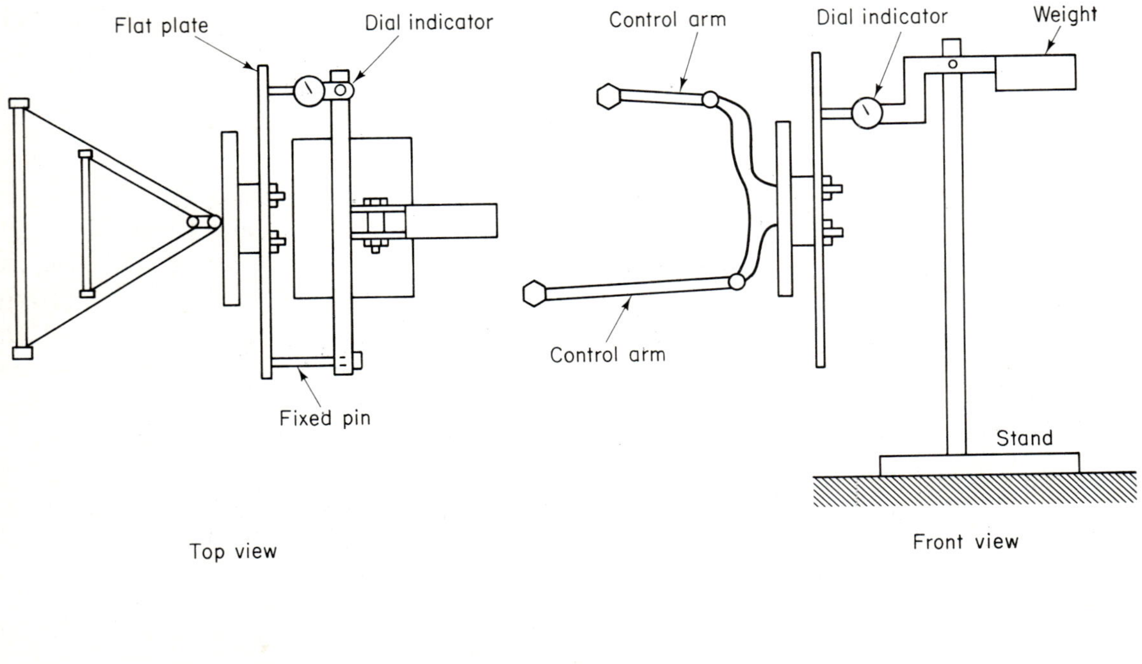

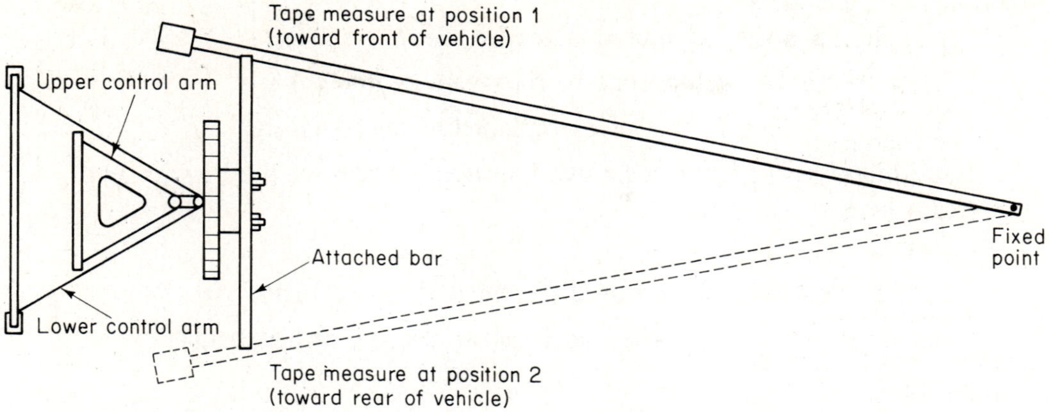

FIGURE 12-18 Top view of measuring roll steer using a tape measure and a bar bolted in place of the wheel.

ends and these can be shimmed up or down for the desired change. The rack and pinion gear can also be shimmed up or down. With parallelogram linkage, the idler arm pivot or the steering gear mounting can usually be moved slightly, within its mounting holes, to provide the desired result. On a race car, the mounting holes can be modified to produce the desired results.

With a steering linkage ahead of the suspension, lowering the rack and pinion gear (or the center link) will reduce the amount of roll understeer. On a steering linkage behind the suspension, the gear or center link must be raised for the same results. On some race cars, the height and length between pivot locations on the rack can be changed to provide adjustment to get the desired reaction. On production vehicles, it is much more difficult to change the pivot points because the tapered tie rod end mountings cannot be shimmed. This limits the adjustment you can make to moving the pivot location of the center link.

12-7 SUMMARY

This chapter covered both normal steering linkage service and steering-linkage-related handling problems. The problems associated with darting (self-steering over bumps) were also covered. In addition, the last section gave some information on steering linkage design considerations used for racing vehicles. It was included to provide you with a greater depth of understanding than is normally available to the technician and, as a result, to better prepare you to diagnose steering and handling problems.

STUDY QUESTIONS

Section 12-1

1. Name three ways a tie rod end can be freed from the taper socket in a steering arm.
2. What is done during disassembly that makes it easier to realign tow and the steering wheel?
3. What special care should be taken before tightening the adjuster sleeve clamps?

Section 12-2

4. When should a tie rod spreader not be used?
5. What joints must be loosened to remove a center link?
6. When is an idler arm and pivot replaced as an assembly?
7. What might result from using two hammers to remove the Pitman arm from the Pitman shaft?

Section 12-3

8. What problem is indicated by different toe-out on turns?
9. What must be replaced when an integral steering arm is bent?

Section 12-4

10. Why is it important that the tie rods are the same length?
11. Why is it important to have the center link on level?

Section 12-5

12. What results from incorrect wheel setback?
13. What is the best way to measure wheel setback?
14. What is the maximum acceptable wheel setback?
15. What causes wheel setback?

Section 12-6

16. What action does bump steer have on most production vehicles?
17. Define: compliance steer.
18. What is the basic principle used when measuring roll steer?
19. What must be done to the suspension before roll steer can be measured?

13

Manual Steering Gears

Current trends in vehicle designs require a precise, responsive steering system. This has evolved from a system that provided relatively slow steering with a large margin for driver error. Older vehicles were heavy, slow responding, and soft riding using bias ply tires. Today's vehicles have undergone major design changes to meet the requirements of the modern driver. In addition to these requirements, the basic automobile has moved from front-engine, rear-wheel drive to front-engine, front-wheel drive. These changes have all had an influence on the steering system design.

The driver controls the direction of the vehicle's front wheels through the steering gear. Modern steering gears are made of two major units: a gear unit and a steering column. The gear unit multiplies the driver's steering effort to provide enough force for steering control. The steering column connects the steering wheel to the gear unit.

13-1 BASIC STEERING GEAR REQUIREMENTS

The steering system provides a means to change the rotation of the steering wheel into spindle rotation about the steering axis. This points the wheels in a new direction. The driver sits behind the steering wheel on one side of the vehicle. The front wheels move up and down while encountering bumps and dips in the road surface. The vehicle must not self-steer while encountering these changes in the road surface. In addition, the system must be simple, require minimum service, and provide maximum reliability.

The mechanical advantage needed in the steering system occurs at two points: between the steering column shaft and the linkage and between the linkage and the wheel pivot. The *steering ratio* is commonly described as the angular rotation (number of turns) of the steering wheel compared to the angular rotation of the wheel spindle.

The driver's turning effort is multiplied by the mechanical advantage through the steering system to turn the front wheels, even when the vehicle is stationary. The gear ratios need to be fairly high to provide this mechanical advantage. For responsive steering, the gear ratios should be fairly low. The front wheels must be able to be steered approximately 30 degrees in each direction to provide a reasonable vehicle turning radius. This means the wheels have a turning arc of about 60 degrees when the steering wheel is turned from one extreme to the other. Most current-design automobiles require about three full turns of the steering wheel to turn the front wheels from one extreme to the other. In this example, the steering wheel would rotate 1080 degrees ($3 \times 360°$) as the wheels turn 60 degrees. The steering system has an overall steering ratio of

FIGURE 13-1 Typical manual rack and pinion steering gear.

1080/60 = 18:1. Most modern automobiles have a steering ratio of about 16:1, while earlier, heavier automobiles had ratios as high as 25 and 30:1.

Manual steering gears have evolved from early worm and sector designs to the more popular rack and pinion type. Manual steering gears are based on the action between a worm gear and a spur gear section or a rack and pinion to provide the required mechanical advantage (gear ratio) between the steering shaft and linkage. With this mechanical advantage, the vehicle can easily be controlled and the front-wheel road shock will not twist the steering wheel from the driver's hands.

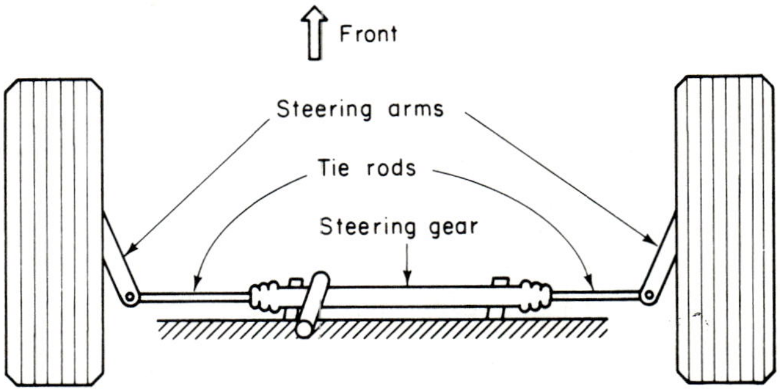

FIGURE 13-2 Top view of a typical rack and pinion steering gear linkage.

13-2 RACK AND PINION STEERING

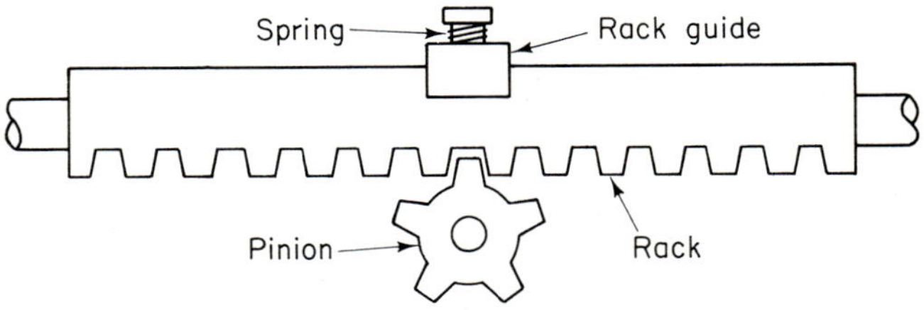

FIGURE 13-3 Operation of a rack and pinion steering gear.

Rack and pinion steering was first introduced in European vehicles. In 1971, Ford was the first domestic manufacturer to use it in high-volume production on the Pinto. Since then, it has become the major type of steering gear for small, light vehicles. The rack and pinion steering gear is simple, light, and responsive (Figure 13-1). It costs less to make and takes less space than the recirculating ball steering gear. There are far less points for friction to dampen steering feedback so the driver has more road feel. Unfortunately, this also increases steering harshness and noise when operating on rough roads. This may cause damage more quickly than it would using a recirculating ball steering gear.

The rack and pinion steering gear takes the place of the center steering link used in parallelogram steering linkages. The steering gear is fastened to the frame or dash bulkhead through rubber-type insulators (Figure 13–2). The rack ends, under a rubber boot, commonly fasten to the tie rods with ball-type ends. Some manufacturers connect the tie rod to the rack center, which is referred to as a center take-off gear. A pinion, controlled by the steering wheel, has teeth that engage with the teeth of the rack. The operation of a rack and pinion system is shown in Figure 13–3. A yoke bearing rack guide puts a preload on the rack directly across from the pinion so there is no free play between the gears.

13-3 WORM AND RECIRCULATING BALL STEERING

The worm and recirculating ball gear, usually referred to as a *recirculating ball gear* (Figure 13-4), has been popular in domestic automobiles for years. With the move toward front-engine, front-wheel-drive and the demand for more precise handling, it has been largely replaced by the rack and pinion gear. Some manufacturers still use recirculating ball steering in larger automobiles. It also remains the popular system for vans and light trucks.

Basic Construction. The recirculating ball gear changes the rotation of the steering wheel to the rotation of an output shaft called a *sector* or *Pitman shaft.* The Pitman arm is splined to the Pitman shaft. One end of the steering center link (Figure 13-5) connects to the Pitman arm. This changes the rotating motion to a back-and-forth motion to move the center link and tie rods.

A nut with exterior teeth fits over the worm in a recirculating ball steering gear (Figure 13-6).

FIGURE 13-4 Section view of a typical manual recirculating ball steering gear.

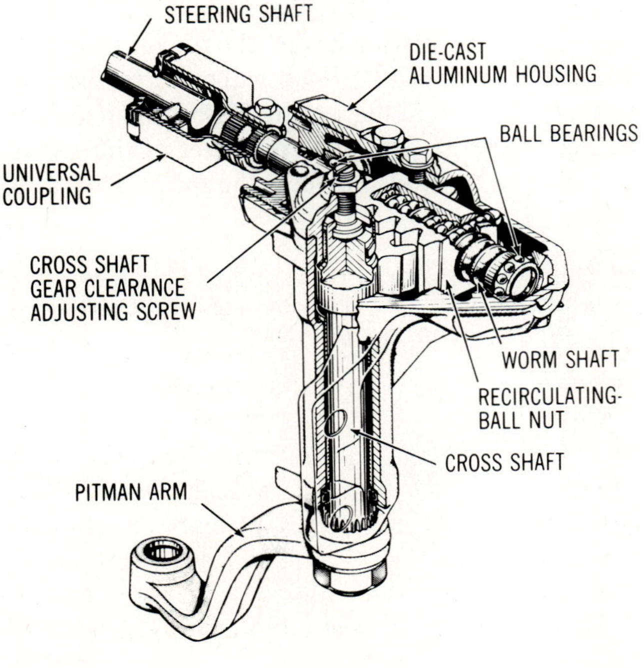

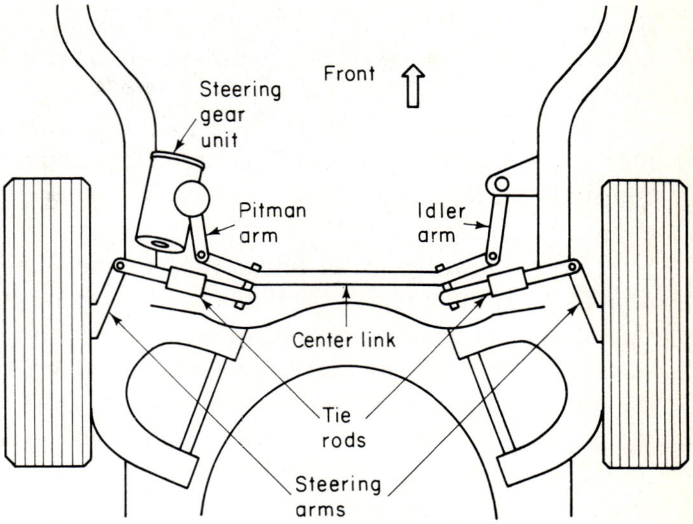

FIGURE 13-5 Top view of a typical parallelogram steering linkage ahead of the control arms.

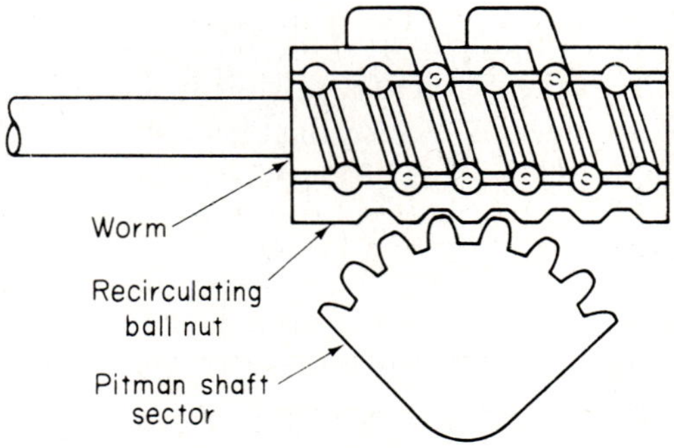

FIGURE 13-6 Operation of a recirculating ball steering gear.

Threads between the nut and worm have grooves that will take ball bearings. Selective-fit balls, very close to the same size, are in grooves so the turning action is as free rolling as a ball bearing. The ball groove may be cut for either right- or left-hand thread directions.

The balls operate in two separate ball races. These races are on an angle so the balls move endways as the worm turns. The balls are redirected by a guide from the end of each slot, across the outside of the nut, and back into the slot entrance. The guide can be identified at the bottom of Figure 1-7.

The gear teeth on the ball nut engage with the sector that, in turn, is part of the cross-shaft, sector shaft, or Pitman shaft. This whole assembly is enclosed in a housing that is bolted to the chassis frame.

Each end of the worm is supported in ball bearings. A worm lash adjustment is used to preload the bearings that hold the worm. The adjustment is on either the upper or lower end of the housing.

Needle bearings or bushings support the Pitman shaft. Both may be located on the Pitman arm side of the sector or one may be on each side of the sector. The ball-nut gear teeth and the sector teeth are cut at an angle. Moving the sector gear into close mesh with the ball-nut teeth reduces the amount of backlash. This adjustment is made on the cover at the upper end of the cross-shaft. You can see how the lash adjuster screw controls backlash in Figure 13-7.

The steering gear teeth are cut to provide a closer fit at the center teeth than at the end teeth as illustrated in Figure 13-8. This reduces gear backlash at the center position where most driving occurs. At the same time, it allows cost-reducing gear tolerance at the off-center positions where the close fits are not needed. Sector backlash adjustments are made with the steering gear centered so the teeth are at the tight center position.

The manual steering gear housing is partly filled with a specified lubricant. A seal is used at the lower end of the Pitman shaft to keep the lubricant from leaking out the bottom of the housing around the Pitman shaft. In most manual steering gears, there is only a dust seal on the upper end of the worm shaft because it is above the lubricant level.

FIGURE 13-8 Typical steering gear sector-to-ball-nut contact. (Courtesy of Buick Motor Division, General Motors Corporation)

FIGURE 13-7 Cross-section of a typical worm ball-nut-type steering gear showing the angled teeth that allow lash adjustment. (Courtesy of Chevrolet Motor Division, General Motors Corporation)

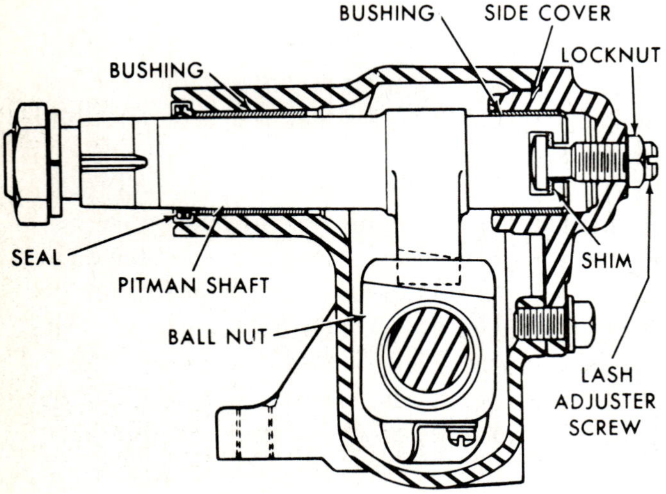

Recirculating Ball Steering Ratios. There is a gear reduction in the steering gear and there is also a gear reduction between the Pitman arm and the steering arms.

The steering gear ratio is between 16:1 and 25:1, depending on manufacturer and vehicle model. As an example, the early Corvette had a 16:1 ratio manual recirculating ball steering gear. Intermediate- and full-size cars of the same vintage used a 24:1 ratio manual steering gear. The ratio is determined at two different points in the gear assembly. If we think of the worm and ball nut as a screw and nut, it is easy to see that the ratio can be changed by changing the angle of the groove between the worm and the ball nut. By increasing the pitch of the ball race on the worm, the ball nut will move farther for the same rotation of the worm. This would decrease the steering ratio. The pitch diameter of the sector gear (Figure 13-9) also affects the ratio. The ratio

FIGURE 13-9 The pitch diameter of the sector affects the steering gear ratio.

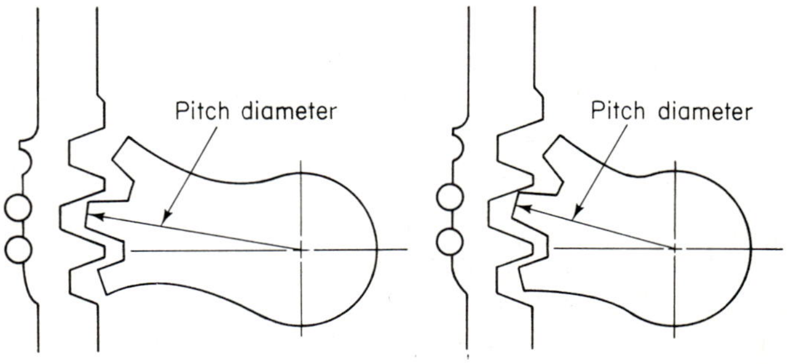

of the gear is increased as the radius of the sector gear is increased. Commonly, within the same gear type, the ratio is changed by the groove pitch between the worm and ball nut.

An overall steering ratio can also be changed by changing the length of the Pitman arm or the steering arms. Some models of the same vehicle line have different ratios by using different length arms. This is commonly used to change the overall ratio as much as 15 percent.

13-4 STEERING GEAR SERVICE

The steering gears are extremely reliable and maintenance-free. In this section, we discuss general service that can be performed on both rack and pinion and on recirculating ball gears. The better you understand the operation of the steering system, the better you can diagnose and solve problems. You will need to use a shop manual when you service a specific gear.

Rack and Pinion Steering Gear Service. Most rack and pinion steering gears are serviceable. Normal service includes boot replacement, inner tie rod end adjustment or replacement, pinion preload adjustment, rack and pinion mesh preload adjustment, bushing replacement, and inspection of the rack for straightness or wear. Some rack and pinion gears are not serviceable and must be replaced when they are faulty.

The steering gear is normally removed before it is serviced. It must be disconnected from the steering linkage and steering column before it can be removed from the chassis. Rack and pinion gears are disconnected from the steering arms at the outer tie rod ends. Remove the retaining nut and use a puller to free the tie rod ends (see Figures 12-1 and 12-2).

Sometimes, it is necessary to service the inner tie rod end on a rack and pinion steering gear. It is easier to service this joint when the steering gear has been removed. It can be serviced on most vehicles using special tools and special service techniques. Do not try to service these joints while the gear is still on the vehicle without the proper tools and intructions.

Loosen the lower clamp bolt at the universal joint and lightly pry the joint collar to free the steering gear from the steering column. Then loosen the gear from the frame or cowl. When the rack and pinion gear is behind the engine, it will have to be removed through the wheel well. On a number of vehicles, it will be necessary to loosen and lower a frame member enough for clearance to remove the gear. The gear is freed from the steering column universal joint as the gear is removed.

The manual rack and pinion steering gear is normally tightly sealed and filled with steering gear lubricant. If a tie rod boot fails, the lubricant will leak out and dirt and water will leak in. Either of these will lead to excessive gear wear. If a boot is cracked, the complete gear should be disassembled and inspected. Clean the outside of the gear to help keep dirt out of the gear parts. Remove the clip from one of the boots and catch the lubricant as you drain it from the gear.

Count the turns as you remove the outer tie rod ends from the rack and pinion gear; then remove the tie rod boots. Wrap a clean shop towel around the end of the gear to keep chips out of the gear as you loosen the inner tie rod end (see Figure 1-20). The inner tie rod end socket is threaded onto the end of the rack (Figure 13-10). The threaded end is locked in place with jam nuts, locking pins in jam nuts, set screws, spiral pins, or staking. You must remove the lock before the tie rod can be removed from the rack. Use an easy-out tool with a slide-hammer puller to remove the spiral pin. Solid pins must be drilled out. Special alignment fixtures should be used to drill out the pin. The hole should be no more than 1/4 inch deep. Staking should be ground

FIGURE 13-10 The inner tie rod end is threaded onto the end of the rack.

FIGURE 13-11 Holding the rack with a wrench across the teeth to keep it from rotating as the tie rod end is unscrewed.

FIGURE 13-13 Removing the rack-bearing yoke.

off. Hold the rack with a wrench across the teeth to keep it from rotating as you unscrew the tie rod from the rack (Figure 13-11). Clean and inspect the threads on the end of the rack. If there are two locking holes in the rack already, the gear must be replaced. The tie rods from the center take-off rack (Figure 13-12) are removed by unbolting the inner tie rod bolts.

Clamp the gear in a vise on one of its mounting tabs. It is best to hold a rack and pinion gear in a mounting fixture.

On most rack and pinion steering gears, the rack-bearing yoke (Figure 13-13) and, possibly, the rack, will have to be removed before the pinion and/or pinion bearing can be removed. Remove the bearing plate or adjuster plug; then lift the pinion assembly from the housing.

After the pinion is removed, the rack will slide from the housing. Inspect the rack housing bushing, rack, and rack teeth for wear and/or damage. Place the rack in small V-blocks and place a dial indicator against it. Rotate the rack and measure the runout (Figure 13-14). If the runout is too great, usually 0.15 inch (0.38 mm), the rack must be replaced. Bent racks are not uncommon and they cause slight binding. This results in a limited wheel return force and wear. The entire steering gear assembly is usually replaced if the rack is bent. Note the condition of the parts and make the required checks as you disassemble the steering gear. Inspect the parts for looseness, binding, scoring, and wear as described in the service manual.

FIGURE 13-14 A dial gauge is placed against the rack as it is rotated in V-blocks.

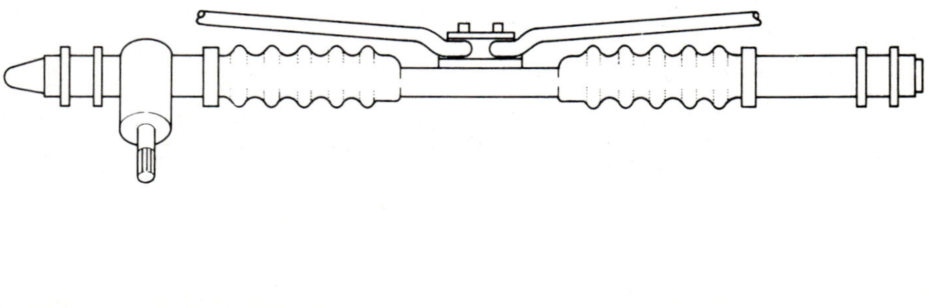

FIGURE 13-12 The inner tie rod ends are unbolted from the center-take-off rack.

Pinion damage can result from contaminants or other problems, such as wheel shimmy. The bearings should be inspected before assembly. Damaged bearings should be replaced. Make all necessary repairs. Replace all seals and lubricate them with steering gear lubricant. The lubricant specified by some manufacturers is gear lubricant while other manufacturers specify lithium grease. Make certain you use the correct lubricant.

Repair kits are available to rebuild rack and pinion gears. With some gear designs, special tools are required to service them. Shops often specialize in one or two gear designs and have the special tools for them. Assembly instructions must be followed carefully to ensure that all parts are put in their proper place with the correct adjustments. The steering worm pinion bearing preload is set during assembly. It is adjusted with shims (Figure 13–15) or an adjuster plug with a lock nut (Figure 13–16; also see Figure 13–13). The preload is specified either as a required turning torque or a required preload distance (shim-type only). This preload keeps the rack the correct distance from the pinion. The yoke on the rack opposite the pinion is spring loaded to hold the rack against the pinion (see Figure 13–13). Run the assembled gear through its full turning range and feel for any roughness or binding. Correct any problems you find. If there are any problems, they can be corrected at this time without having to remove the gear from the vehicle again.

FIGURE 13-15 Shims used to adjust the pinion-bearing preload.

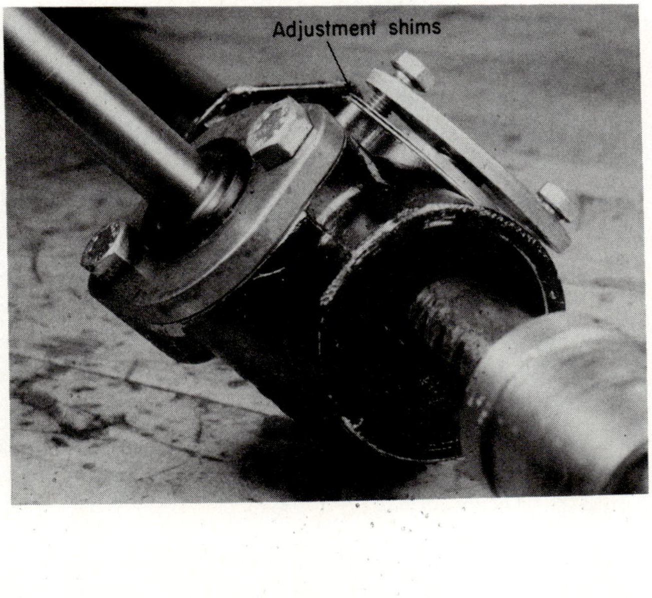

FIGURE 13-16 An adjuster plug and lock nut used to adjust pinion-bearing preload.

Install new inner socket assemblies on the rack. Use a good multipurpose grease in the socket assembly. Tighten the assembly to the correct torque. You will need a crow's-foot wrench on a torque wrench to tighten the socket assembly. Hold the rack with a wrench across the teeth to keep it from rotating as you tighten the socket assembly. Check the socket movement with a spring scale (Figure 13-17). It should take between 1 and 9 pounds on the outer end of the tie rod to move it. Lock the joint socket assembly by staking or installing a new pin. Use a drilling fixture if you need to drill a new hole. Be sure to follow the locking instructions carefully. Install a new boot and new clamps on the end farthest from the pinion gear. Use the correct clamps. Screw clamps tend to cut the boot and cause a leak. Pour the correct amount of gear lubricant (usually 90-weight gear lubricant) in the end closest to the pinion gear. Install the remaining boot and clamps.

Mount the steering gear in the vehicle using the correct bushings and clamps. Insert the pinion in the universal joint on the steering column as you install the steering gear assembly. Make sure you center the rack and pinion gear housing. When three mounts are used, tighten the center mount before tightening the end mounts (Figure 13-18). Make sure the

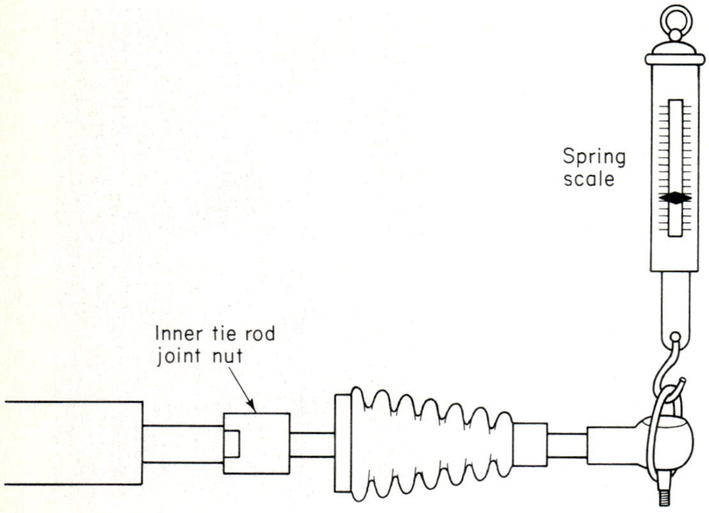

FIGURE 13-17 A spring scale is used to check the correct tightness of the inner tie rod end.

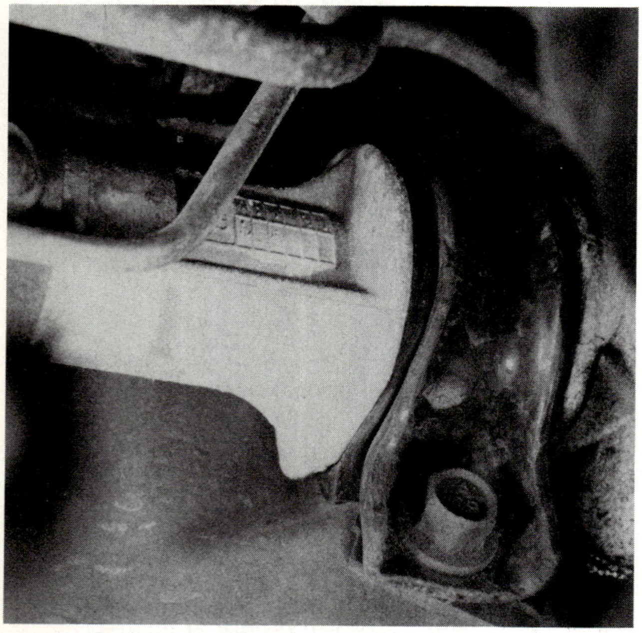

FIGURE 13-18 Typical mounting clamp used to hold a rack and pinion steering gear.

steering operates smoothly through its full range of travel. Complete the assembly by installing the tie rod ends, turning them the same number of turns they had when removed. The tie rods should be the same length as discussed in Section 12-6. The repair is completed when the vehicle toe is set.

Recirculating Ball Steering Gear Service. The recirculating ball steering gear is relatively service-free. If necessary, the worm bearings and sector lash screw can be adjusted. It is easier to make the adjustments with the gear out of the vehicle. Any other repair is done with the gear removed.

To perform the lash adjustment with the gear installed, first disconnect the steering linkage from the steering gear. This will keep the steering linkage friction from affecting the gear turning torque measurements. The horn ring assembly and cover pad are removed from the steering wheel so a torque wrench can be used on the steering wheel-retaining nut. Use an inch-pound torque wrench with the correct size socket on the steering wheel retaining nut to measure the turning torque.

The worm shaft torque should be measured and adjusted first, if necessary. It is measured by rotating the steering wheel to one limit stop. Then, rotate it back approximately one-half turn and measure the turning torque at the point where there is free play between the ball-nut and sector gears. Normally, the torque should be about 5 to 8 inch-pounds. The worm-bearing preload is adjusted by loosening the worm adjuster lock ring and rotating the worm bearing adjuster (at the left in Figure 1-21) to obtain the proper torque. When the proper reading is obtained, the lock ring is tightened. The preload is rechecked at each end of the gear's travel. Check the gear for smoothness as it is moved from limit to limit.

The over-center torque reading is checked by turning the steering gear to its straight-ahead position. The torque reading is taken as the gear is rotated through one full turn (one-half turn each side of center). The reading should increase and then fall off. The sector lash adjusting screw (see Figure 13-7) is turned clockwise to increase the torque measurement and counterclockwise to decrease the torque. Normally, the over-center torque should be about 10 to 15 inch-pounds higher than the worm-bearing preload. A maximum reading in the over-center position should be limited to about 25 inch-pounds. The gear is rotated through its full travel in both directions as the torque reading is checked and any roughness is noted. If the over-center readings are over 5 inch-pounds more in one direction than the other, the sector endplay may be excessive. If any roughness is felt, the gear should be removed and inspected.

Do not adjust a gear that has roughness as it is turned. The steering gear assembly must be removed to correct excess sector end play or roughness. *Note:* The values given are approximate. Check the correct values in the shop manual for the gear being serviced.

Steering Gear Removal. The steering gear can be removed from the vehicle after removing the steering shaft coupling and the Pitman arm. Remove the Pitman arm attaching nut and pull the arm from the Pitman shaft (see Figure 12-6). The position of the arm should be noted because the arm can be placed in any one of three or four positions; however, only one position will allow the gear to function in the vehicle. Remove the bolts that hold the gear on the vehicle frame and remove the gear assembly. With the gear centered, the sector retaining plate bolts (Figure 13-19) are removed, and the sector can be lifted from the case.

The worm shaft is removed by loosening the adjuster plug lock ring and removing the adjuster plug while sliding the worm shaft from the case (Figure 13-20). The gear parts can now be cleaned and inspected. Any worn or damaged parts should be replaced. Before assembly, the sector shaft end play should be checked. This is usually controlled with a selective washer on the lash adjusting screw (Figure 13-21). It is called a *shim* in Figure 13-7. If the endplay is within specifications, the gear is assembled, reversing the disassembly procedure. Replace any damaged sector shaft and worm shaft bearings. Install a new sector shaft oil seal. A thin film of gear oil should be placed on all bearing surfaces. The gear adjustment is performed as described in the procedure for the gear still installed. The gear is filled with the proper lubricant and installed in the vehicle. All fasteners should be properly torqued and secured. Recheck the steering gear and linkage assembly to make sure the steering system will travel smoothly from limit to limit without binding or looseness.

FIGURE 13-19 Removing the sector shaft retaining plate.

FIGURE 13-20 Sliding the worm shaft from the case.

FIGURE 13-21 Selective washer at the head of the lash-adjusting screw to control pinion endplay and steering wheel free play.

STUDY QUESTIONS

Section 13-1

1. What is a typical steering ratio for a modern vehicle?

Section 13-2

2. What are the advantages of a rack and pinion steering gear over a recirculating ball-nut steering gear?

Section 13-3

3. What type of bearings supports the worm shaft in a ball-nut steering gear?
4. What bearing supports the cross- or Pitman shaft in a ball-nut steering gear?
5. How is the ball-nut steering gear made to take up the lash of the gear teeth?
6. Where should the steering gear be set when making sector backlash adjustments?
7. What two things control the gear ratio within the ball-nut steering gear?
8. What steering linkage parts affect the overall steering ratio?

Section 13-4

9. What rack and pinion service can be done while the gear remains on the vehicle?
10. What keeps lubricant in a rack and pinion steering gear?
11. What methods are used to lock the adjustment on the inner tie rod end of a rack and pinion steering gear?
12. What parts must be removed before the rack can be removed?
13. What should be done to a bent rack?
14. How should rack and pinion gears be repaired if you don't have the special tools required?
15. How are the pinion-bearing preloads adjusted?
16. When, during reassembly, is lubricant put in the rack and pinion gear?
17. What adjustment can be made on a recirculating ball steering gear while it remains mounted in the vehicle?
18. How is steering wheel free play adjusted on recirculating ball-nut steering gears?
19. What failure would cause oil to leak from a recirculating ball-nut steering gear?
20. What parts are checked for wear in a recirculating ball-nut steering gear?

14

Power Steering Systems

Power-assisted steering began to appear on automobiles in the early 1950s. It found its place on the heavier automobiles of the time as drivers were demanding more comfort and ease in driving. These first power steering gears were hydraulic-assisted manual steering gears, both link and recirculating ball types. As the popularity of rack and pinion steering grew, power-assisted rack and pinion steering gears were developed. Until very recently, hydraulic-assisted steering gears were the only type available. Electronic vehicle control has brought about the use of electricity to assist power steering systems.

14-1 POWER STEERING REQUIREMENTS

The power steering system is designed to relieve the driver of the majority of the physical effort needed to steer the vehicle. It provides feedback to give the driver a feel of control of the vehicle. The power steering gear returns the vehicle to the straight-ahead position if the steering wheel is released after a turn. This is called *neutral returnability*. The power steering gear reduces feedback from rough roads. In addition, the system uses a minimum of power while assisting steering. It reduces the power requirements to the absolute minimum while driving straight ahead. Careful design has minimized the power used for steering so it does little to reduce fuel mileage.

A careful balance between the amount of assist and the amount of road feel is designed into power steering systems. Today's vehicles require greater steering effort over some of the earlier vehicles. Power steering improves control of the responsive-handling vehicles. The amount of driver effort compared to the amount of assist is an indication of steering and handling feel.

Power-assisted recirculating ball steering gears usually have lower gear ratios than their manual steering counterparts. Power-assisted rack and pinion gears usually have the same ratios as their manual counterparts with considerably reduced steering effort.

14-2 POWER STEERING GEARS

Power steering is basically power-assisted manual steering. The driver supplies part of the steering effort. The power-assist part of the steering system supplies the remaining steering effort. Reduction in driver steering effort needed with power steering allows the power steering gear ratio to be reduced on some systems to about two-thirds of the manual steering ratio. Power steering in these systems provides much faster steering response than manual steering.

The power steering system has an engine-driven oil pump with reservoir, control valve, and pressure surfaces to assist steering effort. A typical power steering system is illustrated in Figure 1-22. The belt-driven pump is on the front of the engine. In some applications, a direct-drive pump is mounted at the front of the crankshaft. Pressure surfaces in most power steering systems are located in the steering gear housing. This type of power steering gear is referred to as *integral*. Some power steering pressure surfaces are located in a separate power cylinder connected between the steering linkages and vehicle frame. These are called *link-type power steering gears*. A power rack and pinion is similar to a link-type power steering gear. This is shown in Figure 14-1.

Most hydraulic power steering systems use a belt to drive a pump. The pump forces hydraulic fluid through lines into pressure chambers to assist steering. Belts wear and fluid leaks. An electric-motor-driven power steering pump was developed for the 1988 Fiero. Electronic power steering, introduced in 1988, is simpler. It uses electricity from the vehicle's electrical system to supply the power to an electric-motor-driven power steering pump. There is one less belt and no hydraulic system leaks. Energy is only used when power is being used to assist steering. Unlike hydraulic steering, electronic power steering will still provide steering assist if the engine should stop. It can be used on rear- and mid-engine vehicles.

One design is being developed by TRW. It uses a ball nut and worm on the rack in place of the hydraulic pressure surfaces used in a hydraulic power rack and pinion steering gear. The ball nut and worm operate like the ball nut in a recirculating ball-nut steering gear. A hollow armature electric motor is connected to the worm nut. Figure 14-2 is a simple drawing of this design. The ball nut is held in the housing with a ball bearing. The armature rotates the ball nut to assist in moving the rack. A torque sensor in the pinion housing provides a signal to the electronic controls to operate the motor. The motor supplies the boost required for the turning effort needed at vehicle speed. The electronics can be designed to give different amounts of boost under different operating conditions. The amount of assist can be made to change with changes in vehicle speed, road conditions, or the preference of the driver. Adjustable steering feel may be the next step in electronic power steering.

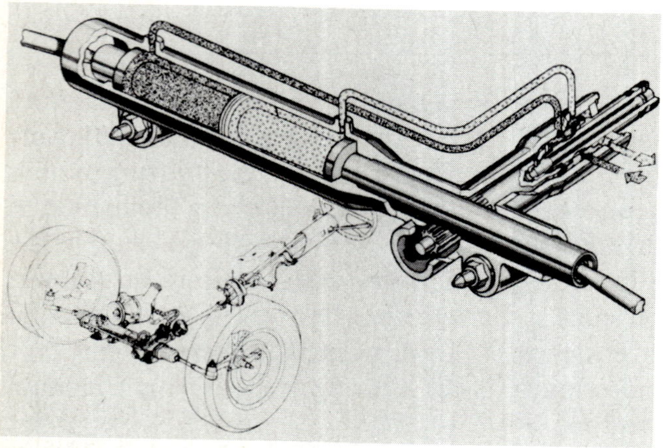

FIGURE 14-1 Section view of a typical power rack and pinion steering gear. (Courtesy of Ford Motor Company)

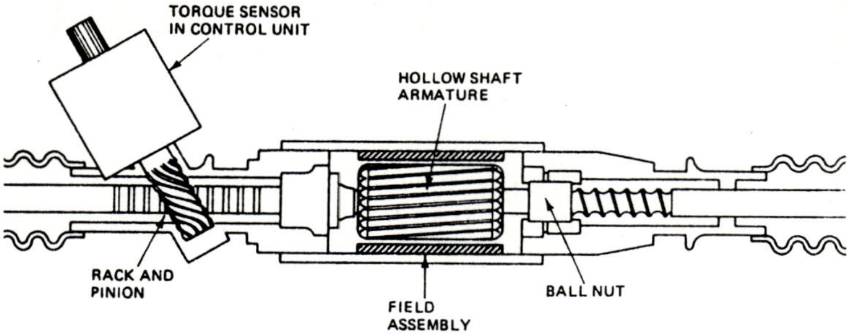

FIGURE 14-2 Electronic power steering gear.

If the electrical system or control system should fail, the gear can be steered like a manual steering gear. You know that you can rotate the steering wheel from one front tire when both front tires are off the ground on a vehicle with a hydraulic power ball-nut steering gear. This forces the ball nut to rotate the worm. It works in a similar way with electronic power steering. With no electricity, the pinion forces the rack endways and this movement forces the ball nut and armature to rotate. It actually takes less force than moving an unpowered hydraulic power ball-nut steering gear because there is no fluid to force through hoses and valves.

The discussion in this chapter covers hydraulic-assist power steering systems as they are currently used on passenger vehicles.

Control Valve. When the engine is running, power steering fluid flows through the system from the pump to the control valve, then back to the reservoir. The same amount of system pressure is applied to both pressure surfaces when the control valve is centered so the pressure has no effect on steering. The power steering fluid flows freely through this open-center fluid circuit. This can be seen in Figure

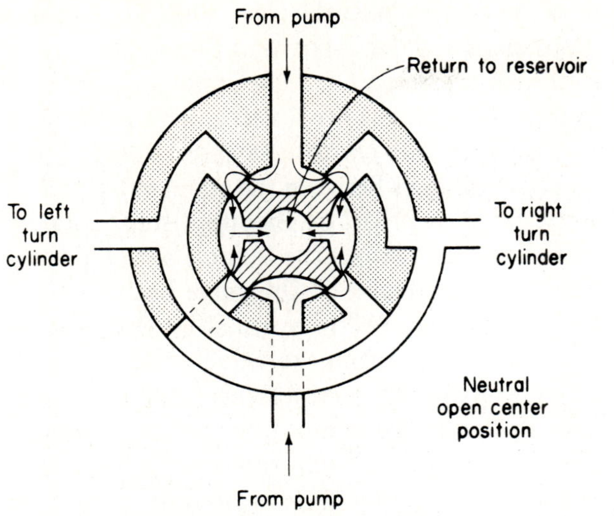

FIGURE 14-3 Centered control valve with all passages open while the steering gear is not being turned.

14-3. When steering effort is applied to the steering wheel, the control valve shifts to block one fluid return passage. This directs fluid to one pressure surface. At the same time, it increases the size of the return passage between the opposite pressure surface and the reservoir (Figure 14-4). The amount of valve shift is proportional to the effort applied to the steering wheel. This, in turn, proportionally increases power assist.

The power steering control valve is either located inside or is attached to the outside of the integral-type power steering gear. In link-type power steering, the control valve may be built into the end of the power cylinder or it may be a separate unit. The control valve on a rack and pinion power steering gear is built into the pinion housing.

FIGURE 14-4 Valve rotated to close the pressure flow going to the right turn cylinder and opening the flow to the left turn passage. This builds pressure in the left turn passage at the same time the right turn cylinder passage is opened to return the fluid to the reservoir.

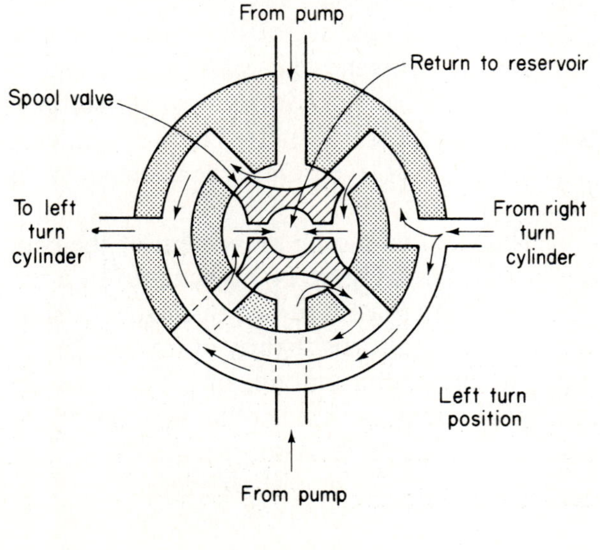

The control valve is balanced between the mechanical input force applied by the steering wheel and the mechanical-hydraulic resistance of the steering linkage and tires on the road. As the steering wheel is turned, its mechanical input force moves the control valve toward the linkage resistance. This is shown in Figure 14-4. In this position, the control valve restricts the pressure area return flow so pressure will build up on one pressure surface. This assists the steering wheel input force to move against the resistance of the steering linkage. During steering wheel movement, the mechanical input force always leads by first moving the control valve. The hydraulic assist force is always trying to catch up with the control valve position by helping to move the steering linkage. When the steering wheel arrives at the desired position, it is held steady and the linkage finally catches up. This centers the control valve to end the power assist.

The control valve has a neutral centering device, usually springs. When no steering effort is being applied to the system, the centering springs center the spool position to balance pressures on both pressure surfaces. Equal pressure holds the steering linkage in place.

If the steering wheel is held in position when the front tires hit an object that tries to turn them, the control valve directs assist pressure to the pressure surface that opposes the turning force. This helps control steering feedback to maintain vehicle control.

Two types of control valves are used: sliding spools and rotating spools. The sliding spool valve may be located between the Pitman arm and steering linkage in the link-type units, mounted concentric with the worm shaft, or placed parallel in a housing outside the steering gear case in integral types. When a rotating spool valve is used, it is mounted concentric with the worm shaft or around the pinion shaft. In all power steering gears, the action of the spool control valve is similar. The rotating spool valve is by far the most popular because of its universal use on rack and pinion steering gears. An example is shown at the top right in Figure 14-1.

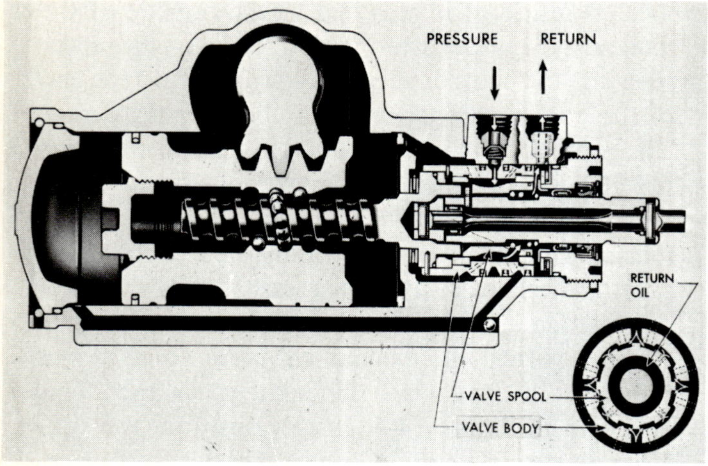

(a)

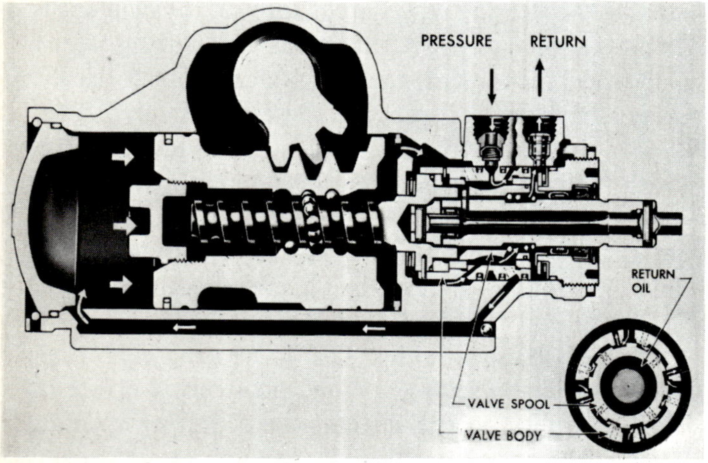

(b)

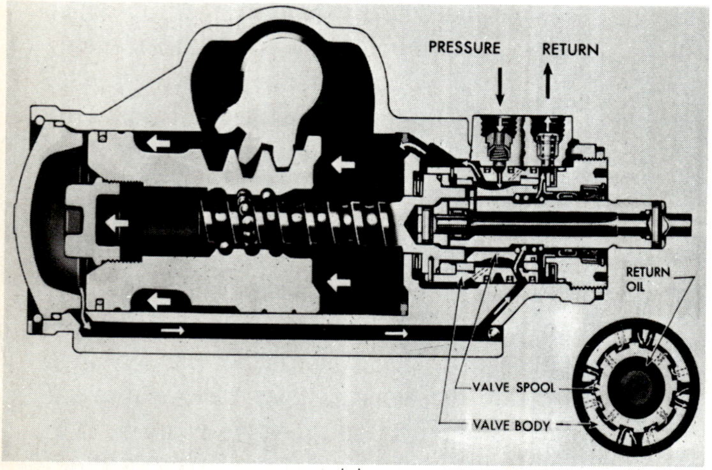

(c)

FIGURE 14-5 Section view of a rotary spool control valve operated by a torsion bar: (a) neutral, (b) right turn, and (c) left turn. (Courtesy of Buick Motor Division, General Motors Corporation)

Rotating Spool Control Valve. One type of rotating spool control valve, shown in Figure 14-5, is operated with a torsion bar. The spool valve body surrounds the spool control valve. The input shaft is connected to one end of the spool valve. The torsion bar connects between the spool valve and valve body. Steering effort on the input shaft rotates the spool valve and twists the torsion bar. This repositions the spool valve in the valve body. The valve body is attached to the worm shaft. Repositioning the spool valve within the valve body directs fluid under pressure to one pressure area and opens a return passage from the other pressure areas.

The power rack and pinion steering gear also uses a rotating spool valve around the pinion shaft. It operates in the same way as the valve just described. You can see this in Figure 14-1. The effort required to turn the steering wheel is carried through a torsion bar to move against the resistance of the steering linkage and wheels. The twist of the torsion bar changes the position of the spool valve in the sleeve ports. This turning movement restricts one of the fluid return passages. The restricted passage causes the pressure to build up in one of the pressure chambers. The pressure in the chamber is applied to the pressure surface of the steering gear in the rack housing to assist the driver to make the turn. The same spool valve turning movement increases the opening to the rotating spool from the other pressure surface chamber.

Sliding Spool Valve. The sliding spool valve in the integral steering gear is controlled by either of two actuating methods. In one valve, movement is caused by a slight endways or axial movement of the worm shaft or steering linkage. The front wheels and steering linkages hold the sector so it resists movement. The worm shaft is moved axially against one of its bearings as the ball nut tries to move the sector. This slight movement is transmitted to the parallel spool valve with a pivot lever. One type is shown in Figure 14-6.

The axial or sliding movement of the concentric spool valve in Figure 14-7 is actuated by a torsion bar. The worm shaft is turned through a torsion bar so the upper end turns slightly more than the lower end. The maximum amount of torsion bar flex is limited by a loosely fitted spline between the input shaft and worm shaft. A short-length helical spline threaded around the input shaft engages an actuator.

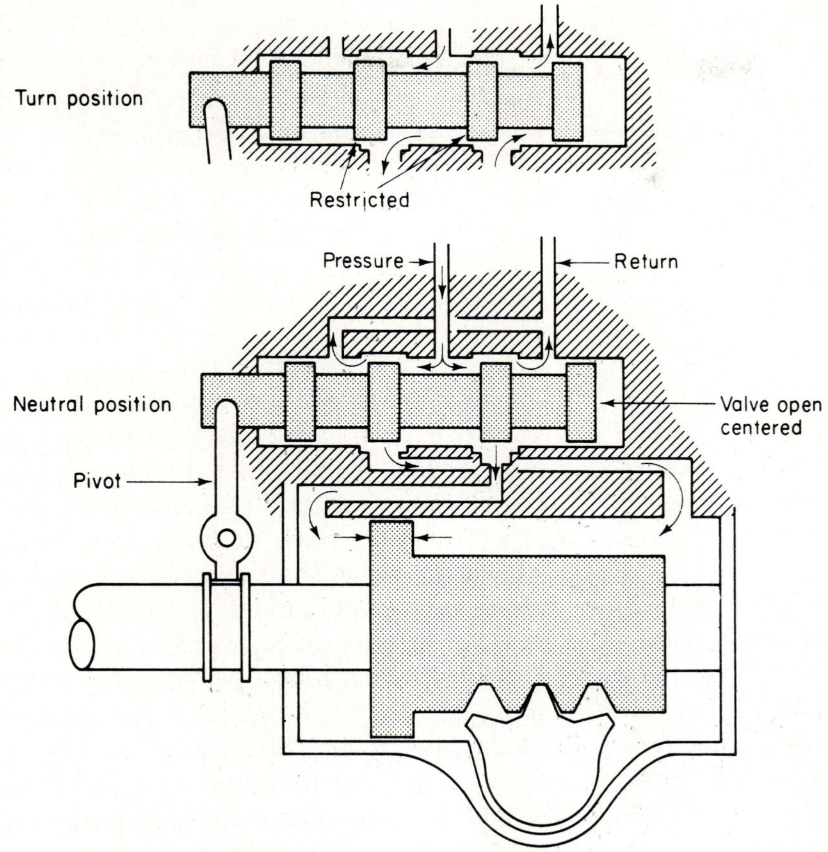

FIGURE 14-6 Principle of operation of one type of sliding spool control valve in neutral (open centered) and during a turn.

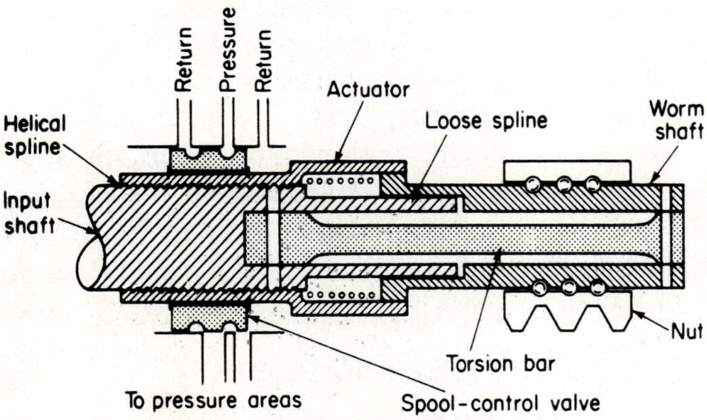

FIGURE 14-7 Section drawing of a torsion bar actuated concentric sliding spool control valve.

The other end of the actuator is splined to the worm shaft.

The steering wheel input effort twists the torsion bar. This twist changes the relative position of the worm shaft actuator assembly in relation to the input shaft position. This action slides the actuator along on the helical splines.

The spool valve is held in position on the actuator with retaining rings so any actuator axial movement also moves the spool valve endways.

Axial movement of the concentric spool valve controls the fluid flow just like the parallel spool type. It opens the fluid return from one pressure surface. At the same time, pressure is built up on the other pressure surface by restricting the return passage. This action provides assist to the steering effort.

Reaction Control. Power steering gears are designed to give the driver some feel for the amount of effort being put into the system. This driver feel is called *reaction control*. The torsion bars provide driver feel when they are used. The pivot-lever-actuated-type of control valve gives driver feel with centering springs and with fluid pressure developed in the steering system.

Power Chambers. The pressure surfaces or areas of the link-type power steering are the surfaces on either side of the piston within the power cylinder. You can see these surfaces in Figures 14-1 and 14-5. The

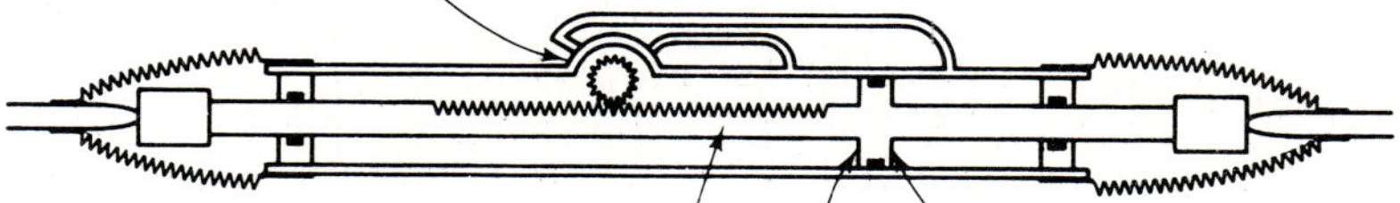

FIGURE 14-8 Pressure surfaces in a power rack and pinion steering gear.

control valve opens a return passage for oil to go to the reservoir from one side of the power piston. At the same time, the return passage from the other side of the piston is closed. This causes the pressure to rise. Fluid is forced into the chamber. This causes the piston to move in the cylinder.

The pressure surface of a link-type power steering gear is located on a double-faced piston within a cylinder. One end of the cylinder is connected to the steering linkage and one end of the piston rod is connected to the frame. Both connections are designed to allow the movement required. Metal lines or hoses are used to connect the sliding spool control valve to the power cylinder. Sometimes, the valve-to-cylinder passages for the fluid are internal. The power piston has two seals on the piston, one facing each end of the cylinder. These seals form the pressure surfaces. Another seal is located around the piston rod where it comes out of the cylinder to keep fluid from leaking out of the system.

The pressure surfaces in the power rack and pinion steering gear are similar to the link type. You saw them in Figure 1-1. Two outward-facing seals are placed on the rack to form the pressure surfaces. In this gear, seals are required on each end of the power rack to prevent external fluid leakage. Metal lines connect the rotating spool control valve to the power chambers. A sectional view of a power rack and pinion steering gear is shown in Figure 14-8.

Ball-Nut Power Piston. In integral power steering gear designs, the exterior of the ball nut acts as a piston and so it is called a *ball-nut power piston*. Each side of this piston is, therefore, a power chamber. The control valve directs fluid pressure to one side of the piston to assist the ball nut in moving the sector in the desired direction.

The ball-nut piston seal ring may be above or below the rack teeth. In either case, the oil pressure can go between the worm shaft and ball nut to pressurize the interior when the ball nut is assisting in the downward direction. Pressure is applied to the ball-nut end and piston flange when assisting in the upward direction. Reaction areas are used to balance the worm shaft and control valve in some power steering applications. Details of these pressure areas can be seen in Figures 14-5 and 14-6.

The ball-nut power piston has a rack, or gear teeth, that rotates a sector in the same way the manual steering does. Steering ratios for power steering gears are usually numerically lower than for manual steering gears, so they have a quicker response to steering wheel movement. A very fast response is not wanted when driving in the straight-ahead position since slight steering wheel movements would cause too much correction. Power steering gears are a compromise when using a constant ratio. Variable-ratio steering gears are used in some automobiles to give a slow ratio when steering straight ahead and a fast ratio on turns. This is done primarily by the way the sector teeth are cut, as illustrated in Figure 14-9.

14-3 POWER STEERING PUMPS

Three types of power steering pumps are used: the vane type, the slipper type, and the roller type. These are shown in Figure 14-10. Their principles of operation and design are very similar. A power steering pump has a rotor within an elliptically shaped cam ring. The pumping spaces on opposite sides of the rotor balance the rotor to minimize bearing loads and wear. Vanes, slippers, or rollers are installed in the rotor slots. Pressure thrust plates on each side of the rotor and cam seal the pump. This assembly is placed in a housing that contains rotor bearings

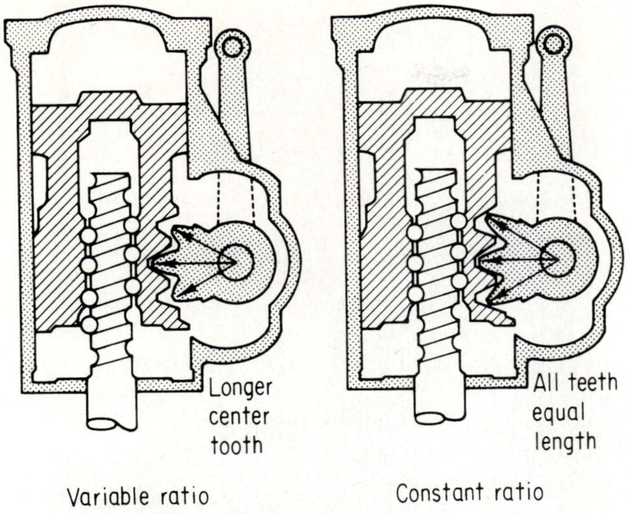

FIGURE 14-9 Constant steering ratio compared to a variable-ratio steering gear. (Courtesy of Buick Motor Division, General Motors Corporation)

and oil passages. The pump housing is usually surrounded by an oil reservoir. Some late-model designs have the reservoir on the side of the pump housing. The pump and reservoir are sealed with O-rings for easily assembled, oil-tight joints (Figure 14-11).

Power Steering Pump Operation. In operation, the rotor spins, causing centrifugal force to throw the vane, slipper, or roller inserts outward so their outer surfaces will rub on the cam. Each slipper has a backup spring to hold it against the cam.

The cam fits the rotor closely at two opposing locations to balance the pump. The spaces or cavity between the inserts gradually expands as the rotor turns them past the close-fitting points. This part of the pump is connected to the inlet passage to bring fluid from the reservoir into these expanding cavities. As the inserts reach the widest part of the cam insert, they pass and close the inlet passage from the reservoir. They then come in contact with the pressure passage. Continued turning now decreases the volume of the cavity between the inserts, forcing the fluid into the pump pressure outlet passage. Each pressure cavity between the inserts works in the same way.

Power steering pumps are positive displacement pumps. Each revolution delivers the same amount of fluid, regardless of the speed at which the pump is rotating. The pump capacity is large enough to supply the fluid volume and pressure required for steering assist during parking while the engine is idling. On some vehicles, it also supplies fluid pressure for the hydroboost power brake.

Power steering assist requirements are very low when driving at highway speeds. At these speeds, the pump will deliver high fluid volume and pressure, unless the pump output is modified. The pump output is modified with a flow control valve and a pressure relief valve.

Pump Control Valve. The greatest amount of power for steering is needed when the steering wheel is turned with the vehicle not moving, usually while parking. The most power-assist pressure is needed at this time. Unfortunately, the engine is idling during this time, so the engine-driven power steering pump is also rotating slowly. Therefore, the power steering pump must be designed to produce high pressure at low pump speeds. Pump speeds are fast when the vehicle is operating at highway speeds,

FIGURE 14-10 Three types of power steering pumps.

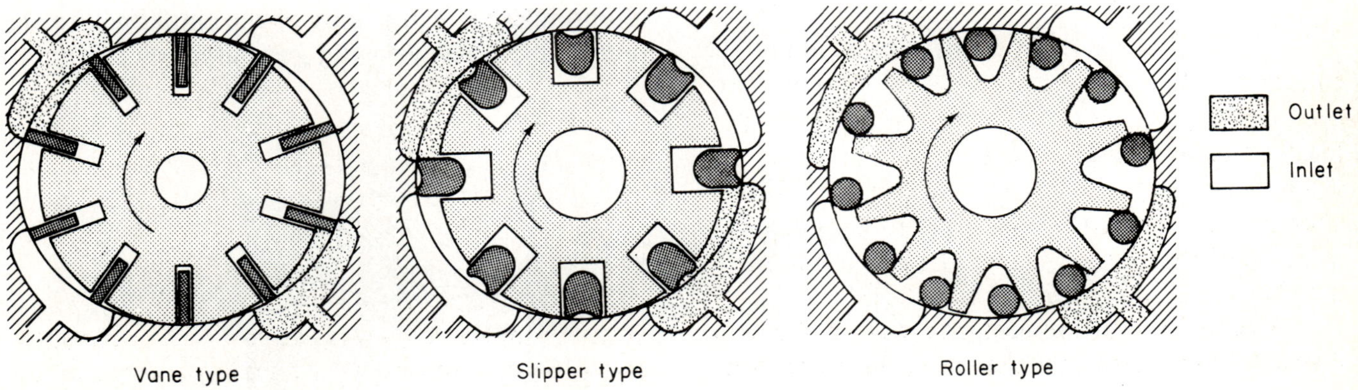

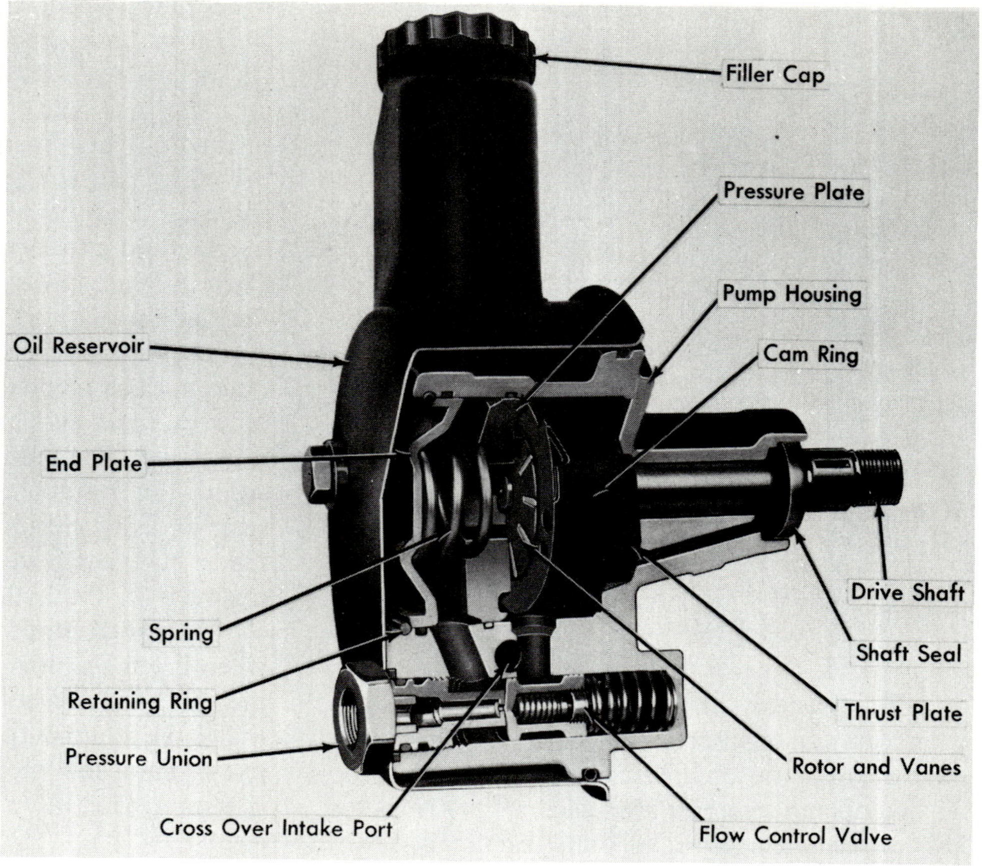

FIGURE 14-11 Section view of a typical vane power steering pump and valves (Courtesy of Cadillac Motor Car Division, General Motors Corporation)

when little or no power assist is needed. A flow control valve is used to reduce high pump volumes produced at highway speeds. The pump control valve also limits maximum pressure.

Flow Control. The flow control valve, illustrated in Figure 14-12, is operated by small differences in pressure and a calibrated spring. The passage from the pump outlet has a restricted opening. Pressure is greater on the pump end of the restricted opening than it is on the steering control valve end of the opening. The pressure drop across the restricted opening increases as the flow of oil through the opening increases.

High oil pressure from the pump side of the restricted opening is directed at one end of the flow control valve. Low oil pressure from the downstream side of the opening is directed at the other end of the flow control valve. A calibrated spring is also on the low-pressure end of the flow control valve.

Flow Control Valve Operation. As the engine speed increases from idle, both the pump fluid flow and pressure increase. This causes a difference in pressure across the restricted opening. The pressure drop causes the flow control valve to move toward its low-pressure end (to the left in Figure 14-13). This movement opens a passage between the pump outlet and pump inlet to bypass part of the fluid back to the pump inlet. The opening gets larger as the pump

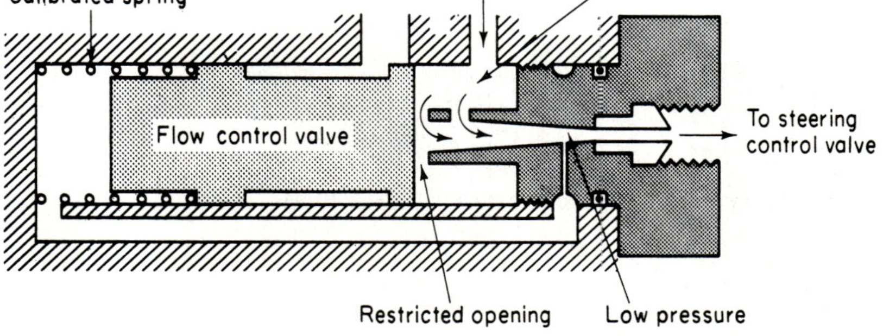

FIGURE 14-12 Operating principles of a flow control valve in position to supply all the fluid to the steering gear.

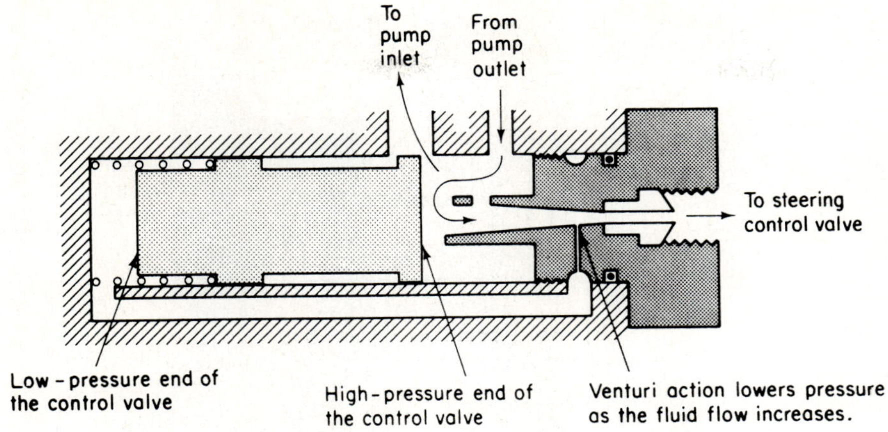

FIGURE 14-13 Operating principle of a flow control valve in position to return most of the fluid to the reservoir.

speed increases to control the amount of fluid going through the power steering system. Recirculating the oil back through the pump reduces the power required by the pump and keeps the oil temperature from increasing too much.

Some control valves have a tapered metering pin that moves in the opening to change its size to match the fluid flow to the steering gear demands. The pin reduces the effective opening size at high pump speeds. Another type of flow control valve reduces the fluid flow by restricting a passage that parallels the opening. This increases the pressure drop to restrict fluid flow. Both types provide precise flow control operation.

Electric motor driven power steering pumps do not need a flow control valve because the pump speed is not controlled by the engine speed. The motor speed is reduced at high vehicle speeds to limit the assist provided.

Pressure Relief. The flow control valve just described operates when there is little restriction and low pressure is required. This occurs when the steering control valve is centered and the pump speed is high. Flow is restricted when the steering control valve is moved off center as it sends fluid to one of the power chambers. This causes the system pressure to rise to assist steering. Consider what happens in the power steering when the driver turns the steering wheel until the steering linkage hits the stop and holds the steering wheel in this full-turn position. Fluid flow is low and the pressure will build to a maximum. Pressure must be limited to avoid damage to the power steering unit seals and hoses. A pressure relief valve limits the maximum pressure by opening a passage between the pump outlet passage and either the pump inlet passage or the pump reservoir. In some power steering pumps, the pressure relief valve is a separate unit. In most pumps, it is built into the flow control valve and acts as a pilot valve as illustrated in Figure 14-14. The pressure relief valve opens when pressure on the low-pressure end of the flow control valve increases to the set maximum pressure. It allows the fluid to flow from the low-pressure end of the flow control valve through the ball check to the pump inlet. This drops the pressure on the low-pressure end of the flow control valve. The flow control valve snaps open (to the

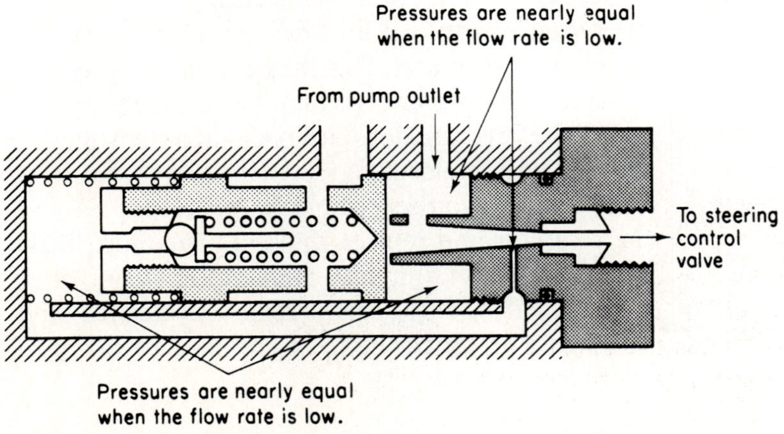

FIGURE 14-14 Operation of a pressure relief valve in the full-pressure position.

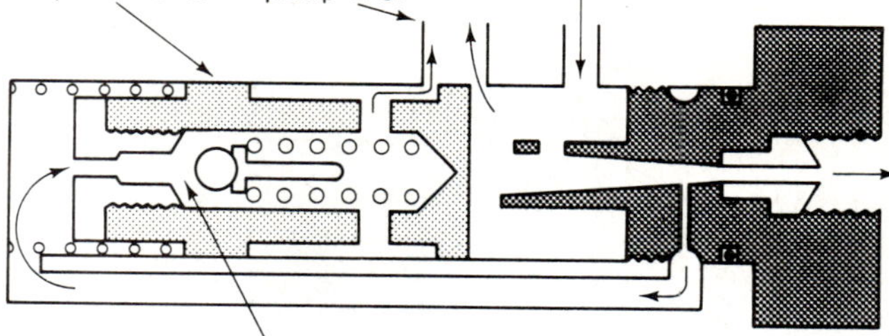

FIGURE 14-15 Operation of a pressure relief valve in the relief position.

left in Figure 14-15). The pump output fluid can then flow freely into the pump inlet. This action lowers the pump output pressure. Pressure and spring force will balance the valve at the pressure for which it is set.

Hoses. The power steering pump is connected to the power steering gear control valve with one high-pressure hose and one low-pressure return hose. The high-pressure hose is usually made in two sizes that will dampen oil pulsations to minimize noise. The ends of the high-pressure hose have screw-type tubing fittings. The low-pressure hose, on the other hand, is connected to the steering gear and reservoir with less expensive hose clamps.

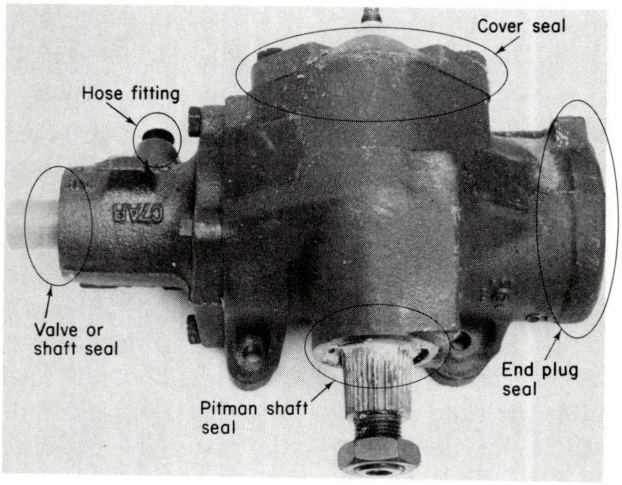

FIGURE 14-16 Location of external leaks from a ball-nut power steering gear.

14-4 POWER STEERING SERVICE

Hydraulic-assist power steering problems are either external fluid leaks or operating problems. Typical external leak points are shown in Figure 14-16 for recirculating-ball-type gears and in Figure 14-17 for rack and pinion gears. External fluid leaks result from damaged oil seals, damaged O-rings, cracked hoses, or cracks in the metal parts. These can readily be seen and corrected by replacing the damaged parts.

Problem Diagnosis. The first thing to do when there is a power steering problem is to see that there is enough fluid in the reservoir and that the drive belt has correct tension. Operation problems causing hard steering or lack of assist involve the pump assembly, the steering gear assembly, or the steering linkage. The steering linkage should be checked for binding before deciding that the pump or steering gear are at fault. A pump pressure test is used to locate the cause of the problem. The pressure test unit consists of a valve with a gauge on the high-pressure side. This unit is connected between the pump and the steering gear, as pictured in Figure 14-18. With the valve open, the pressure should rise when the steering gear is turned against the stops. If it does not rise, the valve is closed. The pump is operating properly if closing the valve causes the pressure to increase. This means that the problem is in the steering gear. The pump is causing the problem if the pressure does not rise when the valve on the test unit is closed.

The reservoir seal can leak. It is easy to remove the reservoir and replace the seal. The most likely cause of pump failure is a sticking flow control or pressure valve. These can be removed, cleaned, and polished with crocus cloth. Care must be taken to

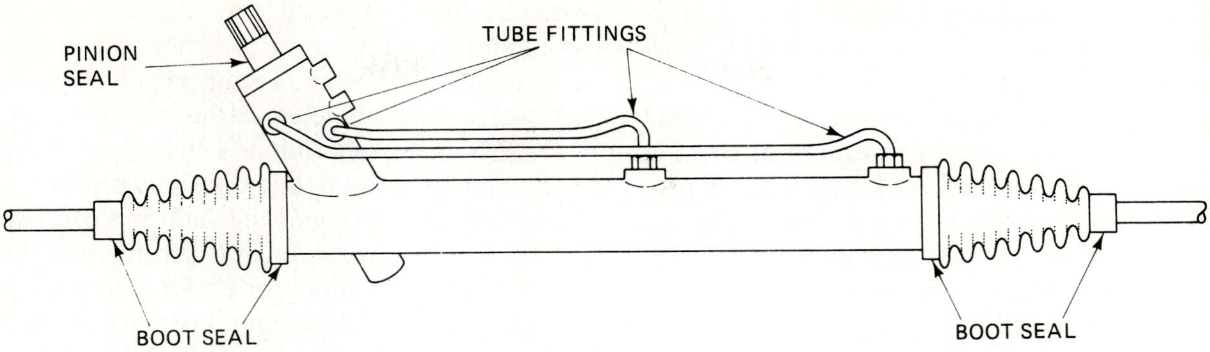

FIGURE 14-17 Location of external leaks from a rack and pinion steering gear.

avoid rounding the edges of these valves. If this does not correct the problem, the pump is usually replaced. The pump is also replaced if the pump shaft seal leaks.

Self-steering or unequal assist are other steering problems. They are usually caused by an improperly centered control valve. Self-steering of the parallel-sliding-type control valve can usually be corrected from the outside of the steering gear. This can be done by changing the position of the valve housing on the steering gear housing. Steering gears using a concentric-type control valve will have to be disassembled and inspected to correct the problem. Unequal assist may also result from internal fluid leaks in the steering gear. The steering gear will have to be disassembled to make either of these repairs. It is general practice to replace all seals any time the gear is disassembled.

Service Procedure. Before disassembly, be sure to have the service manual that applies to the gear being serviced. Usually, special tools are required for alignment and installing seals without damage. These must be used for satisfactory servicing. To avoid

FIGURE 14-18 Test setup used to check the operation of the power steering pump.

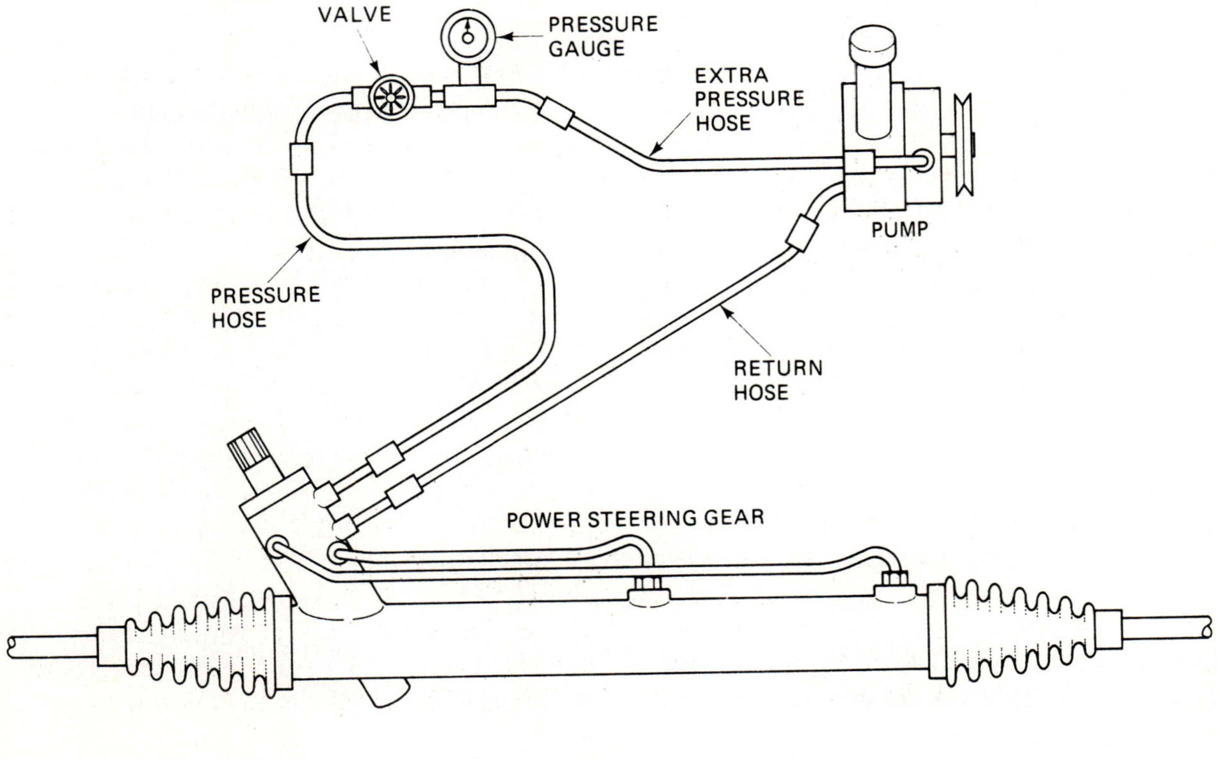

ne disassembly instructions must be care- owed and all precautions observed while us- e special tools.

The power steering gear is removed from the icle in the same way that mechanical steering ears are removed. See Section 13-4 for removal procedures. Clean the outside of the gear externally before disassembly to avoid contaminating the inside of the gear. Fluid is drained and the gear is carefully disassembled. Note the condition of the parts and make the required checks as the parts are disassembled. The parts should be thoroughly cleaned after disassembly, then inspected for looseness, binding, scoring, wear, etc., as described in the service manual.

Any required repairs are made. Replacement parts, including all new seals, are lubricated with power steering fluid as they are assembled. Unfortunately, technicians have been having about a 50 percent comeback rate on rebuilt gears. Often, pinion housing wear prevents control valve seals from seating correctly. Internal rack seals between the pressure surfaces can be damaged during installation, especially if a technician tries to install them without using the correct tools. As a result, it has become common practice to replace faulty power rack and pinion gears with either new or factory-remanufactured gears. Technicians have had good success in repairing recirculating ball steering gears.

Power steering fluid is a clear fluid with properties similar to automatic transmission fluid. Some manufacturers package their fluid as a combination power steering and automatic transmission fluid. A special fluid is required for electro-hydraulic power steering pumps. Assembly instructions must be followed carefully to be sure that all parts are put in their proper place with the correct adjustments. The worm gear preload in recirculating ball steering gears is set as part of the assembly procedure. The Pitman shaft backlash adjustment is done in the same way as on manual steering gears. After assembly, the steering gear unit should be tested before it is installed in the vehicle. This is done by connecting the power steering hoses to the gear. With fluid in the system and the engine running, the gear should be operated throughout its range to purge air from the system. Details for each gear type are given in the applicable service manual. If the steering gear operates correctly, the unit should be installed and retested with the front wheels off the floor.

STUDY QUESTIONS

Section 14-1

1. What does power steering do for the driver?
2. How does the gear ratio differ between manual and power steering gears?

Section 14-2

3. Where are the pressure surfaces in a power steering gear?
4. What are the advantages of electronic steering?
5. What is meant by the term *open-circuit control valve*?
6. What does the control valve do to make a different pressure on each pressure surface within the power steering gear?
7. What force balances the control valve?
8. What force always leads power steering assist?
9. How does power steering help to control steering feedback?
10. Name two types of power steering control valves.
11. Name two types of sliding control valves.
12. What causes reaction control in the power steering control valve?
13. What is the advantage of a variable-ratio steering gear?

Section 14-3

14. What balances the power steering pump rotor?
15. Describe power steering pump operation.
16. What opens and closes the power steering pump inlet and outlet passages?
17. What is meant by the term *positive displacement pump*?
18. Why is it necessary to modify the output of the pump?
19. Under what operating conditions does the flow control valve operate?
20. Under what operating conditions does the pressure-relief valve operate?
21. How do high-pressure and low-pressure power steering hoses differ?

Section 14-4

22. What is the first thing to check when there is a power steering problem?
23. How do you know the pump is operating normally when making a pressure test of the power steering system?
24. What parts are usually replaced on power steering pumps?
25. What precautions should be taken when rebuilding a power steering gear?
26. How do you get air out of a power steering system?

15

Vehicle Dynamic Behavior

The study of the behavior of a vehicle as it is driven at a constant speed, accelerated, stopped, and turned is the topic of this chapter. As any object that has weight is moved, forces must act upon it to change its speed or its direction. The forces required for these changes, and the actions that take place as a result of these forces, are covered in greater detail in this chapter.

Vehicle dynamic behavior is addressed in this text because of the great interest in ride quality and handling by the automobile manufacturers. An understanding of these principles will help you in diagnosing handling and ride problems. This chapter is optional for readers more interested in the service aspects of safe vehicle repair.

15-1 VEHICLE AXIS SYSTEM

To help you understand the motions of a vehicle, the possible motions of a vehicle must be described. We will use common terminology and vectors to describe these motions. They are shown in Figure 15-1.

x-axis. This axis describes straight-line forward or backward motion of the vehicle while operating at constant speed, accelerating, or braking. Movement along the *x*-axis is called *longitudinal motion*.

y-axis. This axis describes the *vertical motion*, up or down, of the vehicle. Vertical motion is used to describe the motion of the body as the vehicle encounters bumps, dips, etc.

z-axis. This axis describes the sideways motion forces, left or right, of the vehicle as it turns. This movement is also called *lateral motion*.

Roll. This describes rotation of the body around the *x*-axis. The tendency of the body to roll occurs as the vehicle is turned.

Pitch. This describes rotation of the body around the *z*-axis. Pitch occurs during acceleration and braking as one end of the vehicle moves up and the other end moves down.

Yaw. This describes rotation of the body around the *y*-axis. Yaw occurs as a vehicle starts to spin out or begins to oversteer.

The three linear motions—*x, y,* and *z*—and the three rotations—roll, pitch, and yaw—describe the six degrees of freedom of a solid body such as a vehicle. A *degree of freedom* is a direction in which a body is free to move. A six-degrees-of-freedom body can have the six motions described. It can be moving around all three axes at the same time.

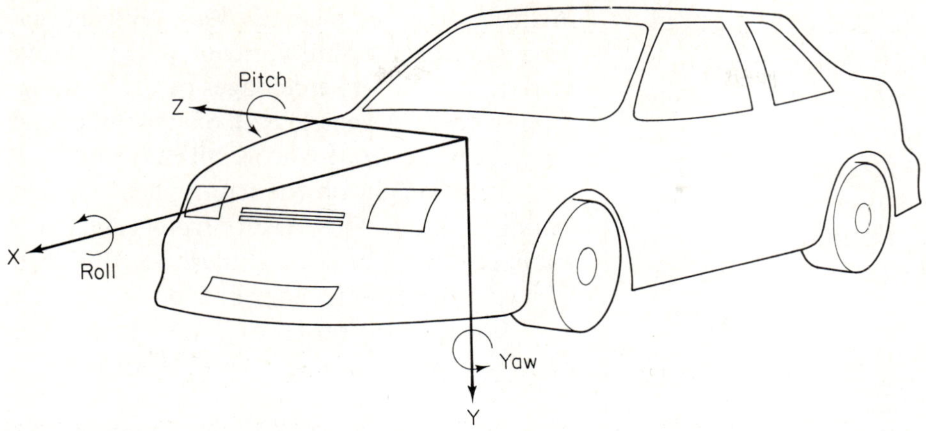

FIGURE 15-1 Six motions or degrees of freedom (free movement) of a vehicle as a result of motion both ways around: roll on the *x* axis, yaw on the *y* axis, and pitch on the *z* axis.

15-2 VEHICLE WEIGHT AND WEIGHT DISTRIBUTION

The term *center of gravity* is very important when studying the behavior of any object in motion. In simple terms, it is the point at which we could lift an object and there would be no rotation about that point, as shown in Figure 15-2.

The center of gravity is important because it is the theoretical point on which all inertia forces act. You will recall that *inertia* is the tendency for an object to stay at rest. Inertia also keeps an object in motion once it is placed in motion. It also keeps the object moving in the same direction once it is moving in that direction. Remember, as well, that Newton's first law of motion states that for every action there is an equal and opposite reaction. In vehicle dynamics, this means that for every force passing through the center of gravity as a vehicle is maneuvered there must be an opposing force. For a vehicle, the opposing forces occur at the tire-to-road contact point in the tire footprint.

FIGURE 15-2 The center of gravity is the point at which the vehicle could be lifted and there would be no motion about that point.

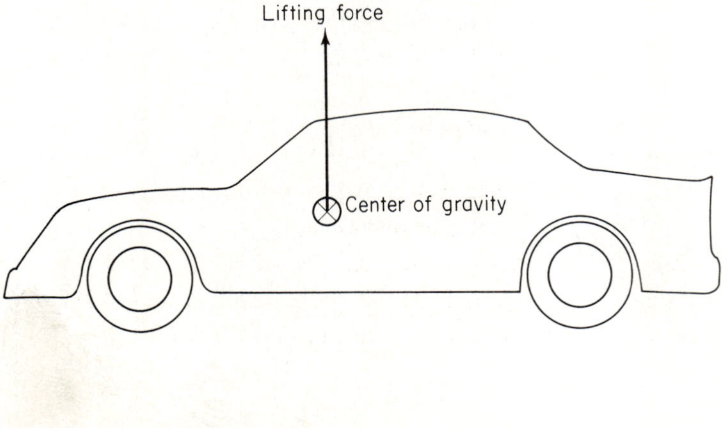

The center of gravity can be found by weighing the vehicle and then, from the weights on the front and rear wheels, the center of gravity, front to rear, can be determined.

Example 1

Left front weight 340 kg
Right front weight 335 kg
Left rear weight 235 kg
Right rear weight 230 kg
Wheelbase length 2.54 meters

$$\text{Total front weight} = 340 + 335 = 675 \text{ kg}$$

$$\text{Total rear weight} = 235 + 230 = 465 \text{ kg}$$

Position of the center of gravity from the front axle

$$X = \frac{W_R \times WB}{W_F + W_R} = \frac{465 \times 2.54}{675 + 465} = 1.036 \text{ meters}$$

The center of gravity is, therefore, 1.036 meters behind the front axle centerline.

The height of the center of gravity can be determined by raising one end of the vehicle and measuring the weight change on the unraised end at a given angle change for the vehicle. The higher the center of gravity, the farther it moves forward as the rear of the vehicle is raised, and the greater the weight change that occurs.

Example 2

A vehicle is weighed and is found to have a total front weight of 675 kg and a total rear weight of 465 kg. The rear of the vehicle is raised 25 cm (0.25 m) and supported under the rear tires. The total front weight is now 685 kg. If the wheelbase length is 2.54 m, find the height of the center of gravity. The rolling radius of the front tires is 30 cm.

$$\text{Angle raised} = \sin^{-1}\frac{\text{Height change}}{\text{Wheel base}}$$

$$= \sin^{-1}\frac{0.25\text{ m}}{2.54\text{ m}} = 5.64°$$

The height of the CG is the rolling radius + height calculated.

Height calculated

$$= \frac{\text{Weight change} \times WB \times \cos(\text{angle})}{W_T \sin(\text{angle})}$$

$$= \frac{(685-675) \times 2.54 \times .995}{(675+465) \times .0983}$$

$$= 0.2153\text{ m}$$

$$\text{CG height} = 0.3\text{ m} + 0.2153\text{ m} = 0.5153\text{m}$$

Weight Distribution. Weight distribution describes how the weight is positioned in reference to the vehicle wheelbase. Commonly, it is described as a percentage of the total vehicle weight on the front and rear wheels. The weight distribution is affected by the placement of the engine and drive line, the frame structure, and the placement of the body panels. In front-engine, front-wheel-drive vehicles, up to two-thirds or 66 percent of the weight is on the front wheels and only one-third or 34 percent is on the rear wheels. With mid-engine vehicles, the weight distribution is commonly about 45 percent on the front and 55 percent on the rear. As the weight distribution approaches 50/50, a vehicle usually has its best cornering ability. Front-weight bias vehicles also demand high percentages of the braking at the front wheels. A vehicle with 66 percent of the static weight on the front wheels will have up to 80 percent of its weight on the front wheels during maximum braking. This results from effective weight transfer during braking (Figure 15-3). Calculation of weight transfer is described later in this chapter.

Locating the center of gravity is relatively easy once the weight at each wheel is known.

$$\text{Distance to front CG} = \frac{W_R \times \text{Wheelbase}}{W_F + W_R}$$

$$\%\text{ Weight on rear} = \frac{W_R}{W_F + W_R}$$

Therefore,

$$\text{Distance from CG to front wheel} = \%W_R \times \text{Wheelbase length}$$

Example 3

A car is measured and has

Front weight W_F = 2000 lbs

Rear weight W_R = 1000 lbs

Wheelbase length = 100 in.

$$\text{Distance from front wheels to CG} = \left(\frac{1000}{1000 + 2000}\right) 100 = 33.3\ \%$$

$$\%\ W_R = \left(\frac{1000}{1000 + 2000}\right) \times 100\% = 33.3\%$$

$$\%\ W_F = (1 - 0.333) \times 100\% = 66.6\%$$

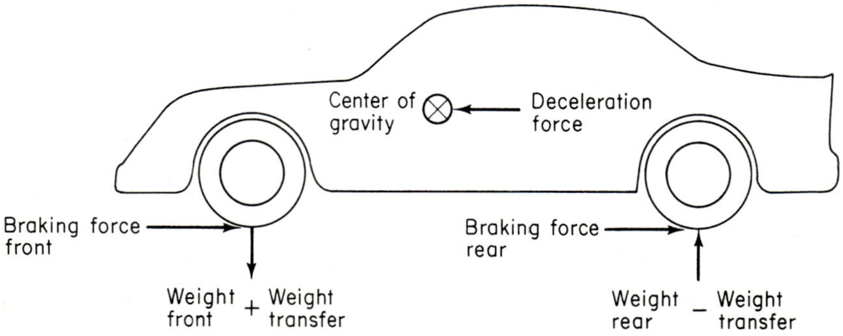

FIGURE 15-3 Effective weight transfer occurs during braking due to deceleration force.

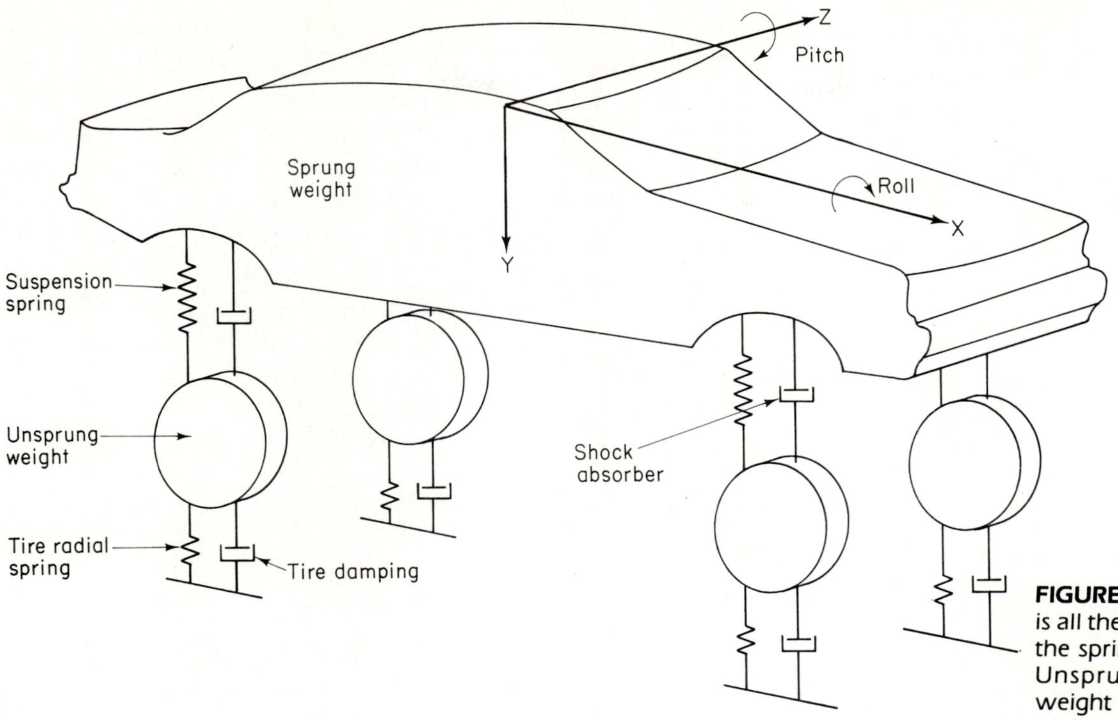

FIGURE 15-4 Sprung weight is all the weight supported by the springs and road surface. Unsprung weight is the weight between the springs and the road surface.

Sprung and Unsprung Weight. The difference between sprung and unsprung weight must be considered when discussing the design of suspension, suspension-control devices, how the vehicle drives, and vehicle handling. *Sprung weight* is all of the weight supported by the vehicle springs. *Unsprung weight* is the weight between the springs and road surface. These can be seen in Figure 15-4. Sprung weight includes the body frame, engine, transmission components, and everything contained therein. Unsprung weight on most vehicles includes tires, wheels, wheel brake mechanisms, steering knuckles, and rear axle assemblies. Suspension control devices, such as shock absorbers, control arms, brake struts, some stabilizer bars, tie rods, and springs, have one end sprung and the other end unsprung. They are not clearly in either category. Their weight is usually considered to be two-thirds sprung weight and one-third unsprung weight. The quality of the vehicle ride diminishes as a higher percentage of the vehicle weight is unsprung. A higher percentage of unsprung weight is especially noticeable while traveling over rough roads. On light vehicles, the ratio of unsprung to sprung weight is kept low for the best possible ride. Usually, the ratio of sprung to unsprung weight is kept to approximately 10:1.

15-3 SPRING GEOMETRY, BOUNCE FREQUENCIES, AND SPRING CENTERS

It is important to understand the important role that suspension springs play in ride quality. As springs are changed, more than just the stiffness of the ride changes. It is also important for the advanced technician to understand how the actual spring rate of the sprung weight (called *ride rate*) is related to the rate of the spring.

Spring Geometry. *Spring geometry* is the relationship that occurs between the spring rate and the wheel rate due to the position of the spring. The *wheel rate* is the force required to cause a deflection divided by the amount of the deflection at the wheel (force/deflection of wheel; usually in lbs/in.). Some suspension designs place the spring on the lower control arm between the frame mounting point and the ball joint (Figure 15-5). The spring on a MacPherson strut suspension acts directly through the tire footprint (Figure 15-6). These two suspensions would have considerably different wheel rates if the

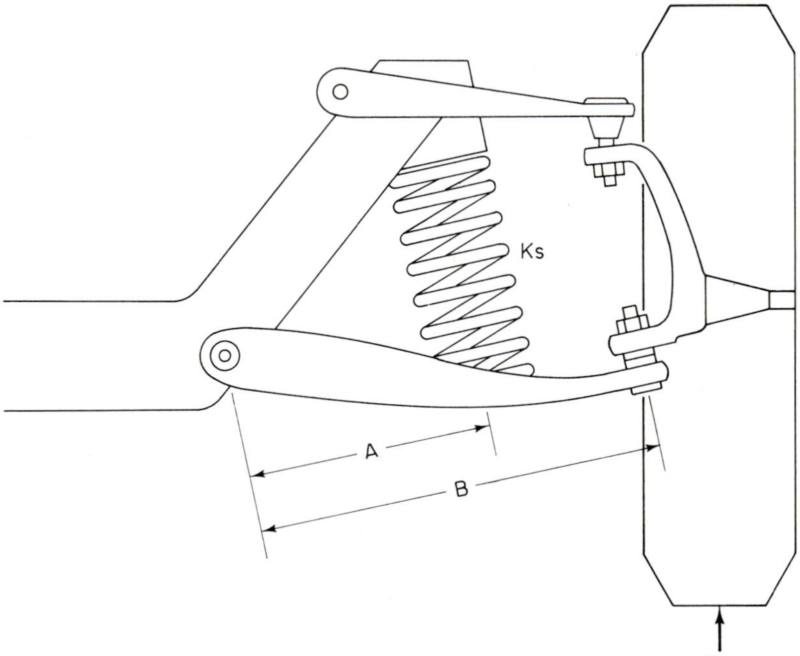

FIGURE 15-5 The spring acts through the control arm to the tire footprint. The spring position on the lower control arm results in a wheel rate lower than the spring rate.

same spring were to be used. Since the wheel rate is actually what affects the ride and handling, the two vehicles would have considerably different rides if everything else were the same. The reason for the difference in ride and handling is the relationship of the wheel movement to the spring movement. This relationship is commonly called the *motion ratio*.

FIGURE 15-6 The spring force on a MacPherson strut acts approximately through the tire footprint.

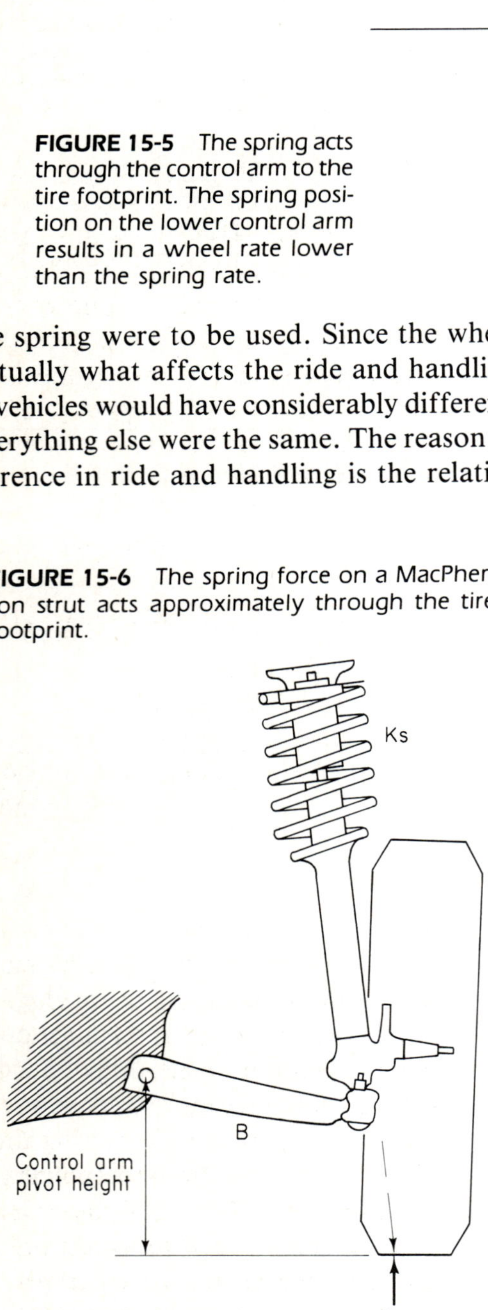

The spring, as you will recall from Chapter 6, is rated in pounds per inch (Newtons/meter). The wheel rate also is measured in lbs/in. To determine the effect the spring rate has on wheel rate, we must look at the mechanical advantage the wheel has on the spring as well as the displacement ratio between the two (Figure 15-7). Consider first the mechanical advantage that the spring has on the ball joint. This ratio is A/B. Next, consider the displacement ratio of C/D. These two must then be multiplied together to get the ratio of wheel rate to spring rate.

$$K_w = \text{Wheel rate}$$

$$K_s = \text{Spring rate}$$

$$K_w = \left(\frac{A}{B} \times \frac{C}{D}\right) \times K_s$$

For example, if the spring is at the 10-inch point on a 20-inch control arm, it is at the half-way point. The A/B ratio is equal to 10/20, or 1/2, and the C/D is in the ratio of 1/2.

$$K_w = (1/2 \times 1/2) \times K_s$$

This is, in fact, always the case. The wheel rate

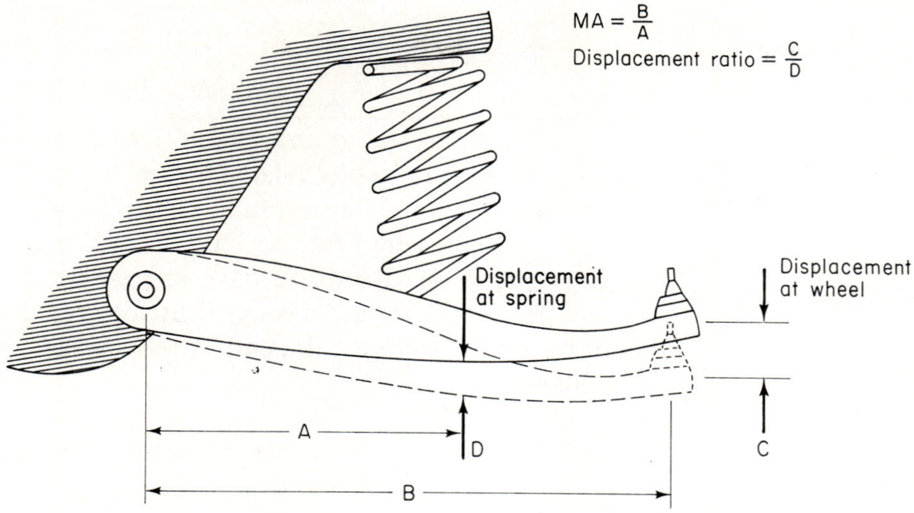

FIGURE 15-7 Wheel rate is affected by the mechanical advantage of the spring position on the control arm (A) and displacement ratio between the wheel (C) and spring (D).

(K_w) is

$$K_w = \left(\frac{A}{B}\right)^2 K_s$$

and the motion ratio is always B/A.

From this point on, we will discuss only wheel rates as they alone affect how a car rides and handles. The spring rates themselves are important only as they are related through the geometry of the suspension.

Frequencies. The cyclic frequency of the suspension and the body of a vehicle are important for both ride and handling. Cyclic frequency can best be understood by examining a simple spring with a weight hanging from it (Figure 15-8). If the weight is set into motion by pulling on it and then releasing it, the weight will move up and down at a given frequency. Frequency is measured in cycles per second. A cycle is completed when the weight moves up and then down to its starting position. The time to complete one cycle is independent of how far the weight is pulled to set it in motion. It is dependent only on the weight (mass) and the spring rate. This frequency is called the *natural frequency* of the spring-mass system. The formula to calculate the natural frequency is

$$F_N = 1/2\pi \sqrt{K/M} \qquad \text{where } M = w/g$$

Using customary units and the acceleration of gravity as

$$g = 386 \text{ in./sec}^2$$

$$F_N = \frac{1}{6.28} \sqrt{\frac{K \times 386}{W}} = 3.13 \sqrt{\frac{K}{W}}$$

where K = lbs/in.
W = lbs

There is one very interesting point about the above formula. The frequency is a function of the spring rate and the weight only. The frequency remains the same whenever the ratio of spring rate to weight remains the same. In automobiles, this is the reason all passenger cars have natural frequencies of the body of about 1.2 cycles/second, independent of the size of the vehicle. It is normal to design the front suspension with a slightly lower natural frequency than the rear suspension. A car built for performance will have ride frequencies above that for

FIGURE 15-8 Cycle frequency demonstrated by the natural frequency of a weight hanging on a simple tension spring.

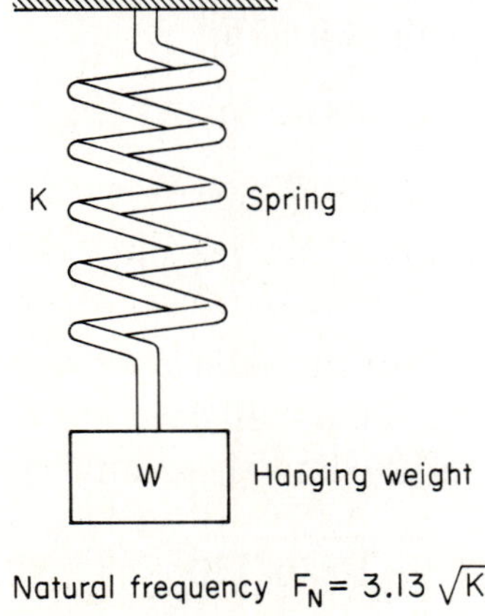

a standard passenger car. A car built to handle well without concern for ride, such as a race car, will have much higher frequencies. In this author's experience, frequencies of about 2.2 cycles/second (Hz) work well for circle track and road race cars that race on low-banked tracks, and with vehicle shapes that apply minimum aerodynamic down force. In this discussion, *ride frequency* means the natural frequency of the sprung mass. In the calculations for an automobile, the spring rate being discussed is the wheel rate. Some simple examples are given to show how these formulas can be used.

Example 4

3000 lb sprung weight car

Front weight = 2000 lb

= 1000 lb/wheel

Front wheel rate = 130 lb/in.

Rear weight = 1000 lb

= 500 lb/wheel

Rear wheel rate = 90 lb/in.

$$F_N \text{ Front} = 3.13\sqrt{130/1000} = 1.13 \text{ Hz}$$

$$F_N \text{ Rear} = 3.13\sqrt{90/500} = 1.33 \text{ Hz}$$

$$\text{Ride frequency} = 3.13\sqrt{K_T/W_t}$$

$$F_N \text{ Ride} = 3.13\sqrt{\frac{130+130+90+90}{3000}} = 1.2 \text{ Hz}$$

Example 5

3000 lb sprung weight car

50/50 weight distribution

Determine wheel rate and spring rate necessary for front

Use: $F_N = 2.2$ Hz $\qquad F_N = 3.13\sqrt{K/W}$

$$K = \frac{(F_N)^2}{(3.13)^2} \times W$$

$$K = \frac{(2.2)^2}{3.13^2} \times \frac{3000}{4} = 370 \text{ lb/in. wheel rate}$$

If the spring is midway between the ball joint and the frame attachment point:

$$K_{\text{wheel}} = (A/B)^2\, K_{spring}$$

$$K_{spring} = (B/A)^2\, K_{wheel} = (2/1)^2\, K_{\text{wheel}}$$

$$= 4 \times K_{\text{wheel}} = 1480 \text{ lb/in.}$$

It is also helpful to understand frequencies of the unsprung mass. These relate to wheel imbalance and wheel tramp. The unsprung spring weight is acted on by the wheel rate and the tire spring rate (see Figure 15-4). Most tires have spring rates between 800 and 1000 pounds per inch. For the unsprung weight, both the tire and wheel rates act on it. Therefore,

$$K = K_{\text{wheel}} + K_{\text{tire}} \qquad W_{\text{us}} = \text{unsprung weight}$$

Example 6

Using $K_w = 130$ lbs/in.

$K_{\text{tire}} = 1000$ lbs/in.

$W_{\text{us}} = 120$ lbs

$$F_N = 3.13\sqrt{K/W} = 3.13\sqrt{(1000+130)/120}$$

$$= 3.13\sqrt{1130/120} = 9.6 \text{ Hz}$$

Ratio of unsprung to sprung frequency at the front wheels:

9.6/1.13 = 8.5:1

An unbalanced tire has a frequency associated with it. Critical speed is reached when the cyclic frequency of the imbalance matches the natural frequency of the unsprung weight. For the previous example,

$$\text{Critical speed (mph)} = .357 \times F_{N\,\text{us}} \times R_{\text{tire}}$$

$F_{N\,\text{us}} = 9.6$ Hz

R_{tire} = Rolling radius

Assume 13 inches (33 cm)

Critical speed = .357 x 9.6 x 13

= 44.5 mph (71 kph)

This means that, for the vehicle we have analyzed, the speed at which an unbalanced tire would be felt on the front is approximately 45 mph. For the rear,

$$F_{N\,\text{us}} = 3.13\sqrt{(90 + 1000)/120}$$

$= 9.43$ Hz

$= 43.8$ mph (70 kph)

assuming the same unsprung weight.

The frequencies for ride, front, rear, sprung, and unsprung weight have now been calculated. The manufacturer has paid careful attention to these. It is important that no two frequencies have the same value. If they were the same, one vibration frequency could excite the motion of another part or the whole vehicle. In most vehicles, the ratio of sprung to unsprung frequencies is close to 10:1. If this were not the case, an unbalanced tire could shake the whole vehicle terribly, not just the suspension and/or steering, as is always the case in current vehicles.

For those interested in further examination of the above calculations, it is possible to obtain the wheel rate of the suspension relatively easily. To do this, place scales under the front and rear tires and measure the static weight at each location. Next, use a jack to raise frame and body exactly 1 inch, measured directly at the fender over the tires at which the wheel rate is to be calculated. Read the scales at the raised end of the vehicle again. The change in weight on the scales is the wheel rate in lbs/in. It is important when doing this to raise both sides equal amounts at the same time so the stabilizer bar does not influence these readings.

The spring center and its location in relation to the center of gravity of the vehicle are also very important. The *spring center* is that point at which a vertical load applied to the body would produce equal displacement of the body, and no pitch or roll would be introduced. In the design of passenger cars, it is important for the spring center to be slightly to the rear of the center of gravity (Figure 15-9). The spring center will naturally move toward the end of the vehicle where a higher wheel rate is being used. The reason for placing the spring center to the rear of the center of gravity is to force the body of the car into a slight amount of pitch as the vehicle encounters bumps. This provides a smoother ride when slightly higher spring rates are used on a vehicle. Whenever the front ride frequency is below the rear ride frequency, the spring center is to the rear of the center of gravity. If the rear suspension is too stiff for the weight on it, an excessive amount of pitch occurs when encountering bumps. If the front stiffness is changed slightly, the spring center may align with the center of gravity and an extremely rough ride can result. As a technician, it is most important for you to understand these relationships. You may be asked to recommend different springs for a vehicle to improve its load-carrying capacity, and the wrong spring choice would ruin ride quality.

The spring center can be located mathematically if the location of the center of gravity, the wheelbase length, and the wheel rates of a vehicle are known.

$$d_{sc} = \frac{-\ [(K_F \times a) - (K_R \times b)]}{K_F + K_R}$$

where K_F and K_R = Front and rear ride rates
a = Distance front wheels to CG
b = Distance CG to rear wheels
d_{sc} = Distance between the CG and the spring center (rearward as shown)

The spring center location is primarily a ride function; however, as we have shown, it is closely tied to the natural frequencies of the vehicle.

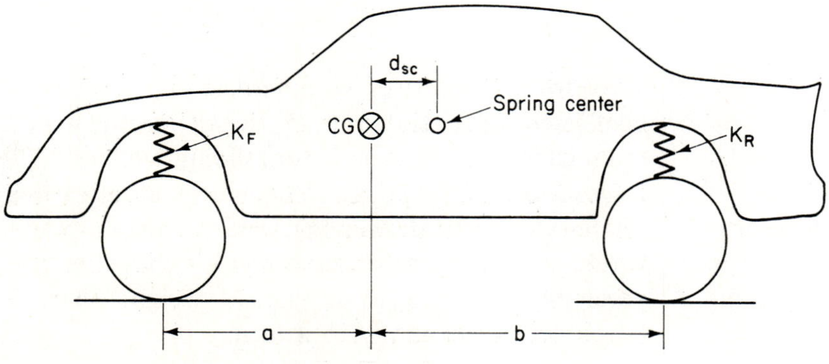

FIGURE 15-9 Typical spring center is located slightly to the rear of the center of gravity.

15-4 ACCELERATION FORCES AND WEIGHT TRANSFER

Forces are developed whenever a moving body that has mass (weight) is accelerated according to Newton's law: $F = ma$ (Force = mass × acceleration). Acceleration increases or decreases straight-line speed or changes vehicle direction. These acceleration forces act through the center of gravity which, on a vehicle, is above the ground. The change produces weight transfer. These forces and resulting weight transfer will be examined in this section.

The lateral (sideways) force that a vehicle develops in a turn depends on the speed the vehicle is traveling, the sharpness of the turn (turn radius), and the weight of the vehicle. Recall that maximum cornering force is limited by the coefficient of friction between the tire and the road. Likewise, maximum deceleration in braking is limited to the tire-to-road coefficient of friction.

It is best to work with the maximum values to understand how the vehicle will behave at the limit of control. We can, however, determine values below the maximum. Only the speed in the turn and the radius of the turn are needed to determine the lateral acceleration occurring in a turn. Since acceleration (in any direction) is the change in velocity, the *lateral acceleration* is the change in direction of the velocity.

$$A_l = V^2/R$$

where V = Velocity in ft/sec
= mph × 1.47
R = Turn radius in ft
A_l = Lateral acceleration ft/sec²

Example 7

A vehicle enters a constant 100 ft radius turn at 30 mph. What is the lateral acceleration?

$$A_l = (30 \text{ mph} \times 1.47)^2/100 = 44^2/100 = 19.36 \text{ ft/sec}^2$$

Commonly, acceleration is related to the acceleration of gravity (g) which, in conventional units, is 32.2 ft/sec² (9.81 m/sec²). The above acceleration in g's would then be

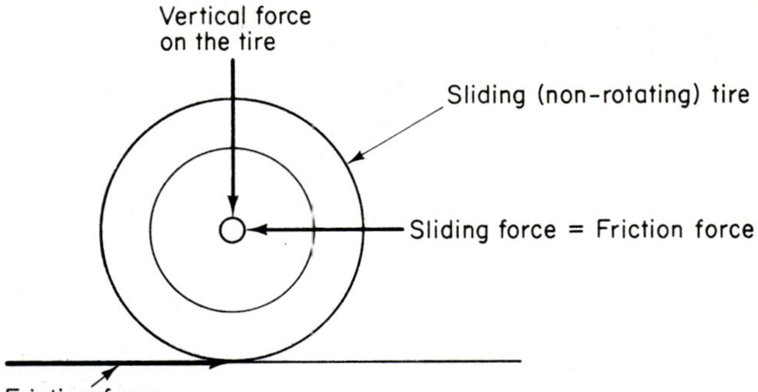

FIGURE 15-10 The coefficient of friction of the tire is determined by dividing the sliding force of the tire by the vertical force (weight) on the tire.

$$19.36/32.2 = 0.6 \text{ g's}$$

The acceleration in g's is a convenient value to use to study the dynamic behavior of a vehicle. Recall that the coefficient of friction of a tire relates the force required to slide a tire to the weight on the tire (Figure 15-10). Since force is always mass x acceleration, the coefficient of friction also indicates directly the percentage of the acceleration of gravity the tire can produce. To help clarify this point,

$$\text{Coefficient of friction} = \frac{\text{Sliding force}}{\text{Normal or vertical force}} = \frac{\text{Mass} \times \text{Acceleration in the plane}}{\text{Mass} \times \text{Acceleration of gravity}} = \frac{w/g \times \text{Acceleration in plane}}{\text{weight}} = \text{Acceleration in plane}/g$$

Therefore, if the tires on a vehicle can produce a coefficient of friction of 0.75, this means the maximum lateral acceleration before sliding is 0.75 x 32.2 ft/sec², or 24.15 ft/sec². Since the acceleration relates the vertical (weight) force to the in-plane force, the lateral acceleration in g's indicates the percentage of the total vehicle weight that is developed horizontally in a turn.

Example 8

Determine the lateral acceleration force on a 3200 lb (1454 kg) car in a 0.7 g turn.

$$F_L = 0.7 \times 3200 \text{ or } (0.7 \times 1454 \text{ kg} \times 9.81 \text{ ft/sec}^2)$$
$$= 2240 \text{ lbs } (9990 \text{ N})$$

Notice that the forces depend on vehicle weight and lateral acceleration. Recall also that lateral acceleration is only a function of the turn radius and the speed in the turn.

Braking and Acceleration Forces. The forces occurring during acceleration can be found in the same way as they were found for lateral acceleration. However, in these conditions, the acceleration or deceleration is in the same direction as the original velocity. The acceleration is just the change in velocity occurring in a given time interval. Also note that acceleration and deceleration act on the moving body in the same manner except that deceleration is negative acceleration.

During braking, all four tires work to produce the deceleration force. Again, this deceleration force is the deceleration force in g's times the vehicle weight.

Example 9

Determine the deceleration force for a 3000 lb vehicle decelerating at 17 ft/sec².

$$F_D = (17/32.2) \times 3000$$
$$= 1584 \text{ lbs } (7063.2 \text{ N})$$

Acceleration forces become slightly more complicated to calculate. This is because, in most vehicles today, the engine drives only two wheels, while the vertical weight is on all four wheels. If we ignore weight transfer (which will be discussed in the next section), the maximum acceleration force available is just the maximum coefficient of friction of the driving tires times the vertical weight on those tires.

Example 10

Calculate the maximum acceleration rate possible for a 3000 lb front-wheel-drive car with 66 percent front weight and a coefficient of friction of 0.7.

$$F_A = 0.7 \times (3000 \times 0.66)$$
$$= 1400 \text{ lbs } (6242 \text{ N})$$
$$\text{Acceleration rate} = \text{Acceleration force/total weight}$$
$$= 1400/3000$$
$$= 0.47 \text{ g's}$$

Notice that the acceleration rate is lower than the coefficient of friction. If we were to examine an all-wheel-drive vehicle, the possible acceleration (assuming the engine were large enough) would be the coefficient of friction. In the above example, the weight transfer would also reduce the percentage of weight on the drive wheels and, as a result, reduce the acceleration rate.

Weight Transfer. Weight transfer is the change in weight at the wheel locations due to cornering and acceleration or braking. These forces act through the center of gravity and, as a result, change the effective weight at each wheel.

Weight transfer is a function of acceleration forces, the height of the center of gravity, and the distance between the wheel locations measured in the direction of the acceleration force. Recall that the acceleration force depends on the magnitude of the acceleration and the mass ($F = ma$). The ratio of the height of the center of gravity to the distance between wheels in the direction of the force is the actual fraction of the force used as weight transfer.

$$WT_L = h/t\ F_L$$

where: WT_L = Lateral weight transfer
h = Height of CG
t = Wheel track
F_L = Lateral acceleration force

$$WT_{\text{long}} = h/l\ (F_a \text{ or } F_D)$$

where WT_{long} = Weight transfer front to rear
h = Height of CG
l = Wheelbase length
F_a or F_D = Acceleration or deceleration forces

Some examples of weight transfer follow.

Wheelbase = 100 in.
Wheel track = 58 in.
Height of center of gravity = 18 in.
Total weight = 2800 lbs

Example 11

The vehicle enters an 80 ft radius turn at 30 mph. Determine the weight transfer to the outside wheels.

$$30 \text{ mph} = 44.0 \text{ ft/sec}^2$$

$$\begin{aligned} F_l &= V^2/R \times 1/g \times W \\ &= (44^2/80) \times (1/32.2) \times 2800 \\ &= 2104 \text{ lbs} \\ A_l &= 44^2/80 = 24.2 \text{ ft/sec}^2 \\ &= 24.2/32.2 = .75 \text{ g's} \end{aligned}$$

Note: If the tires do not have a coefficient of friction of at least 0.75, the vehicle will slide sideways.

$$\begin{aligned} WT_L &= h/t\ F_L \\ &= (18/58) \times 2104 \text{ lbs} \\ WT_L &= 653 \text{ lbs} \end{aligned}$$

This means that prior to the turn, the car had 1400 lbs of weight on both the left and right sides. During the turn, the outside of the vehicle has:

$$W_{out} = (2800/2) + 653 = 2053 \text{ lbs}$$

and the inside of the vehicle has:

$$W_{in} = (2800/2) - 653 = 747 \text{ lbs}$$

$$\text{Total vehicle weight} = 2800 \text{ lbs}$$

Notice that the total vehicle weight must always remain the same. Note also that the fraction of weight transferred is simply

$$h/t \times (A_l \text{ in g's})$$

In the previous case,

$$\begin{aligned} (18/58) \times 0.75 &= .233 \text{ or } 23\% \\ 0.233 \times 2800 &= 653 \text{ lbs} \end{aligned}$$

Example 12

The above vehicle is braking at a maximum rate. Determine the weight on the front and rear wheels if the vehicle has a 65/35 weight distribution. Maximum braking is about 0.75 g's.

$$\begin{aligned} WT_{long} &= h/l\ F_a \\ &= (18/100) \times 0.75 \times 2800 \text{ lbs} \\ &= 378 \text{ lbs} \\ W_F &= (2800 \times .65) + 378 \text{ lbs} \\ &= 2198 \text{ lbs} \\ W_R &= (2800 \times .35) - 378 \text{ lbs} \\ &= 602 \text{ lbs} \end{aligned}$$

Notice the percentage of front weight during braking:

$$\begin{aligned} W_F\% &= 2198/2800 = 0.785 \\ &= 78.5\% \end{aligned}$$

This is why close to 80 percent of the braking comes from the front wheels during hard braking.

Example 13

Determine the lateral acceleration and lateral force needed to cause this vehicle to roll over. *Note:* To roll over, the weight on the inside wheels must be equal to zero.
Therefore,

$$\begin{aligned} WT_l &= 2800/2 = 1400 \text{ lbs} \\ 1400 &= (18/58) \times (2800 \times A_l) \\ A_l &= 1.61 \text{ g's are necessary} \end{aligned}$$

or $1.61 \times 2800 = 4511$ lbs at the road surface

The last example is just used to show how the stability of a vehicle can be determined in a turn. To cause rollover, today's passenger cars require forces at the tire that are higher than is achievable due to friction. Heavy trucks with high loads can obviously roll over much more easily. Observe that, as the center of gravity gets higher and the wheel base or track gets shorter, weight transfer for the same acceleration always increases. Weight transfer during straight-line acceleration increases the rear wheel weights and decreases the front wheel weights.

15-5 SUSPENSION GEOMETRY

Suspension geometry is the positioning and angularity of all of the suspension parts. This positioning has a large effect on how the parts function, the spring rates and stabilizer bars required, and how the car will ride and handle. This section presents these interrelationships and demonstrates how small changes influence vehicle handling and shows how the forces discussed earlier act on suspension geometry.

Roll Center. The *roll center* is the point about which the body rolls while making a turn. The roll center is the only point on the vehicle centerline that does not move up or down during a turn. The front and rear suspensions each have their own roll centers. A line connecting the front and rear roll centers is called the *roll axis*.

The instant center must be determined before the roll center can be located. The instant center is found by extending an imaginary line through both pivots of each control arm to the point where the lines intersect.

The roll center is located at the vehicle center on an extended line connecting the instant center and the center of the tire patch. This can be seen in Figure 15-11.

If Figure 15-11 is examined closely, some very important suspension behavior can be determined. First, the *instant center* is the point in space at which the extended lines of the control arms intersect. It is called an instant center because, for small movements of the suspension linkage, the control arms act as though they are pivoting around this point. The location of the instant center then predicts the camber change that will occur with small motions about the ride height. Notice that when the instant center is toward the center of the vehicle, the wheel produces negative camber as it begins to move up. Also note that, if the instant center is toward the outside, the vehicle initially produces positive camber. Recall that this is an instant center and it is constantly moving as the body moves. The instant center and roll center change position as the ride height changes.

The decision to have positive or negative camber change initially is one the suspension designer must consider. If we examine a vehicle entering a turn (see Figure 7-6), it will always have some body roll. Since the suspension is attached to the body or frame, the body will roll the front suspension and tires on the outside toward positive camber. If, then, the body produces negative camber, it will result in a tire that remains nearly flat on the road surface. If the suspension produces positive camber as the body rolls, this adds to the angle due to body roll and the tire tread ends up at a high angle on the road surface. Negative camber in jounce appears to be a strong choice; however, it is not that clear. Vehicles bounce up and down as they move down the road. This also results in camber change. When one wheel encounters a dip or a bump, camber will also change.

Camber change during jounce and rebound is important. For improved handling in turns, as you would expect from a sports car, it is common to use an instant center toward the inside and negative camber during jounce. For improved ride on heavier, more luxurious cars, it is a commonly accepted procedure to have the instant center toward the outside and positive camber during jounce without real concern for maximum cornering power.

From Figure 15-11, it should be obvious that the instant center predicts the roll center position. The roll center can be above the road surface, at the road surface, or below the road surface. The location of the roll center and the distance between it and the center of gravity influence ride, body roll, and the springs and stabilizer bars used.

Using the vehicle model shown in Figure 15-12, you can see that the roll axis itself is lowered as the roll centers for the front and rear suspensions are lowered. With the center of gravity (CG) remaining at the same height and lowering the roll axis (RA), the distance between the RA and the CG increases. The vehicle will have the same lateral acceleration and the same acceleration force independent of the roll axis position. The acceleration force acts through the CG and tends to rotate the body around the roll axis. For the same lateral acceleration force, a greater distance will increase the torque and, as a result, increase the body roll. Likewise, as the RA is moved toward the CG, the body roll decreases for the same lateral acceleration force.

Roll Couples. A close examination of Figure 15-12 will give you a clearer understanding of body roll, roll moments and roll couples (torque due to the distance between acceleration force and the roll axis), and roll couple distribution. Body roll is usually expressed as an angle and describes the angle between a vertical line to the road surface and a vertical line through the car body. The *roll moment*, or *roll couple*, is the torque about the roll axis due to

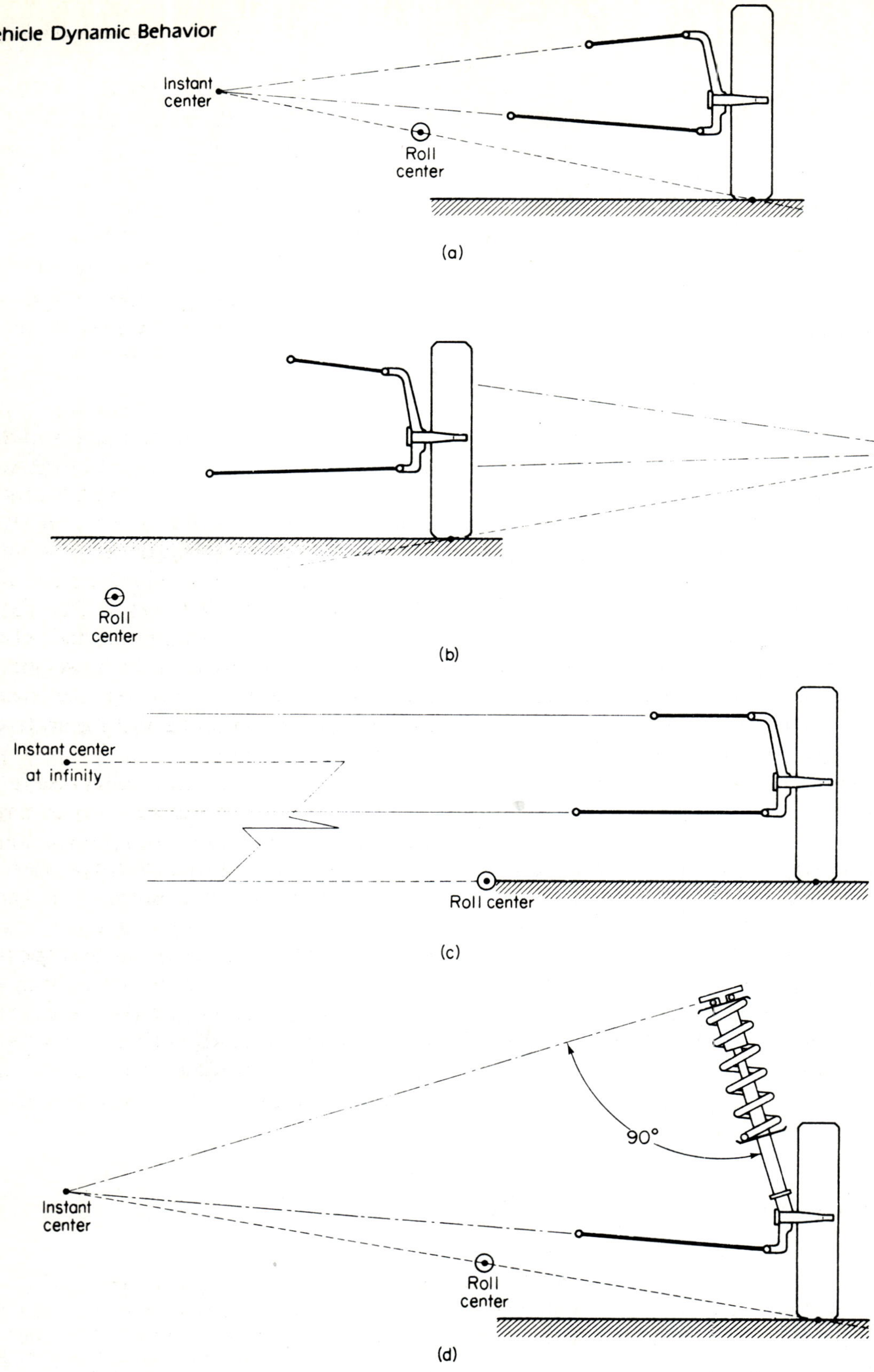

FIGURE 15-11 Illustrations showing how the roll center is located: (a) long-short control arm suspension converging *toward* the vehicle center; (b) long-short control arm suspension converging *away from* the vehicle center; (c) *parallel* control arms make the instant center at infinity and the roll center at the road surface; and (d) the instant center on a MacPherson strut extends at 90 degrees from the top of the strut and through the lower control arm pivot points.

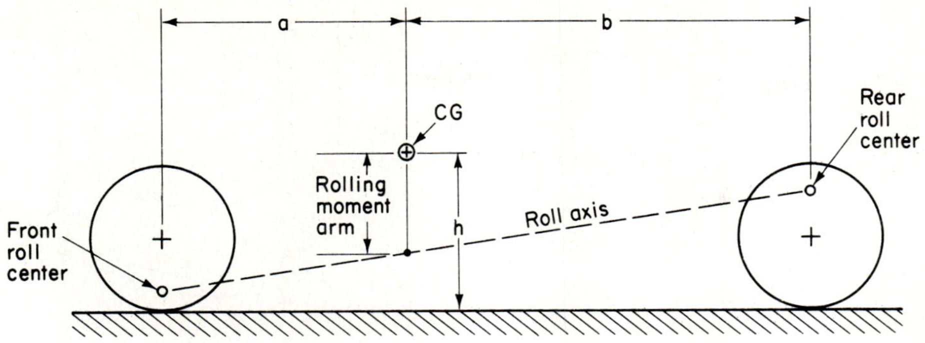

FIGURE 15-12 Schematic diagram of the relationship between roll axis and roll center.

centrifugal force acting on the center of gravity. The *roll couple distribution* describes how the torque (roll moment) is resisted by the front and rear suspension. This distribution affects the division of weight transfer between the front and rear suspensions. As you can see in Figure 15-12, the roll moment is resisted by springs acting at the wheels which really are the wheel rates calculated earlier. If the roll center in the front is lowered, the roll moment (RM) is increased. Since the body acts rigidly, the roll moments must all act about the same RA. This means that, for given wheel rates and a lower front RC, a greater amount of the roll moment must be resisted by the rear suspension.

Lateral Weight Transfer. Let's take some extreme cases to examine this. Figure 15-13 shows a vehicle with a pivoted axle without springs in the front and a sprung axle in the rear. The front offers no roll resistance. Since the same forces and moments are still present, the rear suspension must offer all the resistance to roll. This means that the front, since it is only pivoted, does not transfer any weight from the inside to the outside wheel (a free pivot does not develop torque). Since the amount of weight transfer does not change due to the suspension geometry, all of this weight transfer must be resisted by the rear suspension. The inside rear wheel in a turn would get quite light as the weight transfers to the outside wheel. The outside rear wheel would have considerable weight transferred to it. The front wheels still have the weight they originally had before entering the turn. This high transfer of weight from the rear inside wheel to the rear outside wheel would cause the rear tires to have a high slip angle and cause the vehicle to oversteer.

A second extreme case might also add to our understanding of how sideways weight transfer affects handling. In Figure 15-14, notice the vehicle's single coil spring in the rear which provides a spring for the ride in the rear. However, it has no roll resistance. The two springs out at the front wheels indicate the wheel rates of the front suspension. The front offers all of the resistance to body roll. For a given lateral acceleration, all of the weight transfer control occurs at the front tires. The weight on the outside front tire becomes much greater in a turn; the weight on the inside front tire becomes much less.

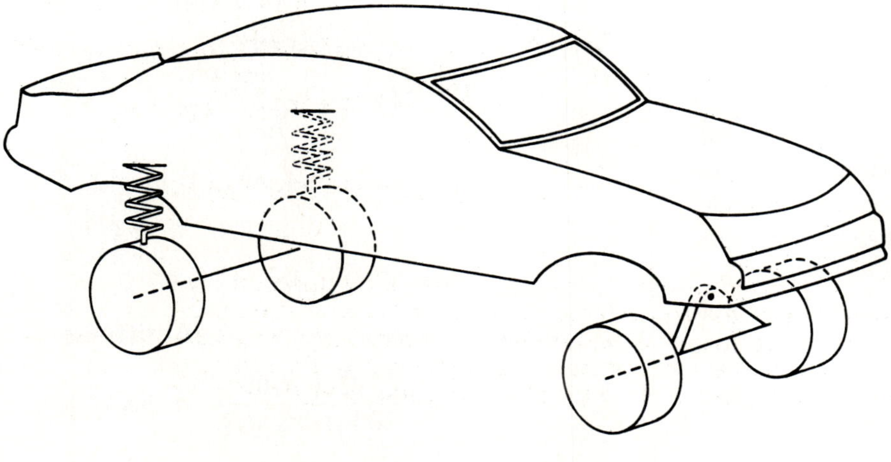

FIGURE 15-13 The rear suspension controls all roll resistance when the front axle pivots at the center and has no roll stiffness.

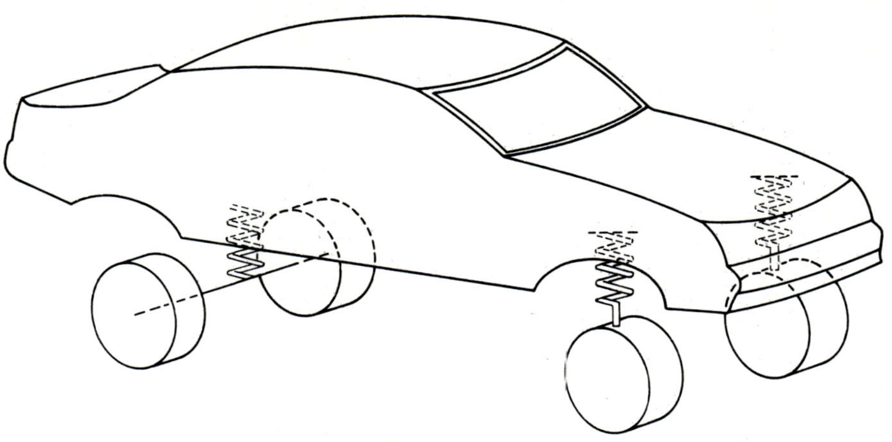

FIGURE 15-4 The front suspension controls all roll resistance when the rear axle has a center spring pivot and zero roll stiffness.

The heavier load on the outside tire will cause a greater slip angle to occur and, as a result, the vehicle will understeer in a turn.

Based on these two extreme examples, you can understand how a normal vehicle, as shown in Figure 15-4, is affected by lateral acceleration and body roll. The front and rear springs offer ride springing as well as resistance to body roll. The actual spring action can be used to determine the wheel rates, as described in Section 15-3. The stabilizer bar, or bars, increases the roll resistance of the body by adding to the roll stiffness without adding to the ride rate. The stabilizer bar resists body roll during a turn by acting to resist the upward motion of the outside wheel and the downward motion of the inside wheel. This added roll resistance increases the weight transfer from the inside to the outside wheel, on the stabilizer end of the vehicle, in the turn. Weight transfer from body roll at one end of the vehicle relative to the other end will increase the slip angle of the tires at that end. If a stabilizer bar is added to the front suspension (or the diameter of one is increased), the vehicle will tend toward roll understeer. Conversely, a stabilizer bar on the rear will tend to decrease the roll understeer.

Understanding how the stabilizer bar and the suspension springs change handling is important to the skilled technician as well as to the performance enthusiast. Changing the load-carrying capacity on a passenger vehicle could cause the vehicle to oversteer in a turn. Any change in the stabilizer bars also affects handling. You have also seen how a change in the ride height on one end of the vehicle can change the roll center position, resulting in a change in the slope of the roll axis and further changing the weight transfer in a turn. Example problems for these cases will show how weight transfer and handling can be determined mathematically.

Example 14 (refer to Figure 15-14)

Vehicle with all the roll stiffness in the rear

Vehicle Data:	
Total weight	2000 lbs
Weight distribution F/R	60/40
Roll center—front (FRC)	4 in.
Roll center—rear (RRC)	10 in.
CG height (H_{CG})	18 in.
Track width	55 in.
Front roll stiffness	0
Wheelbase (WB)	100 in.
Rear roll stiffness	200 ft-lbs/deg
Lateral acceleration	0.4 g

Lateral acceleration force:

$$F_L = 0.4 \times 2000 \text{ lbs} = 800 \text{ lbs}$$

Roll moment arm:

$$\text{Moment arm} = H_{CG} - \left[\frac{(\text{RRC} - \text{FRC})}{\text{WB}} (\text{distance CG to front}) + \text{FRC}\right]$$

$$= 18 - \left(\frac{10 - 4}{100}\right)(40) + 4$$

$$L = 18 - 6.4 = 11.6 \text{ in.}$$

$$\text{Roll moment} = F_L \times L = 800 \times 11.6$$

$$= 9280 \text{ in-lbs} = 773.3 \text{ ft-lbs}$$

$$\text{Roll angle} = \text{Roll moment/roll stiffness}$$

$$= \frac{773.3 \text{ ft-lbs}}{200 \text{ ft-lbs/deg}} = 3.87°$$

Roll couple distribution:

Front = 0 Rear = 200 ft-lbs/deg

$$RCD_r = \frac{\text{Rear}}{\text{Total}} = \frac{200}{200 + 0} = 100\% \text{ rear}$$

Weight transfer:

Front

$$WT_F = \frac{\text{Roll moment front}}{\text{Wheel track}} = \frac{0}{55/12 \text{ ft}} = 0$$

Rear:

$$WT_R = \frac{\text{Roll moment rear}}{\text{Wheel track}} = \frac{773.3 \text{ ft-lbs}}{55/12 \text{ ft}}$$

$$= 168.7 \text{ lbs}$$

In a 0.4 g turn:

Front weights

Inside wheel	*Outside wheel*
600 lbs	600 lbs

Rear weights

Inside wheel	*Outside wheel*
(400 − 168.7 lbs)	(400 + 168.7 lbs)
231.3 lbs	586.7 lbs

Notice that the inside rear tire has only a little more than one-half of its static weight, but the outside rear tire has gained a considerable amount of weight. This suspension design would always tend to oversteer.

Example 15

Refer to Example 14 for data with the front roll stiffness at 300 ft-lb/deg and zero rear roll stiffness

Lateral acceleration force: F_L = 800 lbs (as before)
Roll moment arm: MA = 11.6 in. (as before)
Roll moment = $F_L \times L$ = 773 ft-lbs (as before)
Roll angle = Roll moment/Roll stiffness = 773/300 = 2.58°

Roll couple distribution:
Front = 300 ft-lb/deg Rear = 0
Total = 300 ft-lb/deg

$$RCD_F = \frac{\text{Front}}{\text{Total}} = \frac{300}{300} = 100\% \text{ Front}$$

Weight transfer:

Rear = 0

$$\text{Front } WT_F = \frac{773}{55/12} = 168.7 \text{ lbs (as before)}$$

	Inside wheel	Outside wheel
Front:	600 − 168.7 lbs = 431.3 lbs	600 + 168.7 lbs = 768.7 lbs
Rear:	400 lbs	400 lbs

Notice now that the outside front tire is 178 percent heavier than the inside tire [(768.6/431.3) × 100 = 178%]. The rear tires in a turn retain their static weights. This vehicle design will always understeer in a turn because of the heavily loaded outside front tire.

Example 16

This example is representative of a passenger vehicle.

Vehicle weight	3000 lbs
F/R distribution	65/35
Front wheel rate	100 lbs/in.
Front roll center	2 in.
Front stabilizer bar stiffness	200 ft-lbs/deg
Rear wheel rate	120 lbs/in.
Rear roll center	12 in.
CG height	20 in.
Wheelbase	100 in.
Lateral acceleration	0.5 g's
Track width	58 in.

Lateral acceleration force:

$$F_L = a \times \text{weight}$$
$$= 0.5 \times 3000 = 1500 \text{ lbs}$$

Roll moment arm:

$$MA = H_{CG} - \left[\left(\frac{\text{RRC} - \text{FRC}}{\text{WB}}\right)\right.$$

$$\left.(\text{Distance CG to front}) + \text{FRC}\right]$$

$$= 20 - \left[\left(\frac{12 - 2}{100}\right)(35) + 2\right] = 14.5 \text{ in.}$$

Front roll stiffness:
Due to wheel rate

$$RS_F = \frac{(\text{Wheel rate})\ (\text{Track})^2}{1375}$$

$$= \frac{100 \times 58^2}{1375} = 244.6 \text{ ft-lbs/deg}$$

Total front roll stiffness

$$RS_{FT} = 244.6 + \text{bar stiffness}$$
$$= 244.6 + 200 \cong 445 \text{ ft-lbs/deg}$$

Rear roll stiffness:
Due to wheel rate

$$RS_R = \frac{(\text{Wheel Rate})\ (\text{Track})^2}{1375}$$

$$= \frac{120 \times 58^2}{1375} \cong 294 \text{ ft-lbs/deg}$$

Total roll stiffness:

$$RS_{TOT} = RS_{FT} + RS_R$$
$$= 445 \text{ ft-lbs/deg} + 294 \text{ ft-lbs/deg}$$
$$= 739 \text{ ft-lbs/deg}$$

Body roll angle:

$$RA = \frac{\text{Roll moment}}{\text{Roll stiffness}}$$

$$= \frac{1500 \text{ lbs} \times 14.5/12 \text{ ft}}{739}$$

$$= 2.45° \text{ in a 0.5 g turn}$$

Roll couple distribution:

$$RCD_{Front} = \frac{\text{Front}}{\text{Total}} = \frac{445}{739} \times 100 = 60\%$$

$$RCD_{Rear} = \frac{\text{Rear}}{\text{Total}} = \frac{294}{739} \times 100 = 40\%$$

Roll moment:
Front

$$RM_F = RCD_F \times RM_{Total}$$
$$= 0.6 \times 1812.5 = 1087.5 \text{ ft-lbs}$$

Rear

$$RM_R = RCD_R \times RM_{Total}$$
$$= 0.4 \times 1812.5 = 725 \text{ ft-lbs}$$

Weight transfer:

$$WT_F = \frac{RM_F}{\text{Track}}$$

$$= \frac{1087.5 \text{ ft-lbs}}{58/12 \text{ ft}} = 225 \text{ lbs}$$

$$WT_R = \frac{RM_R}{\text{Track}}$$

$$= \frac{725}{58/12 \text{ ft}} = 150 \text{ lbs}$$

Wheel weights in a turn:

	Inside
Front	975 − 225 = 750 lbs
Rear	525 − 150 = 375 lbs
	Outside
	975 + 225 = 1200 lbs
	525 + 150 = 675 lbs

Notice that the vehicle in Example 16 transfers weight from the inside to the outside of both front and rear tires. A well-designed vehicle will transfer weight in this manner with the distribution of the weight transfer close to the static weight distribution. In most cases, passenger cars will be designed with a higher percentage of the weight transfer on the front. Up to 80 percent front roll couple distribution is not uncommon. This tends to cause some understeer.

15-6 SUMMARY

This chapter has given you an overview of the forces, and the reactions, of a vehicle while it is moving in a straight line as well as in a turn. A technician who understands these relationships will also understand what may happen to ride and handling when springs are replaced with springs of different rates. As passenger car handling has become more precise, it has become more important for service technicians to have an understanding of how replacement parts may affect handling.

STUDY QUESTIONS

Section 15-1

1. When does pitching motion of the vehicle occur?
2. How many degrees of freedom does a vehicle have?

Section 15-2

3. What does the term *center of gravity* describe?
4. Discuss the meaning of the term *weight distribution*.
5. If a vehicle has a 50/50 weight distribution and a 2.5 m wheelbase, where is the center of gravity located?

Section 15-3

6. Is a brake rotor sprung or unsprung weight? Why?
7. If a MacPherson strut suspension system uses a 100 lb/in. spring rate, what is the wheel rate of the suspension?
8. Find the natural frequency of a vehicle with a sprung weight of 2500 lbs and a total ride rate of 550 lbs/in.
9. Why is the spring center positioned to the rear of the center of gravity?
10. If the rear springs are replaced with stiffer springs, what happens to the amount of pitch that occurs due to encountering a bump in the road?

Section 15-4

11. As the radius of a turn is decreased, what happens to lateral acceleration if the speed of the vehicle remains constant?
12. What is the maximum value of lateral acceleration that can be obtained with normal street tires in ft/sec^2, m/sec^2, and g's?
13. If the ride height of a vehicle is increased, what happens to the amount of weight transfer in a turn if the speeds are similar?
14. Will understeer increase or decrease if the front stabilizer bar is replaced with one of a larger diameter?
15. Could oversteer in a turn result if a rear stabilizer bar is fitted to a car that previously had only a front stabilizer bar?

APPENDIX

Basic Computer Programs

The following computer programs have been written for the users of this text. Computers play such an important part in our everyday lives that most educational systems introduce students to their use. Students are also getting more interested and involved in computers earlier in their education. These programs will allow interested students to further their interest in computers while learning the technical topic of the course. No claims are made as to their accuracy or value outside of the educational environment.

The programs were written, using the BASIC language, on an Apple II computer. However, with minimum modification they will run on most microcomputers. These programs were written to enable students to study various aspects of suspension and steering systems by allowing them to change various system parameters and examine the effect on the vehicle by comparing the output. A number of basic assignments can be made using these programs to enhance the students' understanding of steering and suspension.

Although the programs are written to be as friendly as possible, they will be explained briefly here.

FREQUENCY

This program will calculate the natural frequency of the sprung and unsprung mass at each wheel location. The program will ask for the suspension wheel rates (ride rate at the wheel location being analyzed), the sprung weight at the wheel location being analyzed (the weight if a scale is under the tire at that location minus the unsprung weight), the unsprung weight at that wheel location (weight of tire, wheel, brake assembly, and spindle), the tire rolling radius (approximately one-half the tire diameter), and the tire radial spring rate (usually about 1000 lbs/in., although it depends on size and construction). The output includes the sprung mass frequency or ride frequency, the unsprung mass natural frequency, and the critical speed where the vehicle will be sensitive to tire and wheel imbalance. This program will allow the interested user to examine the effect of changing ride rates and tires on the ride frequencies and speed where the vehicle is most sensitive to tire balance. This can also be used to motivate an understanding of basic vibration and forces due to imbalance.

CG LOCATION

Calculations used to predict vehicle behavior and handling usually require that the center of gravity be known. The weight transfer during braking, acceleration, and cornering is affected by the center of gravity location. This program can be used to determine the horizontal and vertical position of the center of gravity.

Required input to the program includes the scale weights of the vehicle at each wheel location (front and rear wheel weights), the rolling radius of the tires, and the wheelbase length. In addition, one end of the vehicle must be raised a measured amount and the change in weight of the unraised end measured. If the above data is input into the program, the center or gravity for the vehicle will be located.

This program can be used to relate static weight distribution to dynamic weight changes that occur on the vehicle. The program can be used to motivate discussion of the effect on handling of cargo and passengers, or other weight placement. This program's greatest value, however, resides in the determination of the specific location of the center of gravity for use in other programs.

WEIGHT TRANSFER

This program can be used to determine the weight transfer that occurs during either braking or acceleration. The weight transfer during braking is directly related to deceleration rate, which is related to tire-to-road coefficient of friction, height of the center of gravity, and wheelbase length. Weight transfer during acceleration is based on the same characteristics as braking; however, maximum acceleration capability, given surplus power, is also limited by the type of drive arrangement and may be limited by the ratio of front to rear roll stiffness of the chassis and the final drive ratio. The latter two effects are based on whether the vehicle has independent or dependent suspension at the driving axle.

This program allows the interested student to study the effects of suspension design on actual dynamic weight distribution. Since the dynamic weight distribution is directly related to maximum acceleration capabilities, this study can be quite interesting to students. On dependent rear-drive vehicles, the effects of front roll stiffness due to increasing stabilizer bar size or increased spring stiffness can be studied. Other relationships of drive arrangements and actual weights during acceleration and braking can be studied. This program can be used with the output from the CG Location program.

CAMBER

This program can be used to approximate the wheel camber changes that occur in long-short arm suspension due to changes in vehicle ride height. In addition, it will determine the scuff travel that occurs due to these same changes and the static roll center height. It can be used to approximate strut-type suspension analysis by simply inserting a value for the upper control arm length that is considerably greater than the lower control arm length and an inner pivot height slightly higher than the lower control arm height. (The strut acts as an infinitely long control arm at 90 degrees to the upper strut mount.)

The program requests as input the location and length of all suspension components. It also requests the ball joint offset (horizontal distance from the ball joint to the centerline of the tread, at the wheel), the front track width, and the travel (jounce and rebound) over which you want to analyze the suspension.

A variety of approximation techniques are used within the program to simplify calculation. Although these approximations do introduce some error at higher changes in ride height, they remain relatively accurate. This program can be used to motivate interest in suspension design and handling. The effects of changing control arm lengths and positions can be shown effectively. The influence of a single change on the camber, scuff, and roll center can be shown. In addition, the program shows how incorrect ride height can affect handling and tire wear even though wheel alignment settings may be correct.

CORNERING WT

This program is designed to realistically determine the actual vertical force acting at each wheel location during high-speed cornering. This force is a result of the static weight plus the weight transfer. This program requires a fairly high degree of knowledge about the suspension system. Depending on what the motivation is, however, much can be learned about elementary vehicle dynamics and the interrelationship of weight transfer, vehicle springing, and roll stiffness.

The program requests information about weight at specific wheel locations, wheelbase length, and wheel track length. It also requests front and rear roll center heights. The roll center height can be determined from the camber program; however, on most automobiles, the front roll center height is in the range of from 0 to 4 inches (0-10 cm) above the road surface. The rear roll center height is commonly 5 to 12 inches (12.5-30.5 cm) above the road surface. (Solid rear axle cars locate it at approximately the height of the axle centerline.) The program also requests front and rear spring rates, which can be determined as discussed in the text. In addition, the front stabilizer bar stiffness is requested. (This is usually between 200 and 800 ft-lbs/degree {275-1100 N-m/degree}.) The cornering acceleration (lateral acceleration) is also requested and this is limited by the coefficient of friction of the tires. The value is requested in g's, which for normal tires is approximately 0.75 at maximum cornering.

The program calculates percentage of weight distribution, roll couple distribution (front-rear), roll couples, and weight transfer. The nonrolling couple is the torque created by the cornering forces acting on the roll axis and, in turn, acting about the centerline of the vehicle at the road surface (roll axis-ground moment). If the roll axis were flat on the road surface, the nonrolling couple would be equal to zero.

This program can be used by those interested in more than a cursory understanding of suspension. This program, accompanied by the Camber and Springs programs, can be used to approximate weight transfer in a flat turn. Although fairly complex in input requirements, much can be learned about handling through the use of this program.

SPRINGS

This program will calculate the deflection constant (spring rate) for a coil spring. It will also determine the stabilizer bar stiffness and the roll stiffness contribution of the front springs, the rear springs, and the stabilizer bar.

This program will allow the interested user to determine values needed to run other programs previously presented. The program can also be used to help students understand the relationships that exist between springs, roll stiffness, and handling.

```
LOAD FREQUENCY
]LIST

10  REM : THIS PROGRAM CAN BE USED TO SHOW NATURAL
20  REM : VIBRATION FREQUENCIES OF A VEHICLE AND
30  REM : SPEEDS WHERE THE VEHICLE IS SENSITIVE TO
40  REM : WHEEL BALANCE.
50  HOME
60  PRINT "THIS PROGRAM WILL CALCULATE THE NATURAL"
70  PRINT "FREQUENCY OF A SUSPENSION SYSTEM"
80  PRINT "AND THE CRITICAL VEHICLE SPEED FOR"
90  PRINT "IMBALANCE EXCITATION OF THE SUSPENSION"
100  PRINT
110  PRINT
120  PRINT "YOU WILL NEED TO KNOW THE FOLLOWING:"
130  PRINT
140  PRINT "    1. SUSPENSION WHEEL RATES"
150  PRINT "    2. SPRUNG WEIGHT AT THE WHEEL"
160  PRINT "    3. UNSPRUNG WEIGHT AT THE WHEEL"
170  PRINT "    4. TIRE ROLLING RADIUS"
180  PRINT "    5. TIRE RADIAL SPRING RATE "
190  PRINT "       (USUALLY ABOUT 1000 LB/IN)
200  PRINT
210  PRINT
220  INPUT "ENTER THE WHEEL RATE IN LBS/IN---";WR
230  PRINT
240  PRINT "ENTER THE SPRUNG WEIGHT AT THE"
250  INPUT "WHEEL LOCATION-------------------";WW
260  PRINT
270  PRINT "ENTER THE WEIGHT OF THE UNSPRUNG"
280  INPUT "COMPONENTS AT THE WHEEL----------";US
290  PRINT
300  INPUT "ENTER THE TIRE RADIAL SPRING RATE";T
310  PRINT
320  INPUT "ENTER THE TIRE ROLLING RADIUS----";RR
330  REM
340  REM : SPRUNG MASS NATURAL FREQUENCIES
350  REM
360 NF = (1 / 6.28) * (WR * 12 * 32.2 / WW) ^ .5
370 NF =  INT (NF * 10 + .5) / 10
380  REM
390  REM : UNSPRUNG MASS NATURAL FREQUENCIES
400  REM
410 NS = (1 / 6.28) * ((WR + T) * 12 * 32.2 / US) ^ .5
420 NS =  INT (NS * 10 + .5) / 10
430  REM
440  REM : CRITICAL SPEED CALCULATION
450  REM
460 E = NS * 3600 * 6.28 * RR / (12 * 5280)
470 E =  INT (E * 10 + .5) / 10
480  HOME
490  PRINT "                    INPUT"
500  PRINT
510  PRINT "WHEEL RATE----------- ";WR;" LBS/IN"
520  PRINT "SPRUNG WEIGHT-------- ";WW;" LBS"
530  PRINT "UNSPRUNG WEIGHT------ ";US;" LBS"
540  PRINT "ROLLING RADIUS------- ";RR;" IN"
550  PRINT "TIRE RATE------------ ";T;" LBS/IN"
560  PRINT
570  PRINT
580  PRINT "                    OUTPUT"
590  PRINT
600  PRINT "NATURAL FREQUENCY (SPRUNG)---- ";NF;" HZ"
610  PRINT "NATURAL FREQUENCY (UNSPRUNG)-- ";NS;" HZ
620  PRINT "CRITICAL IMBALANCE SPEED------ ";E;" MPH
630  PRINT
640  PRINT
650  PRINT
660  PRINT "IF YOU WOULD LIKE TO RERUN THIS"
670  PRINT "PROGRAM TYPE AN R, IF YOU WOULD LIKE"
680  PRINT "TO END TYPE AN E."
690  INPUT A$
```

```
700  HOME
710  IF A$ = "R" THEN 200
720  PRINT
730  PRINT
740  PRINT "YOU SHOULD KNOW THAT:"
750  PRINT
760  PRINT "  1.SPRUNG MASS NATURAL FREQUENCY"
770  PRINT "    IS USUALLY 1-1.3 HZ"
780  PRINT "  2.UNSPRUNG MASS NATURAL FREQUENCY"
790  PRINT "    IS USUALLY 10 TIMES SPRUNG"
800  PRINT "  3.INCREASING SPRING STIFFNESS"
810  PRINT "    INCREASES BOTH NATURAL FREQUENCIES"
820  PRINT "  4.INCREASING SPRUNG MASS DECREASES"
830  PRINT "    SPRUNG NATURAL FREQUENCY"
840  PRINT
850  PRINT "HAVE A GOOD DAY"
860  END
```

```
 RUN FREQUENCY
THIS PROGRAM WILL CALCULATE THE NATURAL
FREQUENCY OF A SUSPENSION SYSTEM
AND THE CRITICAL VEHICLE SPEED FOR
IMBALANCE EXCITATION OF THE SUSPENSION


YOU WILL NEED TO KNOW THE FOLLOWING:

    1. SUSPENSION WHEEL RATES
    2. SPRUNG WEIGHT AT THE WHEEL
    3. UNSPRUNG WEIGHT AT THE WHEEL
    4. TIRE ROLLING RADIUS
    5. TIRE RADIAL SPRING RATE
       (USUALLY ABOUT 1000 LB/IN)


ENTER THE WHEEL RATE IN LBS/IN---130

ENTER THE SPRUNG WEIGHT AT THE
WHEEL LOCATION--------------------900

ENTER THE WEIGHT OF THE UNSPRUNG
COMPONENTS AT THE WHEEL----------75

ENTER THE TIRE RADIAL SPRING RATE1000

ENTER THE TIRE ROLLING RADIUS----12
                        INPUT

WHEEL RATE----------- 130 LBS/IN
SPRUNG WEIGHT-------- 900 LBS
UNSPRUNG WEIGHT------ 75 LBS
ROLLING RADIUS------- 12 IN
TIRE RATE------------ 1000 LBS/IN


                        OUTPUT

NATURAL FREQUENCY (SPRUNG)---- 1.2 HZ
NATURAL FREQUENCY (UNSPRUNG)-- 12.1 HZ
CRITICAL IMBALANCE SPEED------ 51.8 MPH



IF YOU WOULD LIKE TO RERUN THIS
PROGRAM TYPE AN R, IF YOU WOULD LIKE
TO END TYPE AN E.
?E
```

```
YOU SHOULD KNOW THAT:

  1.SPRUNG MASS NATURAL FREQUENCY
    IS USUALLY 1-1.3 HZ
  2.UNSPRUNG MASS NATURAL FREQUENCY
    IS USUALLY 10 TIMES SPRUNG
  3.INCREASING SPRING STIFFNESS
    INCREASES BOTH NATURAL FREQUENCIES
  4.INCREASING SPRUNG MASS DECREASES
    SPRUNG NATURAL FREQUENCY

HAVE A GOOD DAY
```

```
 LOAD CG LOCATION
]LIST

10  REM : THIS PROGRAM IS USED TO LOCATE THE CENTER
20  REM : OF GRAVITY FROM MEASURED SCALE WEIGHTS
30  HOME
40  PRINT "THIS PROGRAM CALCULATES THE CG"
50  PRINT "LOCATION FOR A VEHICLE"
60  PRINT
70  PRINT
80  PRINT "THE REQUIRED INFORMATION IS"
90  PRINT
100  PRINT "    1. FRONT WHEEL WEIGHTS"
110  PRINT "    2. REAR WHEEL WEIGHTS"
120  PRINT "    3. WHEEL BASE LENGTH"
130  PRINT "    4. TIRE ROLLING RADIUS"
140  PRINT "    5. CHANGE IN HEIGHT OF THE"
150  PRINT "       VEHICLE TO PRODUCE THE "
160  PRINT "       MEASURED CHANGE IN WEIGHT"
170  PRINT
180  PRINT "ENTER THE TOTAL WEIGHT ON THE"
190  INPUT "FRONT WHEELS ------------------ ";WF
200  PRINT
210  PRINT "ENTER THE TOTAL WEIGHT ON THE"
220  INPUT "REAR WHEELS ------------------- ";WR
230  PRINT
240  PRINT "ENTER THE WHEEL BASE LENGTH"
250  INPUT "IN INCHES -------------------- ";L
260  PRINT
270  PRINT "ENTER THE TIRE ROLLING RADIUS"
280  INPUT "IN INCHES -------------------- ";R
290  PRINT
300  PRINT "ENTER THE CHANGE IN HEIGHT OF"
310  INPUT "ONE END OF THE VEHICLE -------- ";H2
320  PRINT
330  PRINT "ENTER THE CHANGE IN WEIGHT THAT"
340  PRINT "RESULTED WHEN THE HEIGHT WAS"
350  INPUT "CHANGED ---------------------- ";W2
360 W = WF + WR
370 ANG =  ATN (H2 / L)
380 S =  SIN (ANG)
390 C =  COS (ANG)
400 H = (W2 * L * C) / (W * S)
410 H = H + R
420 X = (WR * L) / W
430 H =  INT (10 * H + .5) / 10
440 X =  INT (10 * X + .5) / 10
450  PRINT
460  PRINT
470  PRINT "THE CG LOCATION IS"
480  PRINT "     ";H;" INCHES FROM THE GROUND"
490  PRINT
500  PRINT "     ";X;" INCHES REAR OF THE FRONT AXLE"
510  END
```

```
]RUN CG LOCATION
THIS PROGRAM CALCULATES THE CG
LOCATION FOR A VEHICLE


THE REQUIRED INFORMATION IS

    1. FRONT WHEEL WEIGHTS
    2. REAR WHEEL WEIGHTS
    3. WHEEL BASE LENGTH
    4. TIRE ROLLING RADIUS
    5. CHANGE IN HEIGHT OF THE
       VEHICLE TO PRODUCE THE
       MEASURED CHANGE IN WEIGHT

ENTER THE TOTAL WEIGHT ON THE
FRONT WHEELS ------------------ 2000

ENTER THE TOTAL WEIGHT ON THE
REAR WHEELS ------------------- 1200

ENTER THE WHEEL BASE LENGTH
IN INCHES --------------------- 100

ENTER THE TIRE ROLLING RADIUS
IN INCHES --------------------- 12

ENTER THE CHANGE IN HEIGHT OF
ONE END OF THE VEHICLE -------- 10

ENTER THE CHANGE IN WEIGHT THAT
RESULTED WHEN THE HEIGHT WAS
CHANGED ---------------------- 25


THE CG LOCATION IS
     19.8 INCHES FROM THE GROUND

     37.5 INCHES REAR OF THE FRONT AXLE
```

```
 LOAD WEIGHT TRANSFER
]LIST

1  REM : THIS PROGRAM IS USED TO DETERMINE ACTUAL
2  REM : WEIGHT DISTRIBUTION DURING ACCELERATION
3  REM : OR BRAKING. IT IS USED TO SHOW HOW
4  REM : SUSPENSION CHANGES CAN AFFECT WEIGHT
5  REM : DISTRIBUTION DURING ACCELERATION
10  HOME : VTAB 10
20  PRINT "          WEIGHT TRANSFER"
30  PRINT "        EVALUATION PROGRAM"
40  PRINT
50  PRINT
60  PRINT "THIS PROGRAM CALCULATES STATIC AND"
70  PRINT "DYNAMIC WEIGHT TRANSFER DURING"
80  PRINT "ACCELERATION AND BRAKING"
90  PRINT
100  PRINT
110  INPUT "PRESS RETURN TO CONTINUE";Z$
120  HOME : VTAB 10
130  PRINT "YOU CAN CALCULATE EITHER ACCELERATION"
140  PRINT "      OR BRAKING CHARACTERISTICS"
150  PRINT
160  PRINT
170  PRINT "IF YOU WOULD LIKE ACCELERATION"
180  INPUT "TYPE -A- ,IF BRAKING TYPE -B-    ";B$
190  IF B$ = "B" THEN 1160
200  HOME
210  PRINT
215  REM : ACCELERATION CALCULATIONS
220  PRINT
230  PRINT "ENTER THE LETTERS CORRESPONDING "
240  PRINT "TO THE DRIVE ARRANGEMENT"
250  PRINT
260  PRINT "  F--FRONT WHEEL DRIVE"
270  PRINT "  R--REAR WHEEL DRIVE W/OPEN DIFF."
280  PRINT "  RL-REAR WHEEL DRIVE W/LIMITED SLIP"
290  PRINT "  RI-REAR WHEEL DRIVE W/INDEPENDENT"
300  PRINT "       SUSPENSION"
310  INPUT A$
320  HOME
330  PRINT "ENTER THE FOLLOWING VALUES"
340  PRINT
350  INPUT " WHEELBASE (IN.)--------------- ";L
360  INPUT " TRACK WIDTH (IN.)------------- ";TW
370  INPUT " TIRE ROLLING RADIUS (IN.)----- ";R
380  INPUT " CG HEIGHT (IN.)--------------- ";H
390  INPUT " FRONT AXLE WEIGHT------------- ";WF
400  INPUT " REAR AXLE WEIGHT-------------- ";WR
410  IF A$ = "RI" THEN 460
420  IF A$ = "F" THEN 460
430  PRINT " FRONT ROLL STIFFNESS "
440  INPUT " (USUALLY 70-90%)-------------- ";RC
450  INPUT " FINAL DRIVE RATIO-------------- ";ND
460  INPUT " TIRE-ROAD FRICTION COEFF.----- ";U

470  HOME : VTAB 10
480  INPUT "PRESS RETURN TO CONTINUE";V$
490 TW = TW / 12
500 RC = RC / 100
510 R = R / 12
520  IF A$ = "F" THEN 550
530  IF A$ = "R" THEN 570
540  GOTO 590
550 PW = (U * WF) / (1 + (H / L) * U)
560  GOTO 600
570 PW = (U * WR) / (1 - (H / L) * U + 2 * U * RC *
    R / (ND * TW))

580  GOTO 600
590 PW = (U * WR) / (1 - (H / L) * U)
600 G = PW / (WF + WR)
610 ACC = G * 32.2
620 WTR = (WF + WR) * (H / L) * G
630 NF = (WF - WTR)
640 NR = (WR + WTR)
650 WPR = WTR / (WF + WR)
660 PF = NF / (WF + WR)
670 PR = NR / (WF + WR)
680 WTR =  INT (WTR + .5)
690 NF =  INT (NF + .5)
700 NR =  INT (NR + .5)
710 G =  INT (G * 100 + .5) / 100
720 ACC =  INT (ACC * 10 + .5) / 10
730 PF =  INT (PF * 1000 + .5) / 10
740 PR =  INT (PR * 1000 + .5) / 10
750 WPR =  INT (WPR * 1000 + .5) / 10
760  HOME
770  PRINT "           INPUT DATA"
780  PRINT
790  PRINT "WHEELBASE (IN.)---------------- ";L
800  PRINT "TRACK WIDTH (IN.)-------------- ";TW * 12
810  PRINT "TIRE ROLLING RADIUS (IN.)------ ";R * 12
820  PRINT "CG HEIGHT (IN.)---------------- ";H
830  PRINT "FRONT AXLE WEIGHT (LBS)-------- ";WF
840  PRINT "REAR AXLE WEIGHT (LBS)--------- ";WR
850  PRINT "TIRE-ROAD COEFFICIENT---------- ";U
860  PRINT
870  PRINT
880  PRINT "           OUTPUT DATA"
890  PRINT
900  PRINT "MAXIMUM ACCELERATION(G'S)------ ";G
910  PRINT "MAXIMUM ACCEL. (FT/(SEC*SEC)--- ";ACC
920  PRINT "MAXIMUM WT. TRANSFER(LBS)------ ";WTR
930  PRINT "FRONT WEIGHT DURING ACCEL.----- ";NF
940  PRINT "REAR WEIGHT DURING ACCEL.------ ";NR
950  PRINT "% OF WEIGHT TRANSFERRED-------- ";WPR
960  PRINT "ACCEL. WT. DISTRIBUTION------ ";PF;"/";PR
970  IF A$ = "F" THEN 1020
980  IF A$ = "RI" THEN 1020
990 WY = (RC * PW * R) / (ND * TW)
1000 WY =  INT (WY + .5)
1010  PRINT "TRANSVERSE WT. TRANSFER (LBS)-- ";WY

1020  PRINT
1030  INPUT "PRESS RETURN TO CONTINUE";Z$
1040  HOME
1045  GOSUB 2000
1050  PRINT "TYPE THE LETTER CORRESPONDING TO"
1060  PRINT "WHAT YOU WOULD LIKE TO DO AT THIS"
1070  PRINT "POINT IN THE PROGRAM."
1080  PRINT
```

```
1090  PRINT "RECALCULATE WEIGHT TRANSFER>>> R"
1100  PRINT "CALCULATE BRAKING WT. TRANS.>> B"
1110  PRINT "    END>>>> E"
1120  INPUT C$
1130  IF C$ = "R" THEN 200
1140  IF C$ = "B" THEN 1160
1150  GOTO 1570
1160  REM  BRAKING WT TRANSFER
1170  HOME : VTAB 5
1180  IF B$ = "A" THEN 1270
1190  PRINT "  ENTER THE FOLLOWING VALUES"
1200  PRINT
1210  INPUT "WHEELBASE (IN.)---------------- ";L
1220  INPUT "TIRE ROLLING RADIUS (IN.)------ ";R
1225 R = R / 12
1230  INPUT "CG HEIGHT (IN.)---------------- ";H
1240  INPUT "FRONT AXLE WEIGHT (LBS)-------- ";WF
1250  INPUT "REAR AXLE WEIGHT (LBS)--------- ";WR
1260  INPUT "TIRE-ROAD FRICTION COEFFICIENT- ";U
1270 BT = (H / L) * (U * (WF + WR))
1280 BF = BT + WF
1290 BR = WR - BT
1300 WDF = BF / (WF + WR)
1310 WDR = BR / (WF + WR)
1320 C = BT / (WF + WR)
1330 BT =  INT (BT + .5)
1340 BF =  INT (BF + .5)
1350 BR =  INT (BR + .5)
1360 WDF =  INT (WDF * 1000 + .5) / 10
1370 WDR =  INT (WDR * 1000 + .5) / 10
1380 C =  INT (C * 1000 + .5) / 10
1390  HOME
1400  PRINT "            INPUT DATA"
1410  PRINT
1420  PRINT "WHEELBASE (IN)----------------- ";L
1430  PRINT "TIRE ROLLING RADIUS (IN)------- ";R * 12
1440  PRINT "CG HEIGHT (IN)----------------- ";H
1450  PRINT "FRONT AXLE WEIGHT (LBS)-------- ";WF
1460  PRINT "REAR AXLE WEIGHT (LBS)--------- ";WR
1470  PRINT "TIRE-ROAD COEFFICIENT---------- ";U
1480  PRINT
1490  PRINT
1500  PRINT "            OUTPUT DATA"
1510  PRINT
1520  PRINT "DECEL. RATE (FT/SEC/SEC)------- ";U * 32.2
1530  PRINT "MAXIMUM WT. TRANSFER----------- ";BT
1540  PRINT "FRONT WT. DURING BRAKING------- ";BF
1550  PRINT "REAR WT. DURING BRAKING-------- ";BR
1560  PRINT "% OF WT. TRANSFERRED BRAKING--- ";C
1570  END
2000  PRINT  TAB( 6);"LF"; TAB( 25);"RF"
2010  PRINT "    ----               ----"
2020  PRINT "    I  I               I  I"
2030  PRINT "    I  I               I  I"
2040  PRINT "    I  I               I  I"
2050  PRINT "    I  I               I  I"
2060  PRINT "    ----               ----"
2070  PRINT  TAB( 5);(NF / 2) - WY; TAB( 24);
     (NF / 2) + WY
2080  PRINT  TAB( 5);"(";WF / 2;")"; TAB( 24);
     "(";WF / 2;")"
2090  PRINT
2100  PRINT
2110  PRINT
2120  PRINT
2130  PRINT
2140  PRINT  TAB( 6);"LR"; TAB( 25);"RR"
2150  PRINT "    ----               ----"
2160  PRINT "    I  I               I  I"
2170  PRINT "    I  I               I  I"
2180  PRINT "    I  I               I  I"
2190  PRINT "    I  I               I  I"
2200  PRINT "    ----               ----"
2210  PRINT  TAB( 5);(NR / 2) + WY; TAB( 24);
     (NR / 2) - WY
2220  PRINT  TAB( 5);"(";WR / 2;")"; TAB( 24);
     "(";WR / 2;")"
2300  INPUT "PRESS RETURN TO CONTINUE";Z$
2310  HOME : VTAB 7
2320 WY = 0
3000  RETURN
```

```
RUN WEIGHT TRANSFER
        WEIGHT TRANSFER
      EVALUATION PROGRAM


THIS PROGRAM CALCULATES STATIC AND
DYNAMIC WEIGHT TRANSFER DURING
ACCELERATION AND BRAKING


PRESS RETURN TO CONTINUE
YOU CAN CALCULATE EITHER ACCELERATION
      OR BRAKING CHARACTERISTICS


IF YOU WOULD LIKE ACCELERATION
TYPE -A- ,IF BRAKING TYPE -B-    A


ENTER THE LETTERS CORRESPONDING
TO THE DRIVE ARRANGEMENT

  F--FRONT WHEEL DRIVE
  R--REAR WHEEL DRIVE W/OPEN DIFF.
  RL-REAR WHEEL DRIVE W/LIMITED SLIP
  RI-REAR WHEEL DRIVE W/INDEPENDENT
     SUSPENSION
?F
ENTER THE FOLLOWING VALUES

 WHEELBASE (IN.)--------------- 100
 TRACK WIDTH (IN.)------------- 58
 TIRE ROLLING RADIUS (IN.)----- 12
 CG HEIGHT (IN.)--------------- 19.8
 FRONT AXLE WEIGHT------------- 2000
 REAR AXLE WEIGHT-------------- 1200
 TIRE-ROAD FRICTION COEFF.----- .75
PRESS RETURN TO CONTINUE
          INPUT DATA

WHEELBASE (IN.)---------------- 100
TRACK WIDTH (IN.)-------------- 58
TIRE ROLLING RADIUS (IN.)------ 12
CG HEIGHT (IN.)---------------- 19.8
FRONT AXLE WEIGHT (LBS)-------- 2000
REAR AXLE WEIGHT (LBS)--------- 1200
TIRE-ROAD COEFFICIENT---------- .75



          OUTPUT DATA

MAXIMUM ACCELERATION(G'S)------ .41
MAXIMUM ACCEL. (FT/(SEC*SEC)--- 13.1
MAXIMUM WT. TRANSFER(LBS)------ 259
FRONT WEIGHT DURING ACCEL.----- 1741
REAR WEIGHT DURING ACCEL.------ 1459
% OF WEIGHT TRANSFERRED-------- 8.1
ACCEL. WT. DISTRIBUTION------ 54.4/45.6

PRESS RETURN TO CONTINUE
      LF                 RF
     ----               ----
     I  I               I  I
     I  I               I  I
     I  I               I  I
     I  I               I  I
     ----               ----
    870.5              870.5
    (1000)             (1000)




      LR                 RR
     ----               ----
     I  I               I  I
     I  I               I  I
     I  I               I  I
     I  I               I  I
     ----               ----
    729.5              729.5
    (600)              (600)
```

```
PRESS RETURN TO CONTINUE
TYPE THE LETTER CORRESPONDING TO
WHAT YOU WOULD LIKE TO DO AT THIS
POINT IN THE PROGRAM.

RECALCULATE WEIGHT TRANSFER>>> R
CALCULATE BRAKING WT. TRANS.>> B
   END>>>> E
?R


ENTER THE LETTERS CORRESPONDING
TO THE DRIVE ARRANGEMENT

  F--FRONT WHEEL DRIVE
  R--REAR WHEEL DRIVE W/OPEN DIFF.
  RL-REAR WHEEL DRIVE W/LIMITED SLIP
  RI-REAR WHEEL DRIVE W/INDEPENDENT
     SUSPENSION
?R


ENTER THE FOLLOWING VALUES

 WHEELBASE (IN.)--------------- 100
 TRACK WIDTH (IN.)------------- 58
 TIRE ROLLING RADIUS (IN.)----- 12
 CG HEIGHT (IN.)--------------- 19.8
 FRONT AXLE WEIGHT------------- 2000
 REAR AXLE WEIGHT-------------- 1200
 FRONT ROLL STIFFNESS
 (USUALLY 70-90%)-------------- 75
 FINAL DRIVE RATIO-------------- 3.23
 TIRE-ROAD FRICTION COEFF.----- .75
PRESS RETURN TO CONTINUE
          INPUT DATA

WHEELBASE (IN.)---------------- 100
TRACK WIDTH (IN.)-------------- 58
TIRE ROLLING RADIUS (IN.)------ 12
CG HEIGHT (IN.)---------------- 19.8
FRONT AXLE WEIGHT (LBS)-------- 2000
REAR AXLE WEIGHT (LBS)--------- 1200
TIRE-ROAD COEFFICIENT---------- .75


          OUTPUT DATA

MAXIMUM ACCELERATION(G'S)------ .3
MAXIMUM ACCEL. (FT/(SEC*SEC)--- 9.8
MAXIMUM WT. TRANSFER(LBS)------ 193
FRONT WEIGHT DURING ACCEL.----- 1807
REAR WEIGHT DURING ACCEL.------ 1393
% OF WEIGHT TRANSFERRED-------- 6
ACCEL. WT. DISTRIBUTION------ 56.5/43.5
TRANSVERSE WT. TRANSFER (LBS)-- 47

PRESS RETURN TO CONTINUE
    LF                  RF
   ----                ----
   I  I                I  I
   I  I                I  I
   I  I                I  I
   I  I                I  I
   ----                ----
   856.5               950.5
   (1000)              (1000)



    LR                  RR
   ----                ----
   I  I                I  I
   I  I                I  I
   I  I                I  I
   I  I                I  I
   ----                ----
   743.5               649.5
   (600)               (600)
```

```
PRESS RETURN TO CONTINUE
TYPE THE LETTER CORRESPONDING TO
WHAT YOU WOULD LIKE TO DO AT THIS
POINT IN THE PROGRAM.

RECALCULATE WEIGHT TRANSFER>>> R
CALCULATE BRAKING WT. TRANS.>> B
   END>>>> E
?B
          INPUT DATA

WHEELBASE (IN)----------------- 100
TIRE ROLLING RADIUS (IN)------- 12
CG HEIGHT (IN)----------------- 19.8
FRONT AXLE WEIGHT (LBS)-------- 2000
REAR AXLE WEIGHT (LBS)--------- 1200
TIRE-ROAD COEFFICIENT---------- .75


          OUTPUT DATA

DECEL. RATE (FT/SEC/SEC)------- 24.15
MAXIMUM WT. TRANSFER----------- 475
FRONT WT. DURING BRAKING------- 2475
REAR WT. DURING BRAKING-------- 725
% OF WT. TRANSFERRED BRAKING--- 14.8
```

```
 LOAD CAMBER
]LIST

10  REM : THIS PROGRAM CAN BE USED TO
20  REM : SHOW HOW CAMBER CHANGES AND
30  REM : SCUFF OCCURS AS THE SUSPENSION
40  REM : HEIGHT IS CHANGED, THE ROLL
50  REM : CENTER HEIGHT IS ALSO DETERMINED
60  HOME : VTAB 5
70  PRINT
80  PRINT "THIS PROGRAM CALCULATES WHEEL CAMBER"
90  PRINT "CHANGE, SCUFF TRAVEL AND ROLL CENTER"
100  PRINT "HEIGHT OF A FRONT SUSPENSION"
110  PRINT
120  PRINT
130  PRINT
140  PRINT "YOU WILL NEED"
150  PRINT "   1. UPPER CONTROL ARM LENGTH"
160  PRINT "   2. LOWER CONTROL ARM LENGTH"
170  PRINT "   3. UPPER BALL JOINT HEIGHT"
180  PRINT "   4. LOWER BALL JOINT HEIGHT"
190  PRINT "   5. UPPER CONTROL ARM PIVOT HEIGHT"
200  PRINT "   6. LOWER CONTROL ARM PIVOT HEIGHT"
210  PRINT "   7. BALL JOINT OFFSET"
220  PRINT "   8. TRACK WIDTH"
230  PRINT
240  PRINT "IF YOU DO NOT HAVE ALL THE NECESSARY"
245  PRINT "INFORMATION TYPE -NO- AND THE PROGRAM"
247  PRINT "WILL END ";
250  INPUT C$
260  IF C$ = "NO" THEN 900
270  PRINT
280  PRINT "PLEASE TYPE IN THE INFORMATION AFTER"
290  PRINT "EACH QUESTION MARK
300  PRINT
310  INPUT "UPPER CONTROL ARM LENGTH-------- ";A
320  PRINT
330  INPUT "LOWER CONTROL ARM LENGTH-------- ";B
340  PRINT
350  INPUT "UPPER BALL JOINT HEIGHT--------- ";C
360  PRINT
370  INPUT "UPPER CONTROL ARM INNER PIVOT HT ";E
380  PRINT
390  INPUT "UPPER BALL JOINT OFFSET--------- ";D1
400  PRINT
```

```
410  INPUT "LOWER BALL JOINT HEIGHT---------  ";D
420  PRINT
430  INPUT "LOWER CONTROL ARM INNER PIVOT HT ";F
440  PRINT
450  INPUT "LOWER BALL JOINT OFFSET---------  ";E1
460  PRINT
470  INPUT "FRONT TRACK WIDTH---------------  ";G
480  PRINT
490  INPUT "SUSPENSION TRAVEL ALLOWED-------  ";X
500  PRINT
510  PRINT
520 H = (C - D)
530 I = (C - E)
540 J = (D - F)
550 Z =  - X
560  HOME
570  REM : SCUFF TRAVEL DETERMINATION
580  PRINT "RIDE HEIGHT"
590  PRINT "(+) IS COMPRESSION      SCUFF TRAVEL (IN.)
600  PRINT " ------------------      ------------"
610 Y = (Z ^ 2 / (2 * H)) * ((C / B) - (D / A)) +
     (Z / H) * (((C * J) /   B) - ((D * I) / A))

620 Y =  INT (Y * 1000)
630 Y = Y / 1000
640  PRINT "          ";Z; TAB( 28);Y
650 Z = Z + .5
660  IF Z > X THEN 690
670  GOTO 610
680  PRINT
690  INPUT "WHEN YOU WANT TO CONTINUE TYPE ANY DIGIT
     ";B7
700 S =  - X
710  HOME
720  REM : CAMBER CHANGE DETERMINATION
730  PRINT "RIDE HEIGHT"
740  PRINT "(+) IS COMPRESSION      CAMBER DEGREES"
750  PRINT "------------------         ------"
760 U = (S ^ 2 / 2 * (1 / A - 1 / B) + S *
     (I / A - J / B)) / (H + (D1 * I / A - E1 * J / B)
     + S * (D1 / A - E1 / B))

770 R = U *  - 57.32
780 R =  INT (R * 1000)
790 R = R / 1000
800  PRINT "          ";S; TAB( 28);R
810 S = S + .5
820  IF S > X THEN 850
830  GOTO 760
840  REM : ROLL CENTER DETERMINATION
850 V =  - (G / (2 * H)) * (((C * J) / B) - ((D * I)
     / A))

860 V =  INT (V * 100)
870 V = V / 100
880  PRINT
890  PRINT "THE ROLL CENTER IS ";V;" INCHES FROM"
892  PRINT "THE ROAD SURFACE (- INDICATES BELOW"
895  PRINT "THE ROAD SURFACE)"
900  END
```

```
 RUN CAMBER

THIS PROGRAM CALCULATES WHEEL CAMBER
CHANGE, SCUFF TRAVEL AND ROLL CENTER
HEIGHT OF A FRONT SUSPENSION


YOU WILL NEED
    1. UPPER CONTROL ARM LENGTH
    2. LOWER CONTROL ARM LENGTH
    3. UPPER BALL JOINT HEIGHT
    4. LOWER BALL JOINT HEIGHT
    5. UPPER CONTROL ARM PIVOT HEIGHT
    6. LOWER CONTROL ARM PIVOT HEIGHT
    7. BALL JOINT OFFSET
    8. TRACK WIDTH

IF YOU DO NOT HAVE ALL THE NECESSARY
INFORMATION TYPE -NO- AND THE PROGRAM
WILL END ?

PLEASE TYPE IN THE INFORMATION AFTER
EACH QUESTION MARK

UPPER CONTROL ARM LENGTH-------- 9.5

LOWER CONTROL ARM LENGTH-------- 16

UPPER BALL JOINT HEIGHT--------- 21.75

UPPER CONTROL ARM INNER PIVOT HT 20.75

UPPER BALL JOINT OFFSET--------- 5

LOWER BALL JOINT HEIGHT--------- 10.75

LOWER CONTROL ARM INNER PIVOT HT 10.25

LOWER BALL JOINT OFFSET--------- 3

FRONT TRACK WIDTH--------------- 62

SUSPENSION TRAVEL ALLOWED------- 2


RIDE HEIGHT
(+) IS COMPRESSION      SCUFF TRAVEL (IN.)
 ------------------      ------------
          -2                .123
          -1.5              .084
          -1                .051
          -.5               .023
          0                 0
          .5                -.018
          1                 -.031
          1.5               -.039
          2                 -.041
WHEN YOU WANT TO CONTINUE TYPE ANY DIGIT 1


RIDE HEIGHT
(+) IS COMPRESSION      CAMBER DEGREES
------------------         ------
          -2                .333
          -1.5              .33
          -1                .271
          -.5               .161
          0                 0
          .5                -.21
          1                 -.465
          1.5               -.764
          2                 -1.106

THE ROLL CENTER IS 1.27 INCHES FROM
THE ROAD SURFACE (- INDICATES BELOW
THE ROAD SURFACE)
```

```
 LOAD CORNERING WT
]LIST

10  REM : THIS PROGRAM ALLOWS YOU TO DETERMINE THE
20  REM : AFFECT OF CHANGING SPRINGS, STABILIZER BARS
30  REM : AND SUSPENSION COMPONENTS ON WEIGHT TRANSFER
40  REM : AND HANDLING.
50  HOME
60  PRINT "THIS PROGRAM CALCULATES ROLL ANGLE"
70  PRINT "ROLL COUPLES, ROLL COUPLE DISTRIBUTION"
80  PRINT "AND WEIGHT TRANSFER."
90  PRINT
100  PRINT "YOU WILL NEED TO KNOW"
110  PRINT "  1. TOTAL VEHICLE WEIGHT AND WEIGHT"
120  PRINT "      DISTRIBUTION"
```

```
130  PRINT "                   OR"
140  PRINT "  2. WEIGHT AT EACH WHEEL LOCATION"
150  PRINT "  3. WHEEL BASE"
160  PRINT "  4. WHEEL TRACK"
170  PRINT "  5. ROLL CENTER HEIGHTS"
180  PRINT "  6. FRONT SPRING RATES"
190  PRINT "                 OR"
200  PRINT "     WHEEL RATES "
210  PRINT "  7. CONTROL ARM LENGTHS"
220  PRINT "  8. REAR SPRING RATES"
230  PRINT "  9. STABILIZER BAR STIFFNESS"
240  PRINT
250  PRINT "DO YOU KNOW THE WEIGHT AT EACH WHEEL"
260  PRINT "LOCATION; TYPE Y IF YES ";
270  INPUT B$
280  IF B$ = "Y" THEN 460
290  HOME
300  PRINT "TYPE IN THE INFORMATION REQUESTED BELOW"
310  PRINT
320  PRINT "ENTER THE TOTAL WEIGHT ON"
330  INPUT "THE FRONT WHEELS---------------- ";WF
340  PRINT
350  PRINT "ENTER THE TOTAL WEIGHT ON"
360  INPUT "THE REAR WHEELS----------------- ";WR
370  PRINT
380 W = WF + WR
390 P1 = WF / (WF + WR)
400  PRINT
410 W1 = (P1 * W) / 2
420 W3 = W1
430 W2 = (W - (W1 + W3)) / 2
440 W4 = W2
450  GOTO 590
460  PRINT
470  HOME
480  INPUT "LEFT FRONT WEIGHT--------------- ";W1
490  PRINT
500  INPUT "RIGHT FRONT WEIGHT-------------- ";W3
510  PRINT
520  INPUT "LEFT REAR WEIGHT---------------- ";W2
530  PRINT
540  INPUT "RIGHT REAR WEIGHT--------------- ";W4
550  PRINT

560  REM : WEIGHT DISTRIBUTION
570  REM
580  LET W = W1 + W2 + W3 + W4
590 W5 = W1 + W3
600 W6 = W2 + W4
610 W7 = W2 + W3
620 P2 = (W5 / W) * 100
630 P3 = (100 - P2)
640 P4 = (W7 / W) * 100
650 P5 = 100 - P4
660 P8 = ((W1 + W2) / W) * 100
670  PRINT
680  PRINT "THE WEIGHT DISTRIBUTION OF"
690  PRINT "THE CHASSIS IS AS SHOWN"
700  PRINT
710 P2 =  INT (P2 * 10 + .5) / 10
720 P3 =  INT (P3 * 10 + .5) / 10
730 P4 =  INT (P4 * 10 + .5) / 10
740 P8 =  INT (P8 * 10 + .5) / 10
750  PRINT "     FRONT    ";P2;"%"
760  PRINT "     REAR     ";P3;"%"
770  PRINT "     DIAGONAL ";P4;"%"
780  PRINT "     LEFT     ";P8;"%"
790  PRINT
800  PRINT "TYPE 1 TO RECYCLE;2 TO CONTINUE ";
810  INPUT B
820  IF B = 1 THEN 250
830  HOME
840  INPUT "CENTER OF GRAVITY HEIGHT-------- ";C1
850  PRINT
860  INPUT "WHEEL BASE LENGTH--------------- ";L1
870  PRINT
880  INPUT "FRONT ROLL CENTER HEIGHT-------- ";H1
890  PRINT
900  INPUT "REAR ROLL CENTER HEIGHT--------- ";H2
910  PRINT
920  IF D5 = 4 THEN 1520
930  PRINT "DO YOU HAVE INFORMATION ON"
940  PRINT "  1. SPRING RATES"
950  PRINT "  2. WHEEL RATES"
960  PRINT "ENTER 1 OR 2 ";
970  INPUT B1
980  IF B1 = 1 THEN 1050
990  PRINT
1000  INPUT "FRONT WHEEL RATE---------------- ";K2
1010  GOTO 1160
1020  PRINT
1030  REM : CALCULATING WHEEL RATE FROM SPRING RATE
1040  PRINT
1050  INPUT "FRONT SPRING RATE--------------- ";K1
1060  PRINT
1070  INPUT "LOWER CONTROL ARM LENGTH-------- ";L2
1080  PRINT
1090  PRINT "ENTER THE DISTANCE BETWEEN "
1100  PRINT "THE SPRING CENTER AND THE"
1110  INPUT "INNER CONTROL ARM PIVOT-------- ";L3
1120  IF B1 = 2 THEN 1190
1130 K2 = K1 * ((L3 / L2) ^ 2)
1140  PRINT
1150  PRINT "THE FRONT WHEEL RATE WITH THE ";K1;"
    LB/INCH SPRING IS ";K2 ;"LB/IN"

1160  PRINT
1170  IF D5 = 2 THEN 1330
1180  PRINT
1190  INPUT "REAR SPRING RATE---------------- ";K3
1200  IF D5 = 3 THEN 1330
1210  PRINT
1220  INPUT "DISTANCE BETWEEN THE REAR SPRINGS ";L4
1230  PRINT
1240  IF D5 = 4 THEN 1330
1250  PRINT
1260  HOME
1270  INPUT "FRONT WHEEL TRACK--------------- ";T1
1280  PRINT
1290  INPUT "REAR WHEEL TRACK---------------- ";T2
1300  PRINT
1310  PRINT "STABILIZER BAR STIFFNESS"
1320  INPUT "  (FT-LBS/DEGREE) -------------- ";K4
1330  PRINT
1340  REM : CALCULATION OF ROLL STIFFNESS
1350  REM
1360 K5 = (K2 * (T1 ^ 2)) / 1375
1370 K6 = K5 + K4
1380  IF A$ = "W" THEN 1400
1390 K7 = (K3 * (L4 ^ 2)) / 1375
1400 K8 = K6 + K7
1410  PRINT
1420  IF D5 <  > 0 THEN 1510
1430  PRINT "LATERAL ACCELERATION EXPECTED"
1440  INPUT "(UP TO 1.1 G'S)----------------- ";A1
1450  REM : DETERMINATION OF ROLL COUPLE
1460  REM
1470  REM : ROLL COUPLE IS THE TORQUE ACTING AROUND
1480  REM : THE ROLL AXIS DUE TO CENTRIFUGAL FORCE
1490  REM : DURING CORNERING
1500  REM
1510 F1 = A1 * W
1520 H3 = (H2 - H1) * P3 / 100
1530 H4 = C1 - (H3 + H1)
1540 M1 = H4 * F1 / 12
1550 A2 = M1 / K8
1560 M2 = M1 * K6 / K8
1570 M3 = M1 * K7 / K8
1580 P6 = (M2 / M1) * 100
1590 P7 = (M3 / M1) * 100
1600 M4 = ((W1 + W3) * A1 * H1) / 12
1610 M5 = ((W2 + W4) * A1 * H2) / 12
1620 M6 = M2 + M4
1630 M7 = M3 + M5
1640 T3 = (M6 / T1) * 12
1650 T4 = (M7 / T2) * 12

1660  PRINT
1670 K5 =  INT (K5 + .5)
1680  PRINT
1690  PRINT
1700  PRINT
1710  PRINT
1720  HOME
1730  PRINT "ROLL STIFFNESS            FT-LBS/DEG"
1740  PRINT
1750  PRINT "   FRONT W/O BAR--------- ";K5
1760  PRINT
1770 K6 =  INT (K6 + .5)
```

```
1780  PRINT "   TOTAL FRONT----------- ";K6
1790  PRINT
1800 K7 =  INT (K7 + .5)
1810  PRINT "   REAR------------------ ";K7
1820  PRINT
1830 K8 =  INT (K8 + .5)
1840  PRINT "   TOTAL----------------- ";K8
1850  PRINT
1860  PRINT
1870 M1 =  INT (M1 + .5)
1880  PRINT "ROLL MOMENT-------------- ";M1
1890  PRINT
1900  PRINT
1910  PRINT
1920  INPUT "ENTER <RETURN> TO CONTINUE";Z$
1930  PRINT
1940  PRINT
1950  HOME
1960  PRINT
1970  PRINT "ROLL ANGLE"
1980 A2 =  INT (A2 * 100) / 100
1990  PRINT "  (IN ";A1;" G TURN)----- ";A2;" DEG"
2000  PRINT
2010  PRINT "THE ROLL COUPLE DISTRIBUTION IS"
2020  PRINT
2030 P6 =  INT (P6 * 10 + .5) / 10
2040 P7 =  INT (P7 * 10 + .5) / 10
2050  PRINT "   FRONT      ";P6;"  %"
2060  PRINT "   REAR       ";P7;"  %"
2070  PRINT
2080  PRINT "ROLL COUPLES              FT-LBS"
2090  PRINT
2100 M4 =  INT (M4 + .5)
2110  PRINT "  FRONT(NON-ROLLING)--- ";M4
2120  PRINT
2130 M5 =  INT (M5 + .5)
2140  PRINT "  REAR(NON-ROLLING)---- ";M5
2150  PRINT
2160 M6 =  INT (M6 + .5)
2170  PRINT "  TOTAL FRONT COUPLE--- ";M6
2180  PRINT
2190 M7 =  INT (M7 + .5)
2200  PRINT "  TOTAL REAR COUPLE---- ";M7
2210  PRINT
```

```
2220  INPUT "ENTER <RETURN> TO CONTINUE";Z$
2230  HOME
2240 T3 =  INT (T3 + .5)
2250 T4 =  INT (T4 + .5)
2260 T5 =  INT (T5 + .5)
2270 T6 =  INT (T6 + .5)
2280 T7 =  INT (T7 + .5)
2290 T8 =  INT (T8 + .5)
2300  PRINT "WEIGHT TRANSFER IN THE ";A1;" G TURN"
2310  PRINT
2320  PRINT "      FRONT---------- ";T3;" LBS"
2330  PRINT "      REAR----------- ";T4;" LBS"
2340  PRINT
2350 T5 = W1 - T3
2360 T6 = W2 - T4
2370 T7 = W3 + T3
2380 T8 = W4 + T4
2390 T5 =  INT (T5 + .5)
2400 T6 =  INT (T6 + .5)
2410 T7 =  INT (T7 + .5)
2420 T8 =  INT (T8 + .5)
2430  PRINT "      RF                RR"
2440  PRINT "  -------           -------"
2450  PRINT "  -     -           -     -"
2460  PRINT "  -     -           -     -"
2470  PRINT "  -------           -------"
2480  PRINT "   ";T7;"              ";T8
2490  PRINT "  (";W3;")            (";W4;")"
2500  PRINT
2510  PRINT
2520  PRINT "     LF              LR"
2530  PRINT "  -------           -------"
2540  PRINT "  -     -           -     -"
2550  PRINT "  -     -           -     -"
2560  PRINT "  -------           -------"
2570  PRINT "   ";T5;"              ";T6
2580  PRINT "  (";W1;")            (";W2;")"
2590  PRINT
2600  INPUT "ENTER <RETURN> TO CONTINUE";Z$
```

```
2610  HOME
2620  HOME
2630  PRINT "YOU CAN CHANGE ANY OF THE FOLLOWING"
2640  PRINT "  1. FRONT BAR"
2650  PRINT "  2. FRONT SPRINGS"
2660  PRINT "  3. REAR SPRINGS"
2670  PRINT "  4. ROLL CENTER HEIGHT"
2680  PRINT "  5. END"
2690  PRINT
2700  INPUT "ENTER THE NUMBER CORRESPONDING TO YOUR
         CHOICE";D5
2710  IF D5 = 1 THEN 1300
2720  IF D5 = 2 THEN 910
2730  IF D5 = 3 THEN 1160
2740  IF D5 = 4 THEN 870

2750  HOME
2760  PRINT "YOU SHOULD KNOW THAT"
2770  PRINT "  1. WITH A NORMAL AUTOMOBILE SLIDING"
2780  PRINT "     OF THE TIRES PREVENTS ROLLOVER"
2790  PRINT "  2. RIDE HEIGHT AFFECTS CG HEIGHT"
2800  PRINT "     WHICH AFFECTS WEIGHT TRANSFER"
2810  PRINT "  3. RIDE HEIGHT AFFECTS ROLL CENTER "
2820  PRINT "     POSITION WHICH AFFECTS THE "
2830  PRINT "     DISTRIBUTION FRONT-REAR OF WEIGHT"
2840  PRINT "     TRANSFER"
2850  PRINT "  4. INCREASED SPRING STIFFNESS OR"
2860  PRINT "     STABILIZER AT EITHER END INCREASES"
2870  PRINT "     THE % OF TOTAL WEIGHT TRANSFER AT"
2880  PRINT "     THAT END OF THE VEHICLE"
2890  END
```

```
RUN CORNERING WT
THIS PROGRAM CALCULATES ROLL ANGLE
ROLL COUPLES, ROLL COUPLE DISTRIBUTION
AND WEIGHT TRANSFER.

YOU WILL NEED TO KNOW
  1. TOTAL VEHICLE WEIGHT AND WEIGHT
     DISTRIBUTION
           OR
  2. WEIGHT AT EACH WHEEL LOCATION
  3. WHEEL BASE
  4. WHEEL TRACK
  5. ROLL CENTER HEIGHTS
  6. FRONT SPRING RATES
           OR
     WHEEL RATES
  7. CONTROL ARM LENGTHS
  8. REAR SPRING RATES
  9. STABILIZER BAR STIFFNESS

DO YOU KNOW THE WEIGHT AT EACH WHEEL
LOCATION; TYPE Y IF YES ?Y

LEFT FRONT WEIGHT--------------- 995

RIGHT FRONT WEIGHT-------------- 1005

LEFT REAR WEIGHT---------------- 455

RIGHT REAR WEIGHT--------------- 445


THE WEIGHT DISTRIBUTION OF
THE CHASSIS IS AS SHOWN

     FRONT    69%
     REAR     31%
     DIAGONAL 50.3%
     LEFT     50%
```

```
TYPE 1 TO RECYCLE;2 TO CONTINUE ?1
DO YOU KNOW THE WEIGHT AT EACH WHEEL
LOCATION; TYPE Y IF YES ?N
TYPE IN THE INFORMATION REQUESTED BELOW

ENTER THE TOTAL WEIGHT ON
THE FRONT WHEELS---------------- 2000

ENTER THE TOTAL WEIGHT ON
THE REAR WHEELS----------------- 1000


THE WEIGHT DISTRIBUTION OF
THE CHASSIS IS AS SHOWN

     FRONT     66.7%
     REAR      33.3%
     DIAGONAL  50%
     LEFT      50%


TYPE 1 TO RECYCLE;2 TO CONTINUE ?2
CENTER OF GRAVITY HEIGHT-------- 19.8

WHEEL BASE LENGTH--------------- 100

FRONT ROLL CENTER HEIGHT-------- 1.27

REAR ROLL CENTER HEIGHT--------- 10

DO YOU HAVE INFORMATION ON
   1. SPRING RATES
   2. WHEEL RATES
ENTER 1 OR 2 ?2

FRONT WHEEL RATE---------------- 130


REAR SPRING RATE---------------- 100

DISTANCE BETWEEN THE REAR SPRINGS 45


FRONT WHEEL TRACK--------------- 58

REAR WHEEL TRACK---------------- 58

STABILIZER BAR STIFFNESS
  (FT-LBS/DEGREE) -------------- 400


LATERAL ACCELERATION EXPECTED
(UP TO 1.1 G'S)----------------- .75



ROLL STIFFNESS              FT-LBS/DEG

   FRONT W/O BAR--------- 318

   TOTAL FRONT----------- 718

   REAR------------------ 147

   TOTAL----------------- 865

ROLL MOMENT-------------- 2929


ENTER <RETURN> TO CONTINUE



ROLL ANGLE
  (IN .75 G TURN)----- 3.38 DEG
```

```
THE ROLL COUPLE DISTRIBUTION IS

   FRONT      83  %
   REAR       17  %

ROLL COUPLES               FT-LBS

  FRONT(NON-ROLLING)--- 159

  REAR(NON-ROLLING)---- 625

  TOTAL FRONT COUPLE--- 2589

  TOTAL REAR COUPLE---- 1124

ENTER <RETURN> TO CONTINUE
WEIGHT TRANSFER IN THE .75 G TURN

     FRONT---------- 536 LBS
     REAR----------- 232 LBS

      RF              RR
  -------         -------
  -     -         -     -
  -     -         -     -
  -------         -------
   1536             732
  (1000)           (500)


      LF              LR
  -------         -------
  -     -         -     -
  -     -         -     -
  -------         -------
   464             268
  (1000)          (500)

ENTER <RETURN> TO CONTINUE
YOU CAN CHANGE ANY OF THE FOLLOWING
  1. FRONT BAR
  2. FRONT SPRINGS
  3. REAR SPRINGS
  4. ROLL CENTER HEIGHT
  5. END

ENTER THE NUMBER CORRESPONDING TO YOUR CHOICE1


STABILIZER BAR STIFFNESS
  (FT-LBS/DEGREE) -------------- 200




ROLL STIFFNESS              FT-LBS/DEG

   FRONT W/O BAR--------- 318

   TOTAL FRONT----------- 518

   REAR------------------ 147

   TOTAL----------------- 665

ROLL MOMENT-------------- 2929


ENTER <RETURN> TO CONTINUE


ROLL ANGLE
  (IN .75 G TURN)----- 4.4 DEG

THE ROLL COUPLE DISTRIBUTION IS

   FRONT      77.9  %
   REAR       22.1  %
```

```
ROLL COUPLES                 FT-LBS

  FRONT(NON-ROLLING)--- 159

  REAR(NON-ROLLING)---- 625

  TOTAL FRONT COUPLE--- 2440

  TOTAL REAR COUPLE---- 1273

ENTER <RETURN> TO CONTINUE
WEIGHT TRANSFER IN THE .75 G TURN

     FRONT---------- 505 LBS
     REAR----------- 263 LBS

      RF              RR
  -------         -------
  -     -         -     -
  -     -         -     -
  -------         -------
   1505            763
  (1000)          (500)


      LF              LR
  -------         -------
  -     -         -     -
  -     -         -     -
  -------         -------
   495             237
  (1000)          (500)

ENTER <RETURN> TO CONTINUE
YOU CAN CHANGE ANY OF THE FOLLOWING
  1. FRONT BAR
  2. FRONT SPRINGS
  3. REAR SPRINGS
  4. ROLL CENTER HEIGHT
  5. END

ENTER THE NUMBER CORRESPONDING TO YOUR CHOICE4

FRONT ROLL CENTER HEIGHT-------- 4

REAR ROLL CENTER HEIGHT--------- 8



ROLL STIFFNESS              FT-LBS/DEG

   FRONT W/O BAR--------- 318

   TOTAL FRONT----------- 518

   REAR------------------ 147

   TOTAL----------------- 665

ROLL MOMENT-------------- 2713


ENTER <RETURN> TO CONTINUE


ROLL ANGLE
  (IN .75 G TURN)----- 4.07 DEG

THE ROLL COUPLE DISTRIBUTION IS

   FRONT      77.9  %
   REAR       22.1  %
```

```
ROLL COUPLES                 FT-LBS

  FRONT(NON-ROLLING)--- 500

  REAR(NON-ROLLING)---- 500

  TOTAL FRONT COUPLE--- 2613

  TOTAL REAR COUPLE---- 1100

ENTER <RETURN> TO CONTINUE
WEIGHT TRANSFER IN THE .75 G TURN

     FRONT---------- 541 LBS
     REAR----------- 228 LBS

      RF              RR
  -------         -------
  -     -         -     -
  -     -         -     -
  -------         -------
   1541            728
  (1000)          (500)

      LF              LR
  -------         -------
  -     -         -     -
  -     -         -     -
  -------         -------
   459             272
  (1000)          (500)

ENTER <RETURN> TO CONTINUE
YOU CAN CHANGE ANY OF THE FOLLOWING
  1. FRONT BAR
  2. FRONT SPRINGS
  3. REAR SPRINGS
  4. ROLL CENTER HEIGHT
  5. END

ENTER THE NUMBER CORRESPONDING TO YOUR CHOICE5
YOU SHOULD KNOW THAT
  1. WITH A NORMAL AUTOMOBILE SLIDING
     OF THE TIRES PREVENTS ROLLOVER
  2. RIDE HEIGHT AFFECTS CG HEIGHT
     WHICH AFFECTS WEIGHT TRANSFER
  3. RIDE HEIGHT AFFECTS ROLL CENTER
     POSITION WHICH AFFECTS THE
     DISTRIBUTION FRONT-REAR OF WEIGHT
     TRANSFER
  4. INCREASED SPRING STIFFNESS OR
     STABILIZER AT EITHER END INCREASES
     THE % OF TOTAL WEIGHT TRANSFER AT
     THAT END OF THE VEHICLE
```

```
 LOAD SPRINGS
]LIST

5  REM :SPRING CALCULATION PROGRAM TO DETERMINE
10  REM :COIL SPRING AND STABILIZER BAR STIFFNESS,
15  REM :RIDE (WHEEL) RATE GIVEN SPRING RATE
20  REM :AND THE ROLL STIFFNESS OF THE VEHICLE
25  REM :GIVEN THE ABOVE PARAMETERS
30  HOME : VTAB 5
40  PRINT "THIS PROGRAM CALCULATES"
50  PRINT "  1. COIL SPRING RATES"
60  PRINT "  2. WHEEL RATES GIVEN SPRING RATES"
70  PRINT "  3. STABILIZER BAR TORSIONAL STIFFNESS"
80  PRINT "  4. ROLL RATES OF THE FRONT AND REAR"
90  PRINT "     SUSPENSION GIVEN THE ABOVE"
100  PRINT
```

```
110  PRINT "ENTER IN THE NUMBER CORRESPONDING TO "
120  PRINT "THE CALCULATION YOU NEED";
130  INPUT A
135  PRINT
140  IF A = 1 THEN 180
150  IF A = 2 THEN 500
160  IF A = 3 THEN 840
170  IF A = 4 THEN 980
172  REM
174  REM : SPRING RATE DETERMINATION
176  REM
180  PRINT "YOU WILL NEED TO KNOW"
190  PRINT "  1. DIAMETER OF THE SPRING WIRE"
200  PRINT "  2. MEAN DIAMETER OF THE SPRING"
210  PRINT "  3. NUMBER OF ACTIVE COILS"
220  PRINT "  4. TYPE OF SPRING ENDS"
230  PRINT "     (PLAIN ENDS = -.5)"
240  PRINT "     (SQUARED & GROUND = -1.75)"
250  PRINT "     (PLAIN ENDS GROUND = -1.0)"
260  PRINT
270  PRINT "ENTER THE SPRING WIRE DIAMETER"
280  INPUT "IN INCHES----------------------- ";D
290  PRINT
300  PRINT "ENTER THE SPRING MEAN DIAMETER"
310  INPUT "IN INCHES----------------------- ";D1
320  PRINT
330  INPUT "ENTER THE NUMBER OF COILS------- ";N
340  PRINT
350  PRINT "ENTER THE CORRECTION FOR THE "
360  PRINT "TYPE OF ENDS (EXAMPLE PLAIN"
370  INPUT "ENDS ENTER -.5)----------------- ";C
380 N1 = N + C
390 K = (1500000 * D ^ 4) / (N1 * D1 ^ 3)
400 K =  INT (K + .5)
410  PRINT
420  PRINT "THE SPRING RATE OF THE ";D;" INCH"
430  PRINT "WIRE DIAMETER SPRING IS "
440  PRINT "-----------------------------";K;" LBS/IN"
450  PRINT
460  PRINT "IF YOU WANT TO EXIT THE PROGRAM TYPE 0;
 IF YOU WANT TO CALCU LATE ANOTHER SPRING TYPE 1;
 IF YOU WANT TO CONTINUE WITH WHEEL RATE
 S TYPE 2 ";
470  INPUT B
480  IF B = 0 THEN 1500
490  IF B = 1 THEN 270
500  PRINT
502  REM : WHEEL RATE FROM SPRING RATE DETERMINATION
505  HOME
510  PRINT "ENTER THE LENGTH OF THE"
520  INPUT "SPRING CARRYING CONTROL ARM----- ";R1
530  PRINT
540  PRINT "ENTER THE DISTANCE BETWEEN THE"
550  PRINT "SPRING CENTER AND THE INNER"
560  INPUT "CONTROL ARM PIVOT--------------- ";R3
570  PRINT
580  PRINT "DO YOU WANT TO USE THE ABOVE CALCULATED"
590  PRINT "SPRING RATE OR ENTER A NEW ONE"
600  PRINT "(TYPE N IF NEW; O IF OLD)";
610  PRINT
620  INPUT A$
630  IF A$ = "O" THEN 650
640  INPUT "ENTER THE NEW SPRING RATE------- ";K
650 K1 = ((R3 / R1) ^ 2) * K
660 K1 =  INT (K1 + .5)
670  PRINT
680  PRINT "THE WHEEL RATE FOR THE ";K;" LB/IN."
690  PRINT "SPRING IS------------- ";K1;" LB/IN"
700  PRINT
710  PRINT "IF YOU WANT TO EXIT TYPE 0"
720  PRINT "IF YOU WANT TO REDO ANY PREVIOUS"
730  PRINT "CALCULATIONS TYPE 1; TYPE 2 TO CONTINUE"
740  INPUT B1
750  IF B1 = 0 THEN 1500
760  IF B1 = 1 THEN 30
770  PRINT
775  REM : STABILIZER BAR DETERMINATION
776  REM
780  PRINT "DO YOU WANT TO CALCULATE STABILIZER BAR"
790  PRINT "STIFFNESS (TYPE Y IF YES; N IF NO) ";
800  INPUT B$
810  PRINT
820  IF B$ = "N" THEN 980
830  GOTO 860
831  REM
832  REM :STABILIZER CALCULATION
833  REM
840  INPUT "ENTER THE LOWER CONTROL ARM LENGTH- ";R1
850  PRINT
860  PRINT "ENTER THE STABILIZER BAR DIAMETER"
870  INPUT "IN INCHES-------------------------- ";D2
880  PRINT
890  PRINT "ENTER THE SPACING BETWEEN THE BAR"
900  INPUT "MOUNTING POINTS ON THE FRAME------- ";L
910  PRINT
920 K2 = (3.142 * (D2 ^ 4) * 11500000) / (32 * L)
930 K2 =  INT (K2 + .5)
940  PRINT "THE STABILIZER BAR STIFFNESS IS"
950  PRINT "---------------------- ";K2;" IN-LBS/RAD"
960  PRINT
970  GOTO 1020
972  REM
974  REM :STABILIZER BAR CONTRIBUTION TO
976  REM :ROLL STIFFNESS
978  REM
980  PRINT "ENTER IN THE BAR STIFFNESS IN"
990  INPUT "IN-LBS/RAD-------------------- ";K2
1000  INPUT "ENTER THE LOWER CONTROL ARM LENGTH---
   ";R1
1010  PRINT
1020  INPUT "ENTER THE FRONT TRACK WIDTH----------
   ";T
1030  PRINT
1040  PRINT "ENTER THE DISTANCE BETWEEN THE "
1050  PRINT "CONTROL ARM INNER PIVOT AND THE"
1060  INPUT "POINT WHERE THE BAR ATTACHES---------
   ";R2
1070  PRINT "ENTER THE LONGITUDINAL DISTANCE"
1080  PRINT "BETWEEN THE BAR AND ITS MOUNTING"
1090  INPUT "POINT ON THE CONTROL ARM------------ ";L1
1100 K3 = K2 * (R2 ^ 2 * T ^ 2) / (688 * R1 ^ 2 * L1
   ^ 2)
1110 K3 =  INT (K3 + .5)
1120  PRINT
1130  PRINT "THE ROLL RATE CONTRIBUTION OF THE"
1140  PRINT "STABILIZER BAR WITH A STIFFNESS OF"
1150  PRINT ;K2;" IN-LBS/RAD IS--- ";K3;" FT-LBS/DEG"
1160  PRINT
1170 K4 = (K1 * T ^ 2) / 1375
1180 K4 =  INT (K4 + .5)
1190 K5 = K3 + K4
1200  PRINT "FRONT SUSPENSION ROLL RATE"
1210  PRINT "    WITHOUT BAR--------- ";K4;"
   FT-LBS/DEG"
1220  PRINT "    TOTAL--------------- ";K5;"
   FT-LBS/DEG"
1230  PRINT
1240  PRINT
1250  PRINT "IF YOU WISH TO END TYPE 0; IF YOU WISH"
1260  PRINT "TO REDO AND CALCULATIONS TYPE 1; TO
   CONTINUE"
1270  PRINT "TYPE 2 ";
1280  INPUT B2
1290  IF B2 = 0 THEN 1500
1300  IF B2 = 1 THEN 30
1310  PRINT
1312  REM : REAR SPRING CONTRIBUTION TO ROLL RATE
1314  REM
1320  PRINT "ENTER THE DISTANCE BETWEEN"
1330  INPUT "THE REAR SPRINGS------------------ ";L2
1340  PRINT
1350  INPUT "ENTER THE REAR TRACK WIDTH-------- ";T1
1360  PRINT
1370  INPUT "ENTER THE REAR SPRING RATE-------- ";K6
1380 K7 = K6 * L2 ^ 2 / 1375
1390 K7 =  INT (K7 + .5)
1400  PRINT
1410  PRINT "THE REAR SUSPENSION ROLL RATE"
1420  PRINT "IS---------------- ";K7;" FT-LBS/DEG
1430  PRINT
1440  PRINT "THE FRONT SUSPENSION ROLL RATE"
1450  PRINT "IS---------------- ";K5;" FT-LBS/DEG""
1460  PRINT
1470 K8 = K3 + K4 + K7
1480  PRINT "THE ROLL RATE OF THE CHASSIS"
1490  PRINT "IS---------------- ";K8;" FT-LBS/DEG"
1500  PRINT
1501  INPUT "TO CONTINUE HIT RETURN";Z$
1502  HOME
```

```
1503  PRINT "THINGS YOU SHOULD KNOW:"
1510  PRINT
1520  PRINT " 1.SPRING RATES INCREASE WITH INCREASED"
1522  PRINT "   SPRING WIRE DIAMETER"
1525  PRINT
1530  PRINT " 2.SPRING RATES DECREASE WITH LARGER"
1532  PRINT "   SPRING DIAMETERS AND INCREASED"
1534  PRINT "   NUMBER OF COILS"
1536  PRINT
1540  PRINT " 3.ROLL STIFFNESS INCREASES WITH"
1542  PRINT "   STABILIZER BAR THICKNESS"
1545  PRINT
1550  PRINT " 4.ROLL STIFFNESS INCREASES WITH SPRING"
1552  PRINT "   STIFFNESS AND SPRING SPACING"
1555  PRINT
1557  PRINT " HAVE A NICE DAY"
1560  END
```

```
 RUN SPRINGS
THIS PROGRAM CALCULATES
  1. COIL SPRING RATES
  2. WHEEL RATES GIVEN SPRING RATES
  3. STABILIZER BAR TORSIONAL STIFFNESS
  4. ROLL RATES OF THE FRONT AND REAR
     SUSPENSION GIVEN THE ABOVE

ENTER IN THE NUMBER CORRESPONDING TO
THE CALCULATION YOU NEED?1

YOU WILL NEED TO KNOW
  1. DIAMETER OF THE SPRING WIRE
  2. MEAN DIAMETER OF THE SPRING
  3. NUMBER OF ACTIVE COILS
  4. TYPE OF SPRING ENDS
     (PLAIN ENDS = -.5)
     (SQUARED & GROUND = -1.75)
     (PLAIN ENDS GROUND = -1.0)

ENTER THE SPRING WIRE DIAMETER
IN INCHES----------------------- .8

ENTER THE SPRING MEAN DIAMETER
IN INCHES----------------------- 5.5

ENTER THE NUMBER OF COILS------- 8

ENTER THE CORRECTION FOR THE
TYPE OF ENDS (EXAMPLE PLAIN
ENDS ENTER -.5)----------------- -1

THE SPRING RATE OF THE .8 INCH
WIRE DIAMETER SPRING IS
-----------------------------528 LBS/IN

IF YOU WANT TO EXIT THE PROGRAM TYPE 0
IF YOU WANT TO CALCULATE ANOTHER SPRING
TYPE 1; IF YOU WANT TO CONTINUE WITH
WHEEL RATES TYPE 2?2

ENTER THE LENGTH OF THE
SPRING CARRYING CONTROL ARM----- 16

ENTER THE DISTANCE BETWEEN THE
SPRING CENTER AND THE INNER
CONTROL ARM PIVOT--------------- 8

DO YOU WANT TO USE THE ABOVE CALCULATED
SPRING RATE OR ENTER A NEW ONE
(TYPE N IF NEW; O IF OLD)
?O


THE WHEEL RATE FOR THE 528 LB/IN.
SPRING IS------------- 132 LB/IN

IF YOU WANT TO EXIT TYPE 0
IF YOU WANT TO REDO ANY PREVIOUS
CALCULATIONS TYPE 1; TYPE 2 TO CONTINUE
?2

DO YOU WANT TO CALCULATE STABILIZER BAR
STIFFNESS (TYPE Y IF YES; N IF NO) ?Y

ENTER THE STABILIZER BAR DIAMETER
IN INCHES-------------------------- 1

ENTER THE SPACING BETWEEN THE BAR
MOUNTING POINTS ON THE FRAME------- 40

THE STABILIZER BAR STIFFNESS IS
---------------------- 28229 IN-LBS/RAD

ENTER THE FRONT TRACK WIDTH---------- 58

ENTER THE DISTANCE BETWEEN THE
CONTROL ARM INNER PIVOT AND THE
POINT WHERE THE BAR ATTACHES--------- 14
ENTER THE LONGITUDINAL DISTANCE
BETWEEN THE BAR AND ITS MOUNTING
POINT ON THE CONTROL ARM------------ 12

THE ROLL RATE CONTRIBUTION OF THE
STABILIZER BAR WITH A STIFFNESS OF
28229 IN-LBS/RAD IS--- 734 FT-LBS/DEG

FRONT SUSPENSION ROLL RATE
    WITHOUT BAR--------- 323 FT-LBS/DEG
    TOTAL--------------- 1057 FT-LBS/DEG


IF YOU WISH TO END TYPE 0; IF YOU WISH
TO REDO AND CALCULATIONS TYPE 1; TO CONTINUE
TYPE 2 ?2

ENTER THE DISTANCE BETWEEN
THE REAR SPRINGS------------------ 45

ENTER THE REAR TRACK WIDTH-------- 58

ENTER THE REAR SPRING RATE-------- 100

THE REAR SUSPENSION ROLL RATE
IS---------------- 147 FT-LBS/DEG

THE FRONT SUSPENSION ROLL RATE
IS---------------- 1057 FT-LBS/DEG

THE ROLL RATE OF THE CHASSIS
IS---------------- 1204 FT-LBS/DEG

TO CONTINUE HIT RETURN



THINGS YOU SHOULD KNOW:

 1.SPRING RATES INCREASE WITH INCREASED
   SPRING WIRE DIAMETER

 2.SPRING RATES DECREASE WITH LARGER
   SPRING DIAMETERS AND INCREASED
   NUMBER OF COILS

 3.ROLL STIFFNESS INCREASES WITH
   STABILIZER BAR THICKNESS

 4.ROLL STIFFNESS INCREASES WITH SPRING
   STIFFNESS AND SPRING SPACING

 HAVE A NICE DAY
```

Glossary

ABRASIVE WEAR: Wear caused by the presence of material between two moving parts, resulting in surface erosion.

AERATION: Air mixed with a liquid; air in the brake fluid or shock absorber.

ALL-WEATHER TIRE: A tire designed to provide good traction on dry, wet, and dirt- and snow-covered roadways.

ANALOG VOLTAGE: An electrical voltage that changes in value but is measureable at any point in time.

ANTISEIZE LUBRICANT: A special material applied to adjoining metal parts to prevent the parts from bonding together due to galvanic action. Used on threaded cap screws into aluminum to prevent the aluminum from adhering to the steel screw and preventing disassembly.

ARCHING: Grinding new brake linings to the same diameter (arc) as that of the brake drum surface.

ASBESTOS: A long, fibrous mineral that has been used as a friction material in brakes. Breathing asbestos dust is hazardous to one's health.

ASPECT RATIO: The ratio of the width to the length. On tires, it is the fully inflated height section divided by the cross-section.

AXIAL: Having the same direction as or being parallel to the axis of rotation.

AXIAL LOAD: A force applied in the direction parallel to the axis of rotation of the part.

AXLE FLANGE: A flat surface on the outboard end of the axle shaft to allow wheel attachment.

AXLE SHAFT: The shaft used to transmit power from the differential to the wheels.

BALANCE: Having equal weight on each side of a supporting point. No change in the position of the center of gravity when the part is in motion.

BALL BEARING: An antifriction bearing where the rolling elements are spherically shaped.

BEAD: A reinforcing ridge. The wire reinforcing around a tire opening where it fits the wheel rim.

BEARING CAGE: Spacer to keep balls and rollers in proper relationship to each other in a bearing.

BEARING CONE: The inner race, roller, and cage assembly of a tapered roller bearing.

BEARING CUP: The outer race of a tapered roller bearing.

BEARING FATIGUE: Failure of a bearing due to the continued flexing of the bearing surface from the applied load.

BEARING RACE: The surface on which a roller or balls or a bearing rotate.

BELLMOUTH: The taper of a brake drum.

BIAS: Lines running at an angle. Cord angle of tire plys. Differences in front and rear vehicle weight and brake system split.

BIAS BELTED: Tires with the carcass ply cords at an angle, overlayed with reinforcing belts.

BIAS PLY: Tires with the carcass ply cords at an angle. No belts are used.

BINDERS: Compounds that hold the friction materials together in brake linings.

BOOST: An added force applied to the driver's input effort.

BRAKE BIAS: The stopping effort of the front wheels compared to that of the rear wheels.

BRAKE CALIPER: The hydraulic cylinder at the wheel used to apply the disc brake linings against the rotor.

BRAKE PAD: The brake friction material and its steel backing used on a disc brake.

BRAKE ROTOR: The brake friction surface that rotates at wheel speed designed for contact with the brake pads on disc brake assemblies.

BRINNELLING: A bearing wear condition that results from impact loading or vibration.

BULKHEAD: The structural part of the vehicle connecting the front of the floor assembly to the roof structure.

BURSTING STRENGTH: The resistance to failure at high pressures of parts that carry the pressure.

BUSHING DRIVER: A tool designed to properly force sleeve-type bushings into their bore without damage.

CAP SCREW: A threaded fastening device with a hexagonal end for turning it into a threaded hole.

CASTELLATED NUT: A nut with six raised portions between which a cotter pin is placed to secure the nut.

CENTER OF GRAVITY: A point around which all weights balance.

CENTRIFUGAL: A force in a direction away from the turning of an object that is moving in a curved path.

CHASSIS: The vehicle frame, suspension, and running gear.

CHECK VALVE: A valve that allows flow in one direction and stops flow in the opposite direction.

CHUNKING: Pieces of rubber coming out of the tire tread.

COEFFICIENT OF FRICTION: An expression that relates the force to move an object to the weight or load produced by the object. Coefficient of friction = force/load.

CONCENTRIC: Located evenly around a common center.

CONTACT PATCH: The part of a tire that is in contact with the road surface.

CONTAMINANT: A foreign or undesirable element found in liquids and solids that will adversely affect the operation.

CORD: A string or thread that makes up the fabric used in tire plys.

CREEP: Slight or slow movement.

CROCUS CLOTH: An extremely fine abrasive cloth that is used for polishing.

CUPPING: Heavy, spotty tire wear.

CURB WEIGHT: The weight of a complete vehicle with its normal load, less driver and passengers but with a full tank of fuel.

DAMPEN: To slow or reduce oscillations or movement.

DEFLECTION: Bending or moving to a new position as the result of an external force.

DIAGONAL SPLIT SYSTEM: A brake system design that will allow application of one front and a diagonally opposite rear brake when part of the brake system fails.

DIFFERENTIAL: The part of a drive line that allows the wheels to rotate at different speeds as the vehicle is driven through a turn.

DIVE: The front of the vehicle lowering during braking.

DOUBLE FLARE: A reinforced tubing flare used on brake system tubing to help prevent failure.

DRIVE TRAIN: The parts used to transfer engine power to the drive wheels.

DUAL-SERVO BRAKES: A drum brake assembly with both front and rear shoes, self-energized during forward and rearward wheel rotations.

DYNAMIC: Moving, as in dynamic loading, when the vehicle is moving.

DYNAMIC BALANCE: The center of gravity and the center of rotation remain stationary as the parts move.

DYNAMIC SEAL: A seal to prevent leakage around a moving joint.

EQUALIZER LINK: A common connector in the parking brake system that causes both rear brakes to be applied with the same cable tension.

EXTRUDED: Forced through an opening under great pressure to change the shape of the material.

FADE: A condition that occurs when there is little braking effect with full brake pedal force.

FATIGUE: Natural stress buildup in the structure of the material that leads to failure.

FINAL DRIVE: The final gear reduction between the engine and the drive wheels.

FIXED CALIPER: A disc brake caliper rigidly mounted to the wheel knuckle, requiring at least one hydraulic piston on each side of the rotor.

FLARE NUT: The threaded fitting on the end of a brake tube used to tighten against the tubing flare.

FLUID FILM BEARING: A sleeve-type bearing where an oil film prevents metal-to-metal contact.

FOOT PRINT: The area of contact between the tire and road surface.

FRICTION FORCE: The force that resists slipping or skidding.

FULL-FLOATING AXLE: An axle design usually used on heavy trucks where the veicle weight is carried by bearings in the wheel hubs or the dive wheels and the axle is used only to transfer driving torque.

GALLING: The transfer of material from one part to the contacting part when movement is present.

GASKET: Sealing materials placed in a nonmoving joint that prevent leakage.

GEAR LUBRICANTS: High-viscosity lubricants with extreme pressure additives to provide maximum protection for gears with high torque.

GRAPHITE: A lubricant used to provide protection from galling and seizing when a minimum amount of lubricant is present. Used both dry and as an additive to greases and oils.

GREASE: A semisolid lubricant consisting of a liquid lubricant with a thickening agent used to reduce friction.

HELIX ANGLE: The angle of a thread that causes the part to advance axially as its body is rotated.

HONE: Smoothing a metal surface with a fine stone.

HYDRAULICS: A study of fluids in motion.

HYDROPLANING: High-speed movement over water, causing a tire to be lifted from the road surface and supported on a wedge of water.

HYGROSCOPIC: Having the ability to absorb water.

IMPENDING SKID: The tire traction point at which any increase in side or tractive load will produce tire skid.

INERTIA: The tendency of a body at rest to stay at rest, or when moving, to keep moving.

INTERNATIONAL STANDARDS ORGANIZATION (ISO): The organization responsible for worldwide standardization of measurements.

ISO-FLARE: A brake tube flaring method that upsets the tube and allows a strong, secure attachment when used with the proper tube nut and seal.

LATERAL LOAD: The force on the side of the tire tread.

LATHE-CUT SEAL: A square cross-sectional seal used in dynamic applications where little or no rotation takes place between the two sealing surfaces.

LEADING SHOE: A brake shoe that has the drum rotating from the toe toward the heel.

LEADING-TRAILING BRAKE: A drum brake assembly having one shoe energized in either forward or rearward wheel rotation.

LIMITED SLIP DIFFERENTIAL: A differential allowing unequal torque to be delivered to the axle shafts.

LIP SEAL: A dynamic seal used primarily to seal between rotating assemblies. Positive sealing in one direction only.

LOAD RANGE: The amount of weight that can safely be carried by a tire.

LUG: The flange stud on an axle or hub on which the drum and wheel are fastened.

MACPHERSON STRUT: A simplified suspension design using a strut to replace the upper control arm and the steering knuckle.

MECHANICAL ADVANTAGE: The increased force that the operator is able to apply through a mechanical lever system.

MOLYBDENUM DISULFIDE: An extreme pressure additive found in many greases to reduce friction and prevent galling when minimal lubricant is present.

NEEDLE BEARING: An antifriction bearing with a rolling element that has a length over four times greater than its diameter.

OFFSET: Having parallel centerlines on different planes. The wheel rim center plane does not match the bolt flange center plane.

ORIFICE: A carefully sized opening that controls a flow of fluid.

O-RING SEAL: A circular cross-sectional seal used in dynamic applications where little or no rotational motion occurs. It is also used as a static seal.

OXIDATION PROPERTIES: Refers to the resistance of a liquid or solid to combine with oxygen. High temperatures increase the rate of oxidation, which limits the useful life of some liquids.

PAL NUT: A sheet metal nut placed on a stud on top of a fastening nut to keep it from loosening accidentally.

PARALLELISM: An inspection to ensure that two surfaces operate continuously at the same spacing.

PLY: A layer of cord fabric in a tire carcass.

PLY STEER: The tendency of a tire to always turn in one direction as it rolls. This is the result of the way the tire was constructed.

PORT: An opening or passage through which liquid or gases flow.

PRELOAD: A load caused when assembling a part before the operating load is applied.

PRESS FIT: Forcing a part into an opening that is slightly smaller than the part to make a solid fit.

PREVENTING MAINTENANCE: Performing service and making repairs before failure occurs.

PROPORTIONING: Maintaining the same ratio between input and output pressures throughout the operating range.

RACE: The part of a bearing on which the rollers or balls roll.

RADIAL: The direction that is moving straight out from the center of a circle.

RADIAL LOAD: A load applied through a bearing from the wheel toward the spindle.

RADIAL PLY: Tire cords running directly across the tire carcass from bead to bead.

RIDE: The characteristic feel as one rides in a vehicle.

ROLLER CONTACT BEARING: An antifriction bearing where the rolling element length is less than four times its diameter.

ROTOR RUNOUT: The axial variation of the brake rotor surface as the rotor is rotated.

RTV SILICONE SEALANT: A room-temperature vulcanizing rubber used to seal against leakage at parting surfaces.

SAFETY STANDS: Steel stands used to support a vehicle after it is raised with a jack. They provide safety for the technician when working under a raised vehicle.

SCUFF: Tire slide on the road surface during operation.

SERIES: The designation of a tire aspect ratio.

SERVICE BRAKE SYSTEM: The main braking system of the vehicle which controls braking effort proportional to the driver's demand.

SHOP CLOTH: A wiping cloth especially made for use in service shops. It is usually provided by and cleaned by a laundry service.

SIPES: Slits in the tire tread to produce more blade surface for traction.

SKID: A tire sliding on the road surface.

SMEARING: Shiny polished areas caused by a bearing race slipping in its mount.

SOCIETY OF AUTOMOTIVE ENGINEERS (SAE): A professional engineering society responsible for setting many vehicle standards used in the world today.

SPALLING: A fatigue-related action causing the bearing surface material to break away from the base material.

SPIN BALANCER: An on-car tire balancer which turns the raised tire at a road speed where imbalance can be detected by wheel shake.

SPINDLE: The axis about which nondriving wheels of the vehicle spin.

SQUIRM: The twist of the tire tread in the footprint.

STABILIZER BAR: A torsion bar connected between the left and right suspensions on a vehicle to reduce body roll in a turn and to add stability.

STAR WHEEL: An adjustable link between the primary and secondary brake shoes.

STATIC BALANCE: Balance that causes no movement when the wheel is stopped in any position.

STATIC SEAL: A seal in a joint that has no movement.

STEERING KNUCKLE: The part of the front suspension that rotates to allow the wheels to be steered and which supports the spindle or stub axle.

TACTILE SENSOR: A sensor that allows the vehicle operator to feel when a certain condition is reached. Disc brake pads are made to vibrate when worn to the point where replacement is necessary and this vibration is felt in the brake pedal.

TANG: Projection from a part used for alignment, or for a driving lug.

TAPERED ROLLER BEARING: An antifriction bearing where the rolling element is similar to a roller bearing; however, it has a taper along its axis, allowing it to accept axial loads in one direction.

TIRE CARCASS: The main structural part of the tire to which tread rubber is attached.

TIRE CONTACT PATCH: The part of a tire that contacts the road surface making the footprint.

TIRE FOOTPRINT: The area on the road in contact with the tire.

TIRE SERIES: The groupings of tire sizes having the same aspect ratio.

TORQUE: Twisting force on a shaft.

TRACTIVE FORCE: The friction force in the contact path that causes torque on the wheel.

TRANSAXLE: A drive assembly combining the transmission and final drive assemblies in one casing.

TRI-POT JOINT: A universal joint using bearings on three axes to maintain a constant plane of drive, making it a constant-velocity joint.

UNSPRUNG WEIGHT: The vehicle weight moved by variations in the road surface that is not supported by the suspension springs. Tires, wheels, brake assemblies, and steering knuckles are included in the unsprung weight.

WADDLE: A sideways vehicle shake caused by a faulty radial tire. Most noticeable when a vehicle moves slowly.

WEIGHT TRANSFER: The changes in radial loads on the tires due to the center of gravity location during braking.

WHEEL OFFSET: The distance between the wheel attachment flange and the wheel rim center plane.

WHEEL-PAC BEARING: A preassembled self-lubricated bearing assembly used on the drive wheels with independent suspension.

Index